I0796538

A History of Sautee Nacoochee

Support for the research and writing of this book was funded by the Gregory New and Dr. Tom and India Lumsden endowments to the Sautee Nacoochee Center History Museum. Publication of this book was made possible, in part, by generous gifts from John and Laura Hardman, Shell and Wyck Knox, and the Dot and Lam Hardman Family Foundation.

This book is a project of the Sautee Nacoochee Community Association.

A HISTORY OF Sautee Nacoochee

Tommy Hart Jones

The University of Georgia Press *Athens*

Athens, Georgia 30602
www.ugapress.org

Designed by Erin Kirk
Set in Minion Pro
Printed and bound by Versa Press
The paper in this book meets the guidelines for permanence and durability of the Committee on Production Guidelines for Book Longevity of the Council on Library Resources.

Most University of Georgia Press titles are available from popular e-book vendors.

Printed in the United States of America
29 28 27 26 25 C 5 4 3 2 1

EU Authorized Representative
Easy Access System Europe—Mustamäe tee 50, 10621
Tallinn, Estonia, gpsr.requests@easproject.com

Library of Congress Cataloging-in-Publication Data

Names: Jones, Tommy, author.
Title: A history of Sautee Nacoochee / Tommy Hart Jones.
Description: Athens : The University of Georgia Press, [2025] | Includes bibliographical references and index. | Summary: "A History of Sautee Nacoochee is a complete local Georgia history that's intended to be a "readily accessible compendium of useful information about the valleys." Indeed, Jones has compiled a vast amount of information into a "narrative that identifies historical contexts; documents and incorporates site-specific information; and strives to illuminate the lives of the people who, over many centuries of human occupation and in many different ways, contributed to making Sautee Nacoochee what it is today." The manuscript traces thousands of years of history in the valleys, covering culture and environment all the way up to the current push for preservation and is a comprehensive collection designed for locals, historians, genealogists, and anyone interested in Georgia's natural, political, and economic past"— Provided by publisher.
Identifiers: LCCN 2024034849 (print) | LCCN 2024034850 (ebook) | ISBN 9780820369587 (hardback) | ISBN 9780820369600 (epub) | ISBN 9780820369594 (pdf)
Subjects: LCSH: Nacoochee Valley (Ga.)—History. | Nacoochee Indian Mound (Ga.) | White County (Ga.)—History.
Classification: LCC F292.W48 J66 2025 (print) | LCC F292.W48 (ebook) | DDC 975.8/277—dc23/eng/20241115
LC record available at https://lccn.loc.gov/2024034849
LC ebook record available at https://lccn.loc.gov/20

Vehiculating—as Carlyle would express it—near the sunset hour of a summer day, on my route from Rabun county . . . a quick turn in the road, very unexpectedly opened to my delighted vision the whole stretch of the charming valley of Nacoochee.

—T. ADDISON RICHARDS, *The Orion*, 1843

The valley is perhaps a mile wide, and, as the surrounding hills are not lofty, it is distinguished more for its beauty than any other quality; and this characteristic is greatly enhanced by the fact, that while the surrounding country remains in its original wilderness the valley itself is highly cultivated, and the eye is occasionally gratified by cottage scenes which suggest the ideas of contentment and peace.

—CHARLES LANMAN, *Letters from the Alleghany Mountains*, 1849

Pitner
Englands Gold
B Smith shop
Grist and Saw Mill
Long Hungry
Welch
Massey
Kiker
Meeting House
Capps
C L Williams
Store
D Brown
Indian Mound
Maj Edward Williams
Talbot
Chattahoochee River
Black
J L Richardson
School House
Owen
Moffat
Dukes Creek
Kelly Lot
J Richardson
Russel
Littlejohn
Mill
Russel Lot
Smith
Harris
Mount Yonah
Catty

Contents

AMERICAN RAILWAY

Preface

Sautee Nacoochee is a rural community located southeast of Helen in White County, Georgia. It is centered around two large valleys: Sautee, drained by Sautee Creek, and Nacoochee, drained by the Chattahoochee River. The latter valley is the larger of the two and, historically, gave its name to a wider area, bounded generally by the Blue Ridge Mountains to the northwest, Mount Yonah to the south, and Alec Mountain and Amos Creek on the east. The old General Militia District 427 has been much subdivided, but what remains is an approximation of the boundaries of Sautee Nacoochee. The boundaries of the two districts listed in the National Register of Historic Places that comprise the two valleys are defined by topographical contour lines. Perhaps the postal zip code 30571 is the most accessible boundary for the modern Sautee Nacoochee community.

Virtually every landscape is a palimpsest, created by the imprint, however ephemeral, that people have left everywhere they have been. As one peels back layers of history, the true nature of a place can be better understood, and this study is meant to do that for Sautee Nacoochee. In the broadest terms, I have sought to synthesize an enormous amount of information from many disparate sources into a narrative that identifies historical contexts, that documents and incorporates site-specific information, and that strives to illuminate the lives of the people who, over many centuries of human occupation and in many different ways, contributed to making Sautee Nacoochee what it is today. In essence it is intended as a readily accessible compendium of useful historical information about the valleys.

A variety of scientific reports encompassing the valleys have been published by the State of Georgia, beginning in the 1890s. These include extensive documentation of the area's geological features and mineral resources, including reports on deposits of gold, asbestos, and talc in northeastern Georgia. More recently, Pamela Gore and William Witherspoon's *Roadside Geology of Georgia* has provided an excellent introduction to the foundation of the palimpsest of Sautee Nacoochee.

Archaeologists and pot hunters have taken an interest in Nacoochee since the mid-nineteenth century, and some of them documented what they carried off, beginning with historian Charles C. Jones Jr. (1831–93) in the 1870s. He was followed most famously by George Heye (1874–1947), whose excavation of Nacoochee Mound in 1915 produced some of the first artifacts in the Museum of the American Indian, which he established in New York City in 1916. In the 1930s a federally funded survey of archaeological resources in Georgia mapped numerous sites in and around Sautee Nacoochee. More recent archaeology has continued documentation of some of those earliest layers of the valleys' palimpsest.

A number of people left written accounts of Nacoochee in the eighteenth and nineteenth centuries. The first was Col. George Chicken (1685–1727), in his journal of an expedition out of Charleston into northeastern Georgia in 1715–16. Benjamin Hawkins recorded a trip through the valleys in 1796, and British-born geographer George Featherstonhaugh (1780–1866) saw "Nahcóochay" in 1836 and thought it "an extremely sweet place."[1] Most who came after him agreed, even as they were dispossessing the native people of their land. The story of the Cherokee in northeast Georgia and the western Carolinas in the late eighteenth and early nineteenth centuries is a sorry prelude to the infamous Trail of Tears in the late 1830s.

Research in public records was not exhaustive for the present project, and much could still be gleaned through additional work. The state's records for the 1820 land lottery, which includes lists of "fortunate drawers" and plats of land lots made in 1821, provide important documentation for the valleys. Likewise, the number of public records for Habersham and White Counties that could be searched more thoroughly is very large, especially those documenting the sale and taxation of real estate and probate of estates.

A variety of federal records continues to be an incomparable resource, especially since facsimile copies of many are now readily accessible online. The decennial censuses include population schedules from 1790 to 1950, agriculture and manufacturing schedules from 1840 to 1920, and slave schedules from 1850 to 1860. The records of the Freedmen's Bureau, especially the monthly reports by the district school superintendents, help establish a context for understanding the early history of the Bean Creek community.

Numerous historical maps, the earliest dating to the 1720s, locate what we now know as Sautee Nacoochee in a general way. In the nineteenth century there were land-lot maps, but they often did not depict roadways. The first reliable mapping of the locale was done by the State Highway Board in the 1930s and depicts much of the nineteenth-century road system, including "primitive" roads, many of which were abandoned by the 1940s. The highway maps created after World War II also record much of the built environment, including roads, houses, stores, churches, district courthouses, and schools.

For the particulars of the history of the valleys, three residents left incomparable resources when they mapped the valleys. The first was Reuben Curtis Moffat (1818–94), who in 1891 drew a surprisingly accurate map of Nacoochee as he remembered it from the time that he lived in the community in the 1830s. Victor Hollis (1883–1963), who with his wife, Marnie, ran the George W. Williams orphanage at Nacoochee in the 1920s and 1930s, added to that legacy when he drew his own rendition of Moffat's map in 1922. Finally, Walter B. Lumsden Jr. (1919–2011), whose father was one of the first rural-free-delivery mail carriers in the county, drew a third map of the community as it existed in 1948. Beyond those primary sources, recently published books greatly inform our understanding of the valleys prior to 1819. William McLoughlin's *Cherokee Renascence in the New Republic* and Don Shadburn's *Upon Our Ruins* are especially important. The latter includes historical documentation not published elsewhere and suggests several potential avenues of research that would certainly shed additional light on the late Cherokee period in the history of the valleys.

For general historical documentation for the period after 1819, this project has relied heavily on several modern local histories, especially Dr. Thomas Lumsden's *Nacoochee Valley*, Garrison Baker's *In the Shadow of Yonah*, Matt Gedney's *Living on the Unicoi Road* and *The Story of Helen, Georgia*, and Chris Brooks and David Greear's *Images of America: Helen*. Much work by other local historians, especially the late Bill Huff, has been preserved by the White County Historical Society and provides a rich resource for studying the history of Sautee Nacoochee.

Most important perhaps to preserving the history of Sautee Nacoochee was Walter Lumsden's brother, Dr. Thomas Newton Lumsden (1925–2015), whose *Nacoochee Valley* still provides the most succinct overview of the valleys' history. His papers, which were uncataloged until recently, are a likely treasure trove of additional historical documentation, since his work formed the foundation for successful nomination of the valleys to the National Register of Historic Places.[2] He and his wife, India Dyer Lumsden (1925–2018), remained staunch supporters of the Sautee Nacoochee Community Association (SNCA) from its inception. No one wrote more lovingly of the charms of the valleys than did George Walton Williams (1820–1903), who moved with his parents from North Carolina to Nacoochee when he was still a small child. His published work is an often-sentimental personal memoir, but it contains valuable nuggets of information about the early years of white resettlement of the valleys.

The present project is the result of recommendations contained in the "General Interpretive Plan" prepared for the SNCA in 1991, which called for, among other things, development of a general history of the valleys. After a series of meetings between the author and staff and volunteers of SNCA in 2014, a scope of work was established and work commenced in the spring of 2015. Research and writing were more or less complete two years later.

For their role in bringing me into this project, special thanks go to Chris Brooks, former director of the Folk Pottery Museum of Northeast Georgia, and Judy Barber, former executive director of the SNCA. David Greear, professional photographer, former mayor of Helen, and chair of SNCA's History Museum Committee when the project commenced, and Mary Geidel, committee member and daughter of Dr. Lumsden, have been a critical part of this project from its inception. Without the kind assistance of Joyce Etheridge in the later stages of the work, it might never have been completed. And without Laura Paquette the UGA Press might never have found this work. Finally, without an endowment by the late Gregory New (1928–2019) for the SNCA History Museum and his generous ongoing support for this and other special projects, this history of Sautee Nacoochee would not have been possible. Most of all, I am personally indebted to the SNCA for their decision to publish my work. I trust that in the years to come this history will be useful for SNCA's museum management and interpretative programs and as a guide for further research into the history of this wonderful part of the world.

Tommy Hart Jones
Atlanta, Georgia
14 June 2021

A History of Sautee Nacoochee

1

The Natural Environment

The special character of the natural environment of Sautee Nacoochee, as elsewhere, has largely determined the shape and extent of human use over many thousands of years of inhabitation of the valleys. Over hundreds of millions of years before that, geological forces were at work forming and reforming, building and disintegrating, the great folds and thrusts of ancient rock that give rise to the topography we experience today. And over that land the vagaries of climate and ice ages brought a changing cast of flora and fauna long before the first humans followed ancient trails worn in the landscape over eons of use by mastodon, buffalo, and other Pleistocene megafauna.

Sautee Nacoochee is at the southern end of the great chain of ancient mountains, ridges, hills, and valleys known as the Appalachians. Emerging from the outwash of the coastal plain in central Alabama, they run in a northeasterly direction over 1,500 miles before disappearing into the Atlantic Ocean, off Newfoundland. The south and east sides of White County, including Sautee Nacoochee, are part of Georgia's Upper Piedmont (sometimes called Inner Piedmont or Upland Piedmont), an area characterized by rolling foothills and narrow valleys and punctuated by dramatic monadnocks such as Mount Yonah, three or four miles southwest of the valleys.

Two geophysical subregions cross Sautee Nacoochee: the Hightower-Jasper Ridges District in the northwest and the Central Uplands District in the southeast. Both are notable for their linear ridge-and-valley topography oriented on a northeast-southwest axis, similar to the Valley and Ridge section of northwest Georgia but geologically unrelated. On the north side of White County is the dramatic rise of the Blue Ridge Mountains, which stretch from southern

Aerial view of northeastern White County, ranging from Unicoi Gap (*left of center at top*) to Mount Yonah (*left of center below*) and Lake Burton (*upper right*). At the center is the arc of valleys along the Chattahoochee River and Sautee Creek. U.S. Geological Survey, "National Map Viewer."

Pennsylvania to northern Georgia. The term "Blue Ridge" is typically reserved for the eastern or front range of the Appalachians, but geologically the Blue Ridge also encompasses the Great Smoky Mountains, Brushy Mountains, and the rest of the Unaka Range in North Carolina and Tennessee. In addition to dramatic scenery and waterfalls, the Blue Ridge is noted for having the highest peaks in eastern North America, including Brasstown Bald, located twenty miles northwest of the valleys and, at 4,784 feet above sea level, the highest peak in Georgia.

Geomorphology

All natural landscapes are initially interpreted and described using the terms of "geomorphology" and "orogeny." The former refers to the physical and geological topography of an area, and the latter refers to the dynamic geological processes that lead to mountains and valleys. The 4.55 billion years in which the earth is thought to have existed present a time scale unimaginably long, a scale that, if correlated to a twenty-four-hour day, would find the entire history of humankind beginning four seconds before midnight. As author John McPhee marveled, "If geologic time could somehow be seen in the perspective of human time, . . . [the] sea level would be rising and falling hundreds of feet, ice would come pouring over continents and as quickly go away. . . . Continents would crawl like amoebae, rivers would arrive and disappear like rainstreaks down an umbrella, lakes would go away like puddles after rain, and volcanoes would light the earth as if it were a garden full of fireflies."[1]

Modern geology identifies twenty or so tectonic plates that make up the earth's lithosphere or crust. Generally around sixty miles deep but varying greatly in size, the plates drift slowly and erratically across the hot, viscous asthenosphere of the earth's mantle. Most of the world's mountains, including the Blue Ridge, were built up through contact between these plates. The geology of the Piedmont and Blue Ridge bears the deformed imprint of each of the series of great transformations, known as orogenies, that brought the Appalachians into existence.

Geologists began surveying, recording, and trying to understand the Appalachians in the first decades of the nineteenth century, at first to identify suitable lands for agriculture and mining, later surveying for roads and canals. Creation of the U.S. Geological Survey in 1879 formalized and legitimized the tremendous field effort necessary in the days before satellites and lasers to accurately delineate what lies beneath our feet. The agency produced the first generally accurate maps of northern Georgia, including Sautee Nacoochee, in the 1880s.[2]

OROGENIES AND GOLD

While there have been numerous orogenic events in the earth's history, a series that began nearly a half billion years ago was the genesis of the Southern Appalachians we see today. The force of that movement sent the North American plate dipping beneath the African plate and created a new subduction zone, where the earth's crust was recycling itself back into the mantle. Along that zone a chain of volcanic islands erupted that precipitated northern Georgia's most spectacular resource: gold. Through a complex reaction in hot springs associated with the volcanoes in deep water, deposits of gold-rich quartzite were covered with sand and clay as the volcanoes eroded away.

Successive waves of volcanic islands came and went over tens of millions of years, until all were swept up against the Georgia coast as a combination of continents formed the supercontinent Pangaea, which concluded with the Alleghenian Orogeny around three hundred million years ago. Buried seven miles deep, the gold-bearing quartzite underwent metamorphosis, and the gold again precipitated into hot water with a high silica content that circulated up through faults to leave behind rich, gold-bearing quartz veins. Erosion and uplift brought these near to the surface in time for nineteenth-century miners and geologists to discover them.[3]

As a result of all this motion over hundreds of millions of years, the stratigraphy or layering of the Appalachians has none of the order found at the Grand Canyon, for example; the rocks of the Blue Ridge and the Piedmont were actually formed several hundred miles east of where they lie today. Instead, there is left what McPhee has described as "a compressed, chaotic, ropy enigma four thousand kilometres from end to apparent end, full of overturned strata and recycled rock, of steep faults and horizontal thrust sheets, of folds so tight that what had once stretched twenty miles might now fit into five."[4]

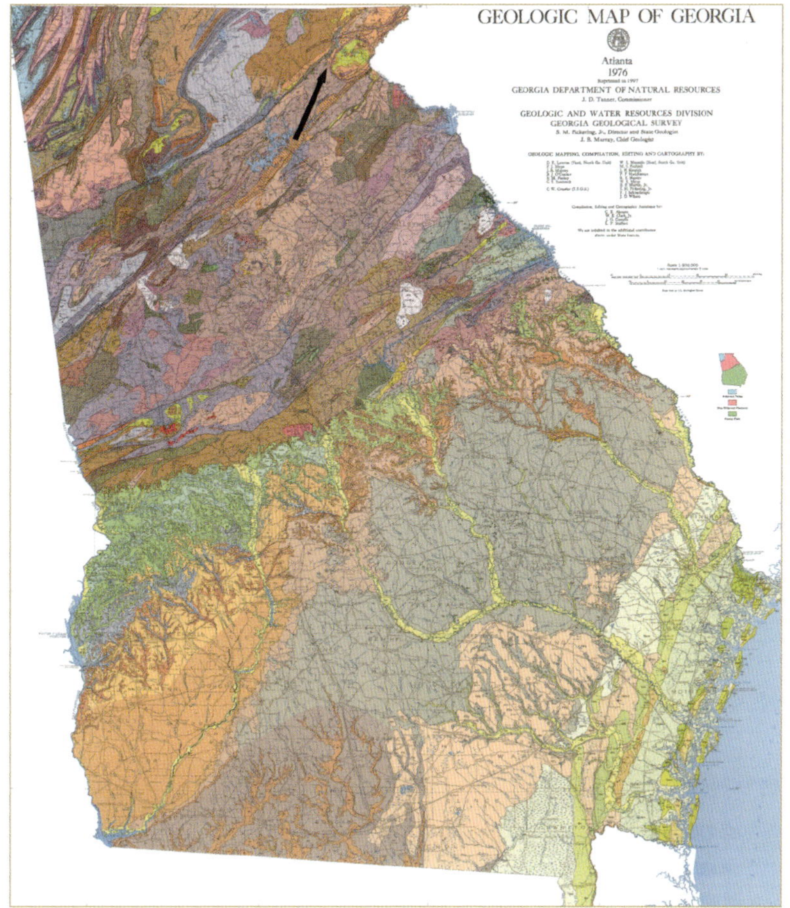

Pickering and Murray's *Geologic Map of Georgia*, 1950, annotated with an arrow to locate Sautee Nacoochee.

Modern geologists find the boundaries between "tectonostratigraphic terranes," or simply terranes, the most significant boundaries in Georgia (and many other places); each has a different stratigraphy, structure, and geological history. Each tells a different story of the evolution of our landscape. Terrane boundaries are generally marked by major fault lines, some of which may represent former plate boundaries.[5] Most of Sautee Nacoochee is part of the Tugaloo Terrane, a large area underlain by metasedimentary rocks, created out of layers of sediment that have undergone metamorphosis under great heat and pressure. This terrane, which encompasses much of the Georgia Piedmont, includes large intrusions of igneous rock—that is, rock formed by

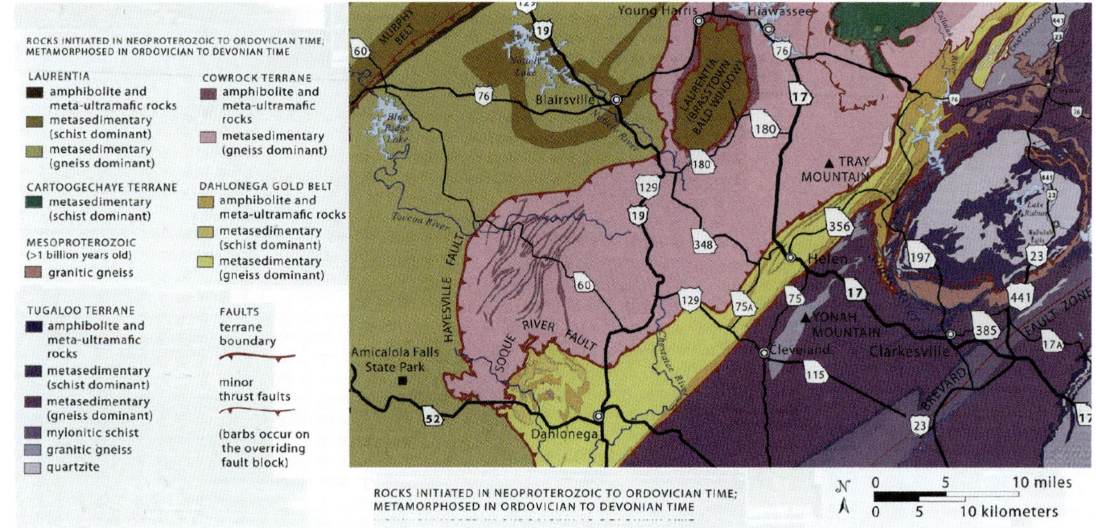

Author's composite of Gore and Witherspoon maps, illustrating the terranes of Sautee Nacoochee and its environs. Shades of purple represent the Tugaloo Terrane; yellow represents the Dahlonega Gold Belt. Gore and Witherspoon, *Roadside Geology*, 218, 224, 228.

magma or lava hundreds of millions of years ago. One of those intrusions is the granitic gneiss that forms Yonah, Sal, and Lynch Mountains to the south and east of the valleys.

In northeastern Georgia the Tugaloo Terrane is thrust over the Dahlonega Gold Belt, a narrow terrane of metasedimentary and metavolcanic rock laced with gold-bearing veins of quartz. The Gold Belt is only three or four miles wide crossing White County but widens west of Dahlonega, as it continues southwest across the Georgia Piedmont. In the northeastern part of the county, the routes of State Route (SR) 356 and 75 Alt form a sort of spine down the wider belt.

CHATTAHOOCHEE FAULT

The Chattahoochee Fault marks the boundary between the Tugaloo Terrane and the Gold Belt. Dukes Creek crosses the fault a mile or so northwest of SR 75, and the intersection of that highway and SR 17 at the western end of Nacoochee Valley is just a few hundred yards away from the fault. It crosses Bean Creek Road and the eponymous creek north of the Bean Creek church. Along the zone of this fault as it runs through northeastern White County were some of the most lucrative gold mines in northeastern Georgia.[6]

BREVARD FAULT ZONE

Early in the study of plate tectonics, it was thought that a single collision of tectonic plates created the Appalachian chain, and that the "suture" for the collision was the Brevard "Fault," a hundred-mile-long band of small, parallel faults that follows the same northeasterly trend as the mountains. The fault passes ten or twelve miles southeast of Sautee Nacoochee and is often described as the boundary between the Upper Piedmont and the Blue Ridge (which is not precisely accurate). Visible in topographical maps, the zone is most notable for the sharp turn it forces on the Chattahoochee River in southern White County and its effect on the river's subsequent course across the Piedmont.[7]

A much more complex chronology has been developed as new data has accumulated. The fault is no suture from a continental collision, which is not to say such a collision did not occur. Additional investigation has revealed that the underlying basement rock is the same on both sides of (and under) the Brevard Fault Zone. As a result, the zone is now interpreted as an instance of thrust faulting that was probably created in two stages, beginning with the area on the southeast side of the zone slipping a few dozen miles to the southwest, followed by that same area thrusting to the northwest over the fault zone as the result of another tectonic collision.[8]

Soils, Hydrology, and Creation of the Valleys

As fast as they are built up, mountains are torn down. Physical and chemical weathering combined with gravity and thousands of years act as powerful agents of change. Through those processes can be described the genesis of the Sautee and Nacoochee Valleys.

SOILS

Rock is broken down by chemical and physical weathering into smaller particles that transform it into sand, silt, and clay. With the addition of organic matter, soil is created. Over thirty distinct soil types have been identified and mapped in and around Sautee Nacoochee, mostly loams of one sort or another. According to a 2015 soil survey, the predominant soil type, found on mountainsides with slopes up to 25 percent, is Hayesville sandy loam. On steeper slopes, where much of the finer sands and clay have washed away, Ashe and Edneyville stony loam predominates. Both soils are classified as residuum, meaning they are created in situ as the underlying granite, gneiss, and mica schist weather and erode. Some of the sandy loams may lie as many as eighty inches deep over the underlying rock, while as little as twenty inches of the stony loams cover steeper slopes.[9]

On the valley floors, the soils are alluvial, meaning they are composed of material brought down from the mountains and deposited by the streams that flow through and occasionally flood the valleys. In these areas the sand, silt, and clay combine in roughly equal proportions with humus to form a variety of loamy soils. Over a quarter of the soil in the valleys is classified as Congaree and Starr soils, deep, well-drained loams, perfectly suited to agriculture. Another third consists of Starr fine and Toccoa sandy loams, also good for agriculture, and about 15 percent is Masada fine sandy loam over a heavy clay base. Finally, 15 percent is Cartecay Complex soil, a rich sandy loam six to eighteen inches above the water table and poorly drained. In most of the valleys, the water table ranges from thirty to more than eighty inches below the surface.

RIVERS AND STREAMS

The primary agent of topographical change is water, a necessary component of most weathering processes and, in the case of running water, the most powerful nongeological force shaping the landscape. Tallulah Gorge is one of the more spectacular local examples of what running water can accomplish, given a few eons. In addition to water's ability to chemically dissolve rock, flowing water carries small particles in suspension that abrade and erode and, given enough volume and speed, especially when the stream is in flood, can move even small boulders downstream. The steep slope of many mountain streams ensures fast-moving water that carves a deep, V-shaped valley. As the slope decreases and water flow begins to slow, larger rocks and pebbles settle out first, followed by sand, silt, and clay as the water slows further still. Through these processes mountains are brought down and valleys built up.

The Chattahoochee River is the largest stream in the area, but the watersheds of several of its larger tributaries ring the valleys and have helped to shape the landscape, most notably Sautee, Chickamauga, Bean, Dukes, Smith, and Spoilcane Creeks. The Chattahoochee originates ten or twelve miles northwest of Sautee Nacoochee in springs on the south side of Jacks Knob (elevation 3,779 feet above sea level), only a few yards south of the Tennessee River Divide. With an elevation of around 3,500 feet at its source, the river drops over 2,000 feet to the floor of Nacoochee Valley, picking up water along the way from Spoilcane Creek, Smith Creek, and over a dozen smaller streams before it reaches the valley.

From its source the river flows southeasterly, but when it reaches Nacoochee Valley, it takes a quick turn to the east as it negotiates the north flank of Sal Mountain. Less than three miles further on, after an equally quick turn to the south, it exits the eastern end of the valley between Sal and Lynch Mountains and winds its way another fifteen miles to the southeast before turning southwesterly into Hall County. About three-quarters of a mile into Nacoochee Valley, the river is joined by Dukes Creek, a large stream renowned for its waterfalls—and its gold. The creek originates eight or nine miles northwest of Sautee Nacoochee in springs that are around 3,200 feet above sea level and also quite near the Tennessee River Divide. Dukes Creek flows generally eastward until its course, like that of the river, is blocked by the granite of Sal Mountain and makes a sharp turn north for the final half mile of its run to the river.

A half mile or so east of the confluence of Dukes Creek and the Chattahoochee River, Sautee Creek flows into the river from the north. Its sources are springs on the southeast flank of Mack Mountain (elevation 2,260 feet), five miles northeast of Sautee Nacoochee, although the actual course of the creek is nearly twice that long. Its main tributary is Chickamauga Creek, which drains an area stretching north of SR 356 and enters Sautee Creek as the latter rounds the northwest flank of Lynch Mountain. A half mile or so downstream from the mouth of Chickamauga Creek, Bean Creek and Ben Creek join Sautee, which continues another half mile before cutting through the western end of Lynch Mountain and joining the river.

The metamorphic rock underlying much of the Blue Ridge and Upper Piedmont is not easily eroded, and rivers and streams are typically narrow and, in the steeper terrain of the Blue Ridge, fast flowing. The size and location of valleys depends, of course, on the underlying geology. The exceptionally rich bottomlands began, the state geologists wrote, with "a flooded condition of the streams during a period of subsidence" millions of years ago.[10]

At a few locations, a peculiarity of the bedrock forces the river or stream into a narrow channel, and, in times of flood, flow can be severely restricted. As the water slowly drains, it leaves behind deposits of sand, silt, and clay that, over millions of years, fill deep canyons to form the rich alluvial soil that has attracted animals and humans for tens of thousands of years. The rocky channel through which the Chattahoochee River flows at Nora Mill slows floods and has led to formation of the small valley around Helen. Similarly, the sharp turn of Dukes Creek a half mile from its mouth helped form the fertile valley floor along that creek, across which SR 75 runs, south of Nacoochee. When in flood, the combined flow of Chickamauga and Sautee Creeks is slowed by passage through the western end of Lynch Mountain. The more open topography at the confluence of the two creeks allowed formation of the broad, irregular Sautee Valley, two and a half miles long and a half mile wide.

Nacoochee Valley was created behind the river's sharp turn, where it is joined by Sautee Creek as they pass between the granite of Sal and Lynch Mountains. It is also around two and a half miles long, but the terrain has produced a more uniformly rectilinear floodplain that, in contrast with the sharp rise of the hills immediately flanking the valley on all sides, is part of Nacoochee's special character.

Climate Change

Ocean-floor coring has shown that, over the last 2.5 million years, the earth has undergone fifty periods of glacial expansion that covered large swaths of North America and northern Europe with ice sheets up to two miles thick. A dozen such periods have occurred over the last hundred thousand years as the ice has retreated and advanced, with accompanying climatic variation.[11]

The Wisconsin Glaciation, the most recent glacial period in North America, began some eighty-five thousand years ago and reached its maximum around twenty-five thousand years ago. A massive ice sheet covered most of Canada and pushed southward to envelope what is now the United States as far south as New York City, central Pennsylvania, and the Ohio and Missouri Rivers. These glaciers were part of a global ice age that locked up so much water that sea levels dropped as much as four hundred feet below what they are today. Much of the continental shelf was exposed, creating huge expanses of new habitat for flora, fauna, and people. Temperatures worldwide averaged eleven degrees cooler than today, and there was a corresponding shift to a much drier climate. Sautee Nacoochee was never glaciated, but higher elevations in the Southern Appalachians were subject to near-glacial conditions and areas of permafrost during the depths of the ice age. Twenty thousand years ago, the local climate would have been more like that of southern Canada today, with average temperatures at least ten degrees colder and as little as half as much rainfall.[12]

By about 16,000 years ago, the ice sheet had begun a rapid retreat, interrupted only by a period between 12,900 and 11,700 years ago, when global temperatures cooled enough to allow glaciers to advance for a few centuries. Then, in the space of a few decades, the climate warmed, ushering in the present interglacial Holocene Epoch.[13] Today the climate of Sautee Nacoochee is warmer than it has been for the past thousand years. The average annual temperature is now around fifty-six degrees Fahrenheit, with an average winter temperature of forty-five and summer of seventy. An average of seventy inches of rain falls each year, along with an average of two or three inches of snow. The average date of the first frost is 20 October and of the last freeze 12 April.

FLORA

Since the Southern Appalachians were not glaciated, it has been suggested that the mountains helped create a "thermal enclave" in extreme southeastern North America, allowing a number of species to continue to thrive, especially in the Coastal Plain and the Florida peninsula. In spite of any thermal enclave, the top of Tray Mountain and other mountaintops higher than about four thousand feet may have been above the tree line and had a landscape of alpine tundra during the depths of the Wisconsin Glaciation, which began some seventy-five thousand years ago. At lower elevations, including much of the Upper Piedmont of Georgia, open coniferous forests of pine, spruce, and larch were typical. Oak, poplar, hickory, and other species adapted to a warmer climate were driven out entirely.[14]

By about eighteen thousand years ago, the glaciers were in retreat, and plant habitats slowly shifted location and composition, eventually leading to a dramatically different landscape. The spruce and pine forests were slowly replaced by northern hardwoods, such as aspen, birch, beech, maple, and elm, and by about ten thousand or eleven thousand years ago, most of the modern plant communities had taken hold. Some of the northern species survived in cooler microclimates at higher elevations and on north-facing slopes, but the southern red oak, poplar, hickory, pine, and, somewhat later, chestnut dominated the forest.[15]

Mature trees often reached a height of two hundred feet, towering over an understory of shade-tolerant species such as redbud, dogwood, rhododendron, mountain laurel, and an even lower understory of smaller shrubs and groundcovers. The great naturalist William Bartram was stunned by the primeval forest he found when he traveled through northern Georgia in the early 1770s:

> Leaving the pleasant town of Wrightsborough, we continued eight or nine miles through a fertile plain and high forest, to the north branch of the Little River, being the largest of the two, crossing which, we entered an extensive fertile plain, bordering on the river, and shaded by trees of vast growth, which at once spoke of its fertility. Continuing some time through these shady groves, the scene opens, and discloses to view the most magnificent forest I had ever seen. We rise gradually a sloping bank of twenty or thirty feet in elevation, and immediately entered this sublime forest; the ground is a perfectly level green plain, thinly planted by nature with the most stately forest trees.

Describing enormous black oak, poplar, black walnut, sycamore, shell bark hickory, beech, elm, and sweet gum, Bartram continued,

> To keep within the bounds of truth and reality, in describing the magnitude and grandeur of these trees, would, I fear, fail of credibility; yet, I think I can assert, that many of the black oaks measured, eight, nine, ten, and eleven feet in diameter five feet above the ground, as we measured several that were above thirty feet girt, and from whence they ascend perfectly strait, with a gradual taper, forty or fifty feet to the limbs; but, below five or six feet, these trunks would measure a third more in circumference, on account of the projecting jambs, or supports, which are more or less, according to the number of horizontal roots, that they arise from.[16]

While many parts of the landscape had been manipulated by humans by the time Bartram saw it, there is no doubt that the forests of eastern North America, when first seen by Europeans, were simply spectacular.

FAUNA

Equally spectacular were the now-extinct megafauna (animals weighing over one hundred pounds) that dominated the animal world in the Late Pleistocene. Mastodon, mammoth, bison, and horses, as well as giant species of sloth, tortoise, beaver, peccary, and armadillo, roamed southeastern North America. Preying on them were several large carnivores that included the short-faced bear, the dire wolf, and the saber-toothed tiger. In 1971 fossilized remains of mastodon, mammoth, bison, deer, and several other smaller species were found in Wilkes County at the only Pleistocene fossil site yet located in the Georgia Piedmont.[17]

The largest and best-known of these animals were the mammoths and mastodons. Both were herbivores, but the mammoths evolved as grazers in open grasslands and were more common on the Piedmont and Coastal Plain. In contrast, mastodons were browsers who favored wooded areas such as the ice-age spruce forests and bogs of eastern North America, which at times included northeastern Georgia. Mastodon and buffalo were herding animals constantly on the move in search of fresh forage, salt, and water, and routes of seasonal migration could be hundreds of miles long. They moved single file through the dense forests of the Southern Appalachians and over eons wore trails that would be followed by the first humans and become the foundation of many roads we still travel today, including, most likely, Unicoi Road through Sautee Nacoochee.

Extinction Events

The biosphere has undergone six major mass extinctions over the history of life on earth, with the Quaternary Extinction the most recent. Occurring at the end of the last ice age, it may have taken place over the course of just a few thousand years. Most of the large megafauna disappeared, leaving smaller species relatively unscathed. There is widespread disagreement about the causes of the extinction. Many recent arguments hold that humans, who were populating North America during this same period, hunted them to extinction. That seems unlikely, given they were hunting with stone-tipped spears, but the entry of humans certainly upset the ecological balance, so much so perhaps that species died off as a secondary effect of the presence of humans. The climate change that accompanied the melting of the ice sheets has also been blamed for the extinction, but it is probable that there was no one single cause.

Except for asteroid impact or other such planetary catastrophe, climate only slowly changes habitats enough to result in migration or extinction. Such natural changes may occur within a thousand or ten thousand years, but often much longer periods are involved. Not so with human activity, which can wipe out a species in a few decades and has been responsible for the ongoing Holocene Extinction. The flora and fauna of the Southern Appalachians have suffered tremendous damage, especially since the late eighteenth century.

Hernando de Soto's pigs began the destruction in the sixteenth century. Brought to America for food and breeding, some were gifted to tribes while others escaped, proliferated, and quickly evolved into fearsome wild hogs. Being omnivorous, hogs are enormously destructive of the landscape, and they remain a tremendous problem in many parts of the South, along with European wild boar introduced by sportsmen in the late nineteenth century.

European trade with the indigenous people began in the late sixteenth century, with deer pelts being the primary commodity. By the early eighteenth century, many tribes had acquired firearms, and the buffalo, beaver, and deer populations went into precipitous decline. William Bartram observed what was happening when he traveled through northeastern Georgia in the 1770s, noting that "the buffalo (Urus) once so very numerous, is not at this day to be seen in this part of the country; a few elk, and those only in the Appalachian mountains."[18]

By 1810 no bison were left east of the Mississippi, the Kentucky long rifles having done their work, and, after the Civil War, the species was intentionally slaughtered to the last few dozen as part of the nation's strategy for conquest of the indigenous people. With bounties offered for their extermination, eastern gray wolves were disappearing as well; the last one was reportedly killed around 1920. The last eastern cougar was killed in 1938.[19] In addition to mammals, many bird species have been decimated over the past two hundred years. The passenger pigeon was one of the most prolific birds in North America, traveling in such numbers that they darkened the skies. With an unfortunate proclivity for nesting in huge flocks and inexplicably remaining in place as hunters blasted away, the birds were shot mostly for sport, and the last one died in the Cincinnati Zoo in 1914. The Carolina parakeet, the Southeast's only true native parrot, was also shot for sport, but mostly for its feathers and to control the damage flocks of them could do to crops and orchards. Last sighted in the wild in 1918, they were declared extinct in 1939.

Plants, too, have gone extinct or nearly so in the Southern Appalachians, including ginseng, which the Cherokee hunted nearly to extinction after they discovered it as a commodity in the eighteenth century. Most notable is the American chestnut, which once made up as much as a quarter of the forest trees in the Southern Appalachians. In the early twentieth century, a fungus was accidentally introduced on imported wood and within a few years had decimated the chestnut. By the 1930s it was virtually extinct. Demonstrating again how all of nature is interrelated, the black bear population, for which chestnuts were the primary food, also went into serious decline and did not recover until they adapted to eating acorns and other nuts.[20]

2

First People

Archaeologists and anthropologists refer to the earliest humans in the Americas as Paleoindians, or Paleo-Indians, not to be confused with the Paleolithic people who lived in other parts of the world for hundreds of thousands of years before coming to the Americas. The first Paleoindians are thought to have reached southeastern North America thirteen or fourteen thousand years ago, several thousand years after their ancestors first migrated across the Bering land bridge between Asia and North America.

These people are thought to have existed as small bands of a few dozen individuals, subsisting as hunters and gatherers. Camping in rock shelters and caves, like those on Mount Yonah and elsewhere in northern Georgia, or temporary shelter provided by hides stretched over branches or poles, they were more or less constantly on the move in search of food and other resources. They left little evidence of their presence in Georgia beyond innumerable stone points from knives and spears.[1]

The preferred material for Paleoindian and most later stone points was chert, a fine-grained sedimentary rock notable for the way it fractures when struck hard enough. Obsidian and quartz also fracture in much the same way, and points and blades that are quite sharp can be made from all three materials. Flint is a high-quality chert that also sparks when struck against iron or iron-bearing rocks and so was used for starting fires from a very early date.

Paleoindian bands were quick to locate known deposits of chert or other rock suitable for toolmaking, and their seasonal movements often centered around those locations. Where good chert was not available, material was often acquired from some distance away. At the Rucker's Bottom archaeological site in Elbert County, as well as elsewhere across northeastern Georgia, early people used local quartz, but chert from other regions was also used, an indication of the beginnings of long-distance trade that eventually resulted in the system of trails that covered eastern North America at the time of European contact.[2]

Perhaps the best known of the Paleoindian people were those of the "Clovis Culture," so called for Clovis, New Mexico, where their first cultural artifacts were discovered in 1932. They were long thought to represent the "first" humans in North America, but it is now generally accepted that humans were present thousands of years earlier. Clovis artifacts have been found in many places, especially the characteristic bifacial, fluted projectile points that were first identified at Clovis. Subsequent research suggests that, while the Clovis complex of tools was developed after humans had gotten past the ice fields of Canada, the points themselves were based on an older, presumably Beringian, model. Regardless of origin, Clovis points were widely used and, in the words of one archaeologist, "a lethal weapon for highly mobile, big-game hunting foragers." Numerous Clovis sites have been identified throughout the mid-latitudes of North America, all dating from about 11,500 to 10,600 BCE.[3]

No major Clovis sites have been found in Georgia, and the Topper Site in Allendale County, South Carolina, on the lower Savannah River, is one of only a few clearly identifiable Paleoindian sites that have been documented anywhere in eastern North America. Numerous Clovis and other Paleoindian stone points have been found, however, including two in Nacoochee Valley, leading to the assumption that Paleoindian hunting parties passed through Sautee Nacoochee as early as 11,000 BCE. Their camps were often located on ridges overlooking game trails, such as the ancient trail that became the Unicoi Road through Nacoochee Valley.[4]

As noted in the previous chapter, the late Paleoindian Period was a time of tremendous change in North America. The forests of spruce and pine that had dominated much of northern Georgia were gradually giving way to what we would now recognize as a northern temperate hardwood forest. At the same time, the megafauna were dying out; by about 10,000 BCE, mammoth, mastodon, saber-toothed tigers, and three-quarters of the other megafaunal genera had gone extinct or, in some cases, were evolving into smaller species. Many have argued that early humans were responsible for the demise of these animals—and there is little doubt that Paleoindians hunted Pleistocene megafauna—but those species were probably already dying out before the first humans arrived.[5]

With extinction of the megafauna, Paleoindian hunting and point technology evolved accordingly, and by the advent of the Archaic Period, ten thousand years ago, the distinctive Clovis points were being replaced by smaller points that were not as finely crafted but were more quickly made and better suited to the hunting of smaller game. By the end of the Paleoindian Period, eight or nine thousand years ago, the climate was warming quickly, and many of the earliest sites of human habitation were being drowned by rising seas, which reached modern levels by about 6000 BCE. The southern temperate hardwood forests of southern oak, hickory, poplar, sweetgum, and chestnut were beginning to evolve as well, although it would be several thousand years before that transition was complete.

Archaic Period (ca. 8000 to ca. 1000 BCE)

During the Archaic Period, regional cultures began to evolve, and some patterns of hunting and gathering developed that would continue into historic times, including a reliance on deer and wild turkey. Early in the Archaic Period, people were already using plant material such as river cane (*Arundinaria gigantea*) to make baskets, mats, and finely woven fabric produced on primitive looms. The *atlatl*, which was a sort of spear thrower; bone fishhooks; awls; and a proliferation of point types, polished stone tools, and other utilitarian objects appeared during the Archaic Period. By the middle of the period, in the Piedmont and mountains of northern Georgia, there was an increasing reliance on quartz for toolmaking.[6]

There were probably hundreds of widely dispersed bands of people across the South by the middle of the Archaic Period, six or seven thousand years ago. Each band consisted of perhaps twenty-five to fifty individuals in three or four families. Archaic people were, like the Paleoindians, migratory hunters and gatherers, following the seasons and game in an area that could range over hundreds of square miles.

Typically, they maintained a base camp in the winter and a series of temporary foraging camps in spring and summer, some of which could well have existed in and around Sautee Nacoochee. While no such sites have been identified in White County, a variety of Archaic points and other tools have been reported, including *atlatl* weights, hammerstones, scrapers, and steatite bowls.[7]

Several large Archaic sites have been identified in Georgia, the closest to Sautee Nacoochee being the Rucker's Bottom site in Elbert County, now flooded by Clarks Hill Reservoir. Those sites have not been as extensively investigated as some sites in Tennessee and Alabama, but what is known about the Archaic Period in Georgia appears to comport with the rest of southeastern North America. While little more than stone tools have been found at most Georgia sites, a rock hearth, pits, and large amounts of stone-working debris have been identified at an Early Archaic site near Cartersville, along with evidence of a "tent-like structure" similar to those found in Tennessee and elsewhere. Although most Archaic dwellings were temporary and insubstantial, evidence for five dwellings, one of which might have been mud-covered, has been documented in the middle Savannah River valley. In Warren County, in the Upper Piedmont, a pit house was excavated. Measuring about twelve by fifteen feet and dug down a foot below grade, it was dated to about 1850 BCE and may have been occupied for a long period.[8]

Around 6000 BCE the climate underwent a rapid warming and, until about 3000 BCE, average temperatures worldwide remained several degrees above what they are today. Known as the Holocene Climate Optimum, the phenomenon's effects were felt most strongly in the Arctic and Siberia. In the midlatitudes the temperature may not have risen much at all, although the first hurricanes in one hundred thousand years began to roil the Caribbean during this period.[9]

The climate was significantly drier, which appears to have led to an increase in natural wildfires, particularly on the drier, leeward slopes of the mountains. The first evidence for intentional burning of local stream bottoms appears from this period, probably to attract big game and make it easier for them to be spotted from campsites on the surrounding hillsides. Whether Archaic people began routinely burning the valleys of Sautee Nacoochee at this early date may never be known, but it is likely that the valleys have been relatively open for a very long time. By the end of the Archaic Period, around 2000 BCE, when evidence for protofarming began to appear, burning of the valley floor might have been a regular, if not routine, event.[10] By about 3000 BCE climatic conditions resembled those of the historic period, and a few scattered bands of people in southern Mexico were beginning to cultivate beans, manioc, yams, and some maize, although they still remained primarily hunters and gatherers. It would be another two millennia before protofarmers would appear in southeastern North America, but late in the Archaic Period people were gathering seasonally available seeds, nuts, goosefoot, and other plants.

SOAPSTONE AND PETROGLYPHS

Steatite, or soapstone, is plentiful in northern Georgia, and significant deposits have been reported around Sautee Nacoochee.[11] From an early date, soapstone was widely used for *atlatl* weights and other objects, but around 2000 BCE people began using quartz picks and chisels to make soapstone bowls and cooking slabs. Because of soapstone's unique thermal qualities, these items were often widely traded. Artifacts from the ancient soapstone quarries at Soapstone Ridge in DeKalb County, Georgia, for example, have been found as far away as the Great Lakes and Mexico.[12]

Since it was easily worked, people also used soapstone when creating petroglyphs, which circumstantial evidence suggests began to appear across northern Georgia in the Late Archaic Period. None of them have been surely dated, but archaeologists have speculated that, since Archaic soapstone quarries have been documented, people may have started creating petroglyphs around the same time. The practice continued into the Mississippian Period, with older symbols sometimes reworked to create newer symbols. Petroglyphs typically appear alongside trails and often at changes in elevation, such as mountain gaps. Some of the best-documented petroglyphs in the vicinity of Sautee Nacoochee are eight boulders at Track Rock Gap, between Blairsville and Brasstown Bald. A significant petroglyph boulder has also been documented on U.S. Forest Service land on the south side of Hickorynut Mountain, or Ridge, northwest of Helen.[13]

At the University of Georgia is a large granite boulder covered with petroglyphs, which was reportedly taken from "the back side of Squirrel

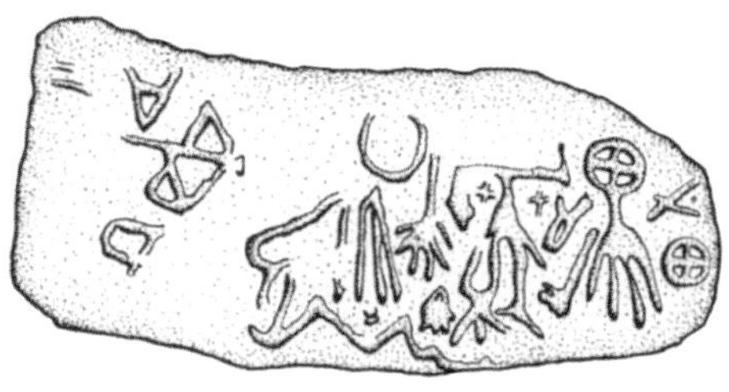

Wauchope's sketch of the petroglyph from Squirrel Mountain in northeastern White County. Wauchope, "Archaeological Survey," 366.

Mountain." No mountain of that name has been identified in maps or other documentation, and the exact location of the boulder was not identified by the Works Progress Administration–funded archaeological survey of North Georgia in the late 1930s. The large boulder came from Forsyth County and was given to the University of Georgia by H. R. Vaughn of Athens in 1963. The meaning and significance of these and most other petroglyphs remain uncertain and were so even to the Cherokee, whose ancestors may have created them. Recent speculation to the contrary, they are almost certainly not evidence for a lost Creek alphabet or an astronomical guide to the heavens. They may simply be ancient graffiti.[14]

LATE ARCHAIC INNOVATION

Around 2500 BCE, along the lower Savannah River or the Saint Johns River in Florida, people produced the first pottery in southeastern North America and some of the first on the continent. Fiber-tempered, the pottery is known as Stallings Island pottery, after the island archaeological site in the Savannah River upstream from Augusta, where it was first discovered. Pottery was used for storage of seeds and nuts but, in and of itself, is not necessarily a sign of agriculture. Pottery allowed for boiling of water and better cooking, which would have reduced food-borne illnesses and parasites, but for unknown reasons pottery was not adopted by the people of the Upper Piedmont, including northeastern Georgia, until much later. The abundance of soapstone—and habit—may have delayed the adoption of a pottery tradition.[15]

In the Late Archaic, purely decorative objects such as pendants and beads began to appear as well. Hafted stone adzes also came into use and made possible the construction of larger, more substantial buildings. They could also be used in making dugout canoes, such as the one found by Delbert Greear in the Chattahoochee River near the mouth of Mauldin Mill Creek in 1974, but most rivers in northeastern Georgia were too rocky and too shallow for much of the year for canoes to be very useful.[16]

The first signs of agriculture in Georgia appear toward the end of the Archaic Period, but plants remained a small part of the typical diet. With many more people competing for resources, the small roving bands of hunter-gatherers began to coalesce into larger groups that could better protect and control resources. Burial practices became more complex, and, very late in the period, the first earthen funerary mounds began to appear at scattered locations.[17]

TRAILS

When the first humans arrived toward the end of the last ice age, they naturally followed the ancient animal tracks that had been worn over eons of use by Pleistocene megafauna. These trails "followed the lines of least resistance," one archaeologist noted, avoiding rough or swampy terrain and dense undergrowth, such as laurel thickets and cane brakes, and often following low ridgelines.[18]

The Late Archaic saw a significant increase in trade in items like soapstone utensils over the trade routes that began to develop during that period. Connecting tribes and regions all across eastern North America, these trails were often used for travel over quite long distances, whether in trade or war or simply to maintain contact with distant family members. These early trails were usually less than two feet wide, as travelers, who would not have horses until after 1700 CE, wound their way single file through the area's hilly and heavily wooded landscape.[19]

The routes of trails were often dictated by the location of convenient river crossings, but nearly all the streams in northeastern Georgia are small enough that the water in and of itself did not usually impede travel, as attested by complaints during the historic period that the Unicoi Trail forded the Chattahoochee River a dozen times in what is now White County. With the advent of agriculture around 1000 BCE, settled villages began to appear, and these too became destinations in their own right.

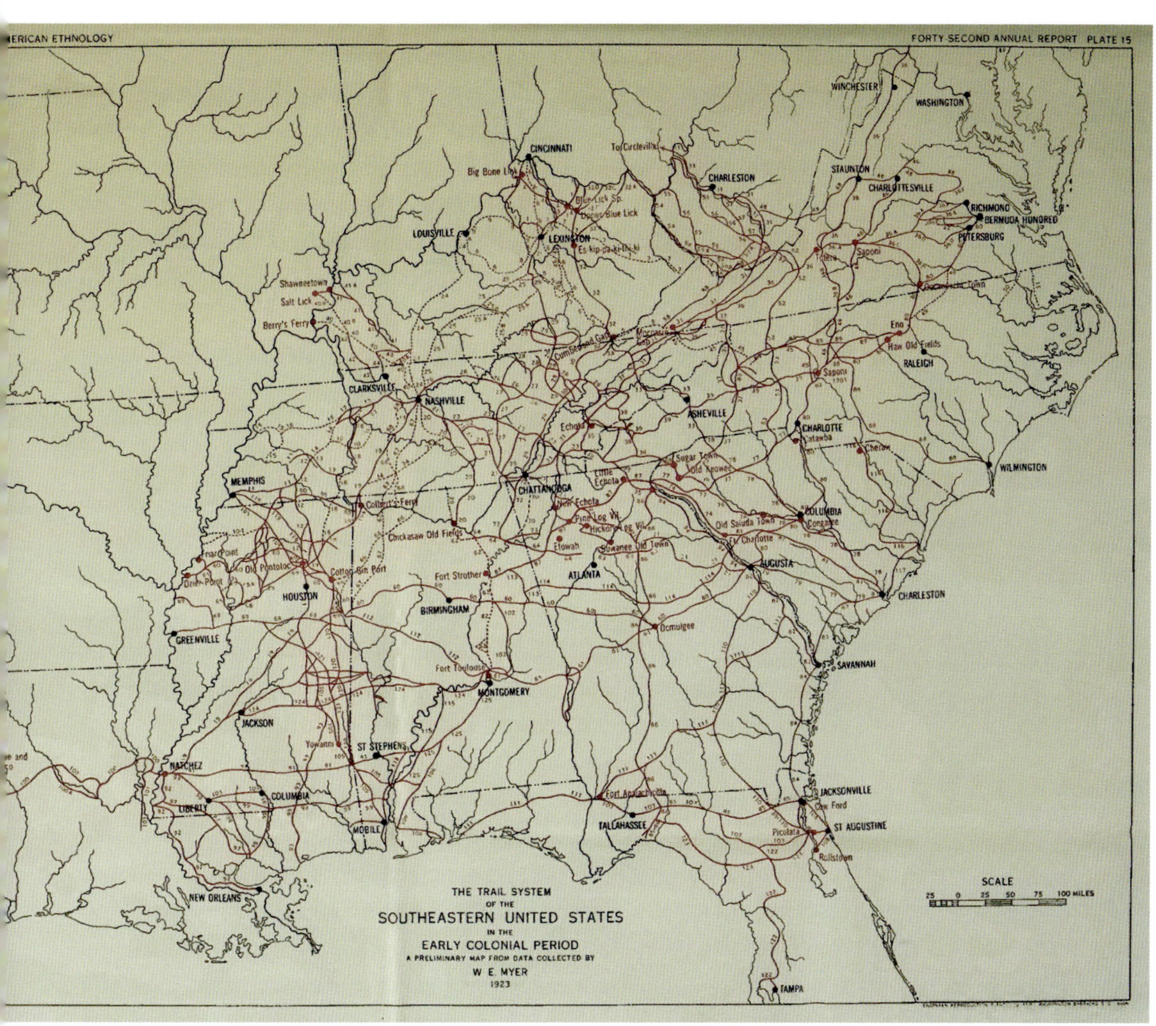

Myer's map "The Trail System of the Southeastern United States in the Early Colonial Period," 1923, showing his interpretation of the earliest trails, each of which is numbered. Myer, "Trail System," 748, plate 15.

The shoals of the Savannah River, at the Fall Line and present-day Augusta, were the site of some of the earliest, more or less permanent human occupation documented in the region, beginning at least four thousand years ago. A series of important, long-distance trails radiated from the shoals and to the west forked and forked again into paths leading to Alabama and Tennessee. Upstream, just at the confluence of the Tugaloo and Chattooga Rivers, which are the headwaters of the Savannah River, were easier fords for trails leading northwest to the Tennessee River Valley and to the west and southwest along the Eastern Continental Divide and the Chattahoochee River to the Gulf Coast.[20]

Woodland Period (ca. 1000 BCE to ca. 1000 CE)

The Woodland Period saw the continuation of trends first evident in the Late Archaic Period, which had included the use of pottery, the beginnings of rudimentary agriculture, and the construction of funerary mounds. Although there was continuity between the Archaic and Woodland Periods, Charles Hudson (1932–2013), one of the foremost authorities on the indigenous peoples of the Southeast, has described the Woodland as representing "a change in ideology and subsistence," which resulted in "the most distinctive, the most completely indigenous culture ever to exist in eastern North America."[21]

POTTERY

Grit- and sand-tempered pottery replaced the fiber-tempered pottery of the Archaic Period, and its use became much more widespread, in some places even before the advent of agriculture. There were numerous local variations in form and design and a constant evolution of style. In northern Georgia the earliest Woodland pottery type, and one documented by archaeologist Robert Wauchope at Nacoochee, is known as Dunlap Fabric Marked, so named for the Dunlap Mound at Ocmulgee National Monument, where it was first identified, and for the way pieces were decorated using fabric-wrapped paddles.

By the beginning of the first millennium CE, production of Cartersville Check Stamped pottery flourished in North Georgia, followed over the next few centuries by Swift Creek Complicated Stamped, as the

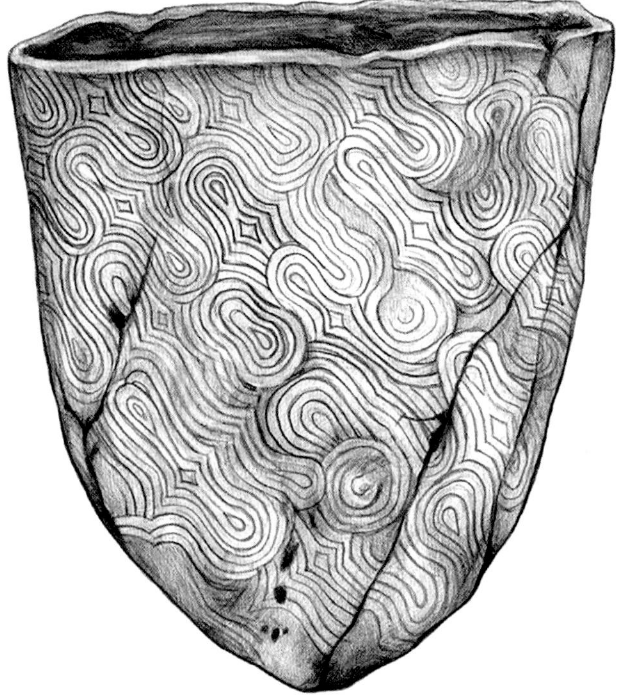

A pottery type common during the Woodland Period across northern Georgia. "Swift Creek Complicated Stamped," Ceramic Type Collection, Laboratory of Archaeology, University of Georgia, Athens. Illustration by Moss Joslin.

influence of the great Hopewell tradition was felt. Originating in the Ohio River valley, that culture is perhaps best known for the Serpent Mound and other effigy mounds in Ohio. In Georgia, Alabama, and South Carolina, the Swift Creek culture is considered to have been part of the Hopewell "interchange," connected by long-distance trade routes and sharing similar beliefs and traditions.

Archaeologists have established many other classifications to help make sense of and establish a chronology for the tremendous increase in cultural artifacts throughout this and later periods. Examples of most of the types common across North Georgia have been documented in and around Sautee Nacoochee, especially during the course of the excavation of Nacoochee Mound in 1915.[22]

MOUNDS AND OTHER BUILDINGS

By about the first century CE, funerary mounds were being built to cover cremated remains and bones all over southeastern North America.

Some of the remains were interred in log or stone tombs. Effigy mounds also began to appear, some of them, like Rock Eagle in Putnam County, Georgia, made entirely of thousands of stones. A variety of stone walls and piles found across northern Georgia likely dates to the Woodland Period as well, although some may have had their origin very late in the Archaic Period. Associated but perhaps not contemporaneous with the petroglyphs at Track Rock Gap are stone walls and stone piles similar to those that have been documented at Mount Yonah, Alec Mountain, Fort Mountain, Stone Mountain, and elsewhere across northern Georgia. Although the Cherokee and other historic tribes continued to create memorial or directional stone piles until they were driven from the state in the nineteenth century, none of the historic tribes claimed responsibility for the ancient walls or for the petroglyphs.[23]

According to James Mooney's great history of the Cherokee, they had several explanations for the origins of the petroglyphs, "the more sensible Indians saying that they were made by hunters for their own amusement while resting in the gap." An ancient tradition also held that they were created at the dawn of time, by birds and animals fleeing some unspecified danger and leaving their tracks in the newly created earth. Mooney reported, too, on "a hole" near Blood Mountain that was like "a small well or chimney, in the ground, from which there came up a warm vapor that heated all the air around." The Cherokee believed that this was because the Immortals had "a townhouse and fire under the mountain."[24]

There was a trend toward more permanent residences, and by late in the period there were hundreds of more or less permanent villages across the Piedmont and Blue Ridge. At least three village sites dating to the Woodland Period have been documented at Sautee Nacoochee, but it is not clear if they were occupied simultaneously or consecutively. Settlements were typically smaller in the Piedmont and mountains than on the coast, and, at least early in the period, they tended to be temporary hunting and gathering stations.

Houses and other structures were round, with wattle-and-daub walls, thatched roofs, and a firepit in the middle of an earthen floor. Built with local materials, they were often located in river floodplains, perhaps because that is where a lot of the edible plants grew. Numerous cylindrical or bell-shaped storage pits have also been documented across North Georgia, but they appear to have been used for the storage of acorns, walnuts, and chestnuts, since there were no cultivated crops until after about 1000 BCE.[25]

THE BOW AND ARROW

For thousands of years, hunters all over the world relied on darts, spears, and *atlatls* to take down their prey. These weapons had the distinct disadvantage of being less than effective at ranges over about twenty-five yards. Documenting the appearance of the bow and arrow has always been hampered by the fact that their organic elements have seldom been preserved, forcing paleontologists and archaeologists to rely on changes in the size and design of points in determining when the new technology was adopted in any given place.

The oldest evidence for the existence of bow-and-arrow technology is in Africa and dates to around 11,000 BCE, but by about 5000 BCE the bow and arrow were found all over Eurasia. The earliest evidence in North America is in the Arctic, dating to about 3000 BCE, but none have been found south of the boreal forests from before the first century CE. By 500 CE, bows and arrows were being used from the Plains westward, and over the next two centuries their use spread across eastern North America.

Around 700 CE, across the Midwest and Southeast, there was a "sudden and dramatic" increase in the incidence of the smaller, specialized points generally associated with the adoption of bow-and-arrow technology, which increased the efficiency of hunting exponentially. By about 800 CE the bow and arrow were in use all across North America, leaving the *atlatl* to disappear from the archaeological record.[26]

PROTOFARMING

Agriculture brought about a revolution in the way people lived, even as traditional hunting and gathering continued. It helped ensure a steady supply of food, and with that came permanent settlements, significant population growth, and an increasingly complex and hierarchical society. The original "neolithic revolution" took place in the Middle East around the eleventh millennium BCE, but similar transitions to agriculture occurred independently in several locations, including the Americas, over several thousand years.

Agriculture in the southern Piedmont had its origins in the Late Archaic, when people began to gather and eat the seeds of certain native plants, including sunflower and sumpweed, both of which may have been cultivated at a very early date. By about 1000 BCE gourd and summer and winter squash appeared in the Lower Mississippi valley. While it has long been assumed that these plants were domesticated in Mexico or Central America, some have argued that they were domesticated independently in southeastern North America. Gourds were particularly useful in dealing with the perennial problem of containers for water and storage needs in general, but they were also used for ornaments and rattles.[27]

About 1000 BCE people began cultivating chenopodium, pigweed, knotweed, giant ragweed, and maygrass, cultivation that might have started by simply weeding out competing species. Over time people began saving and replanting seeds, and conventional agriculture was underway. Around the second century BCE, tropical "flint corn" was introduced but was grown for only a few centuries before disappearing, perhaps because of a cooling climate. Not until after about 700 CE was maize, commonly called corn in North America and Australia, widely grown again in the southern Piedmont. Hunting, fishing, and gathering remained a major component of people's lives through much of the Woodland Period, and it would be important to all the indigenous people well into the historic period.[28]

Mississippian Period (ca. 1000 CE to ca. 1550 CE)

Eleven or twelve hundred years ago, in the middle Mississippi River valley, many of the trends that had developed in the Woodland Period culminated in the Mississippian tradition, which the acclaimed anthropologist Charles Hudson called, "the highest cultural achievement in the Southeast, and for that matter all of North America." It was not, he believed, the result of a Mesoamerican intrusion into eastern North America, as some have argued, an idea that he thought was simply "no longer acceptable."[29]

Earlier definitions of "Mississippian" were rather narrow, focusing on the presence of shell-tempered pottery, rectangular houses, and flat-topped pyramidal mounds. More recently, Hudson has broadened the definition to include "those societies that practiced cleared-field agriculture with maize as the dominant crop, that had hierarchical political organizations with evidence of ascriptive status differentiation, and that shared a set of cult institutions marked by a consistent iconographic complex and pyramidal earthen mounds."[30] This transition happened very unevenly across space and time. It is thought that the people who lived in the mountains of the Southern Appalachians, perhaps including Sautee Nacoochee, made the leap from Middle Woodland to Mississippian in a very short period. Nevertheless, Woodland and even Archaic traditions continued well into the historic period in many locations.[31]

MAIZE, BEANS, AND SQUASH

The foundation of the diet and culture of indigenous people in the centuries before European contact was the cultivation of "the three sisters": maize, beans, and squash, which were almost always planted together. The beans could be supported by the stalks of maize while putting nitrogen back in the soil, and the squash shaded out weeds and helped retain moisture. Significantly, too, the beans and maize, when consumed together, provided a more balanced protein than either could provide on its own. The transition to maize had a significant effect on the environment, since cultivation of maize always entailed widespread clearing of fields, often using fire. The river-bottom landscape would have been, as Hudson put it, "a treeless plain, broken only by wide and imposing canebrakes and occasional fruit and nut trees."[32]

It has been estimated that, at the height of the Mississippian period, as much as an acre of land was needed for every individual in the community for the production of maize. As a result, fields often stretched for miles at places such as Etowah in northwest Georgia and Ocmulgee in Middle Georgia. Almost certainly, the Chattahoochee River valleys of Helen and Nacoochee as well as the creek bottoms of Sautee, Chickamauga, and Dukes Creek would have been similarly planted during much of the Mississippian Period.[33] New gender roles evolved along with the turn to agriculture. Women were the potters, and farming, too, came to be seen as women's work. One historian of the period notes that "female kin groups controlled the fields, the food they produced, and the houses in which those who ate it lived." Men did the hunting and did most of their work in seasonal camps; the forest was for the men, the clearing for the women.[34]

CEREMONIAL CENTERS

With plenty of food, populations increased. Hudson tells us that "towns grew up that were larger, more permanent, and more secularized than the large religious centers of the Woodland tradition." Large, flat-top mounds built as the foundation for temples, mortuaries, chiefs' houses, and other important buildings were the hallmark of the tradition. Typically facing an open plaza, the mounds were usually pyramidal, but some were circular, elliptical, or even long and ridgelike. Whatever the shape, they were the ceremonial center and were enlarged and refreshed over centuries of use.[35]

Always located in a fertile valley, typically on alluvial terraces, towns might be palisaded with vertical poles that were sometimes covered with mud on the outside. Most of the people lived within a few miles of this center, sometimes in smaller villages that could themselves be palisaded and sometimes in unfortified homesteads on the periphery.[36] Mississippian people used prodigious amounts of river cane, as did many tribal people. It was a resource that would also be hugely

Painting by John Kollock (1929–2014), *Nacoochee Mound and the Village of Chota*, as they might have looked prior to European contact. Reproduced with permission from the Kollock family.

important to European settlers as fodder for their horses and cattle. The Mississippian people used it for a variety of purposes, from making toys, whistles, blowguns, and baskets to thatching the roofs of their houses. They also used huge amounts of timber for their buildings and palisades, with one estimate that an average Mississippian village might use thirty thousand small trees over several generations.[37]

Some authorities have identified a South Appalachian Mississippian tradition in Georgia, South Carolina, and western North Carolina that developed late in the Mississippian Period. Hudson notes that "similar artifacts and a strikingly similar series of symbolic forms" developed across the region. Pottery from this period is very distinctive, with complicated stamped surface details that may reflect a local Woodland culture borrowing ideas and designs from Mississippian cultures to the west and northwest. The mounds often contained richly furnished burials that probably reflected the rise of chiefdoms, another hallmark of the Mississippian Period.[38]

CHIEFDOMS

As many as twenty chiefdoms flourished in Georgia in the late Mississippian Period, usually ruled by a group of elders but sometimes by a single chief. The local population and outlying villages might pay tribute in the form of corn, hides, meat, and perhaps prestigious trade items. The chiefdoms were hereditary, mostly through the maternal line, but with power that might stretch no more than a dozen miles in any one direction.[39] These smaller chiefdoms in turn paid tribute to one of several paramount chiefdoms that dominated southeastern North America in the Protohistoric Period. The larger paramount chiefdoms included the Powhatan in the tidewater around Chesapeake Bay, the Cofitachequi in the South Carolina Piedmont, the Guale in coastal Georgia, the Apalachee and Timucua of northern Florida and south Georgia, the Ocute in east Georgia, and the Coosa in northwest Georgia.

The relationship of the chiefdoms at Nacoochee and at Tugaloo to the paramount chiefdoms is not clear. They were certainly outside the sphere of influence of the great chiefdoms at Coosa in northwest Georgia and Cofitachequi in eastern South Carolina, both of which were visited by Hernando de Soto. Most likely, the Nacoochee chiefdom was tributary to the Ocute chiefdom in the Oconee River valley.[40] The concept of "tribes" evolved after European contact with the indigenous people, but, by late in the Mississippian Period, individuated groups were coalescing. Archaeologist Max White, seeing that both the historical record and archaeological discoveries agree, thought there was a considerable continuity from the prehistoric period into the historic period.[41]

Archaeology

In 1849 Charles Lanman (1819–95) passed through the valleys and, like many before and after, marveled at their beauty. He also noted "the artificial memorials of Nacoochee." In *Letters from the Alleghany Mountains*, he wrote, "On the southern side of the valley, and about half a mile apart, are two mounds, which are the wonder of all who see them. They are perhaps forty feet high, and similar in form to a half globe. One of them has been cultivated while the other is covered with grass and bushes, and surmounted, directly on the top, by a large pine tree. Into one of them an excavation has been made, and, as I am informed, pipes, tomahawks, and human bones were found in great numbers."[42] Archaeology is the source of most of what we know about Sautee Nacoochee before the late seventeenth century. Nacoochee Mound was partially excavated in 1915, as were a number of other sites in 1938–39. Limited archaeological investigation of a town or village site near the mound in 2004 provided additional data, but it is clear that only some of the archaeological resources in the valleys have been identified.

CHARLES COLCOCK JONES (1870)

Charles Colcock Jones Jr. (1831–93) was the author of dozens of publications, and many people consider him to have been Georgia's most important nineteenth-century historian. His interest in the history of the indigenous people was deep; by the 1860s he was a serious amateur archaeologist at a time when there were few who knew or cared about archaeology. Jones's *Antiquities of the Southern Indians, Particularly of the Georgia Tribes*, published in 1873, is considered a classic in its field and includes a lengthy report on his excavations at Nacoochee.[43]

In June 1870 James H. Nichols (1834–97), the owner of much of the western end of the Nacoochee Valley, unearthed three stone box graves a few inches below the surface and about thirty feet west of the mound. Jones's narrative does not make clear how much Nichols had dug up on his own before Jones saw the site, but it appears Jones may have been present for at least part of the excavation. The grave at the center of the three was the only one covered by slabs of stones laid over vertical slabs to form a box. In it "a male skeleton, measuring more than six feet, lay extended at full length." He was an old man, Jones wrote, noting that "the few teeth remaining in the lower jaw were much worn, and the alveolar processes had been greatly absorbed."[44] That he was able to identify the sex and age of the skeleton suggests his expertise or, perhaps, the assistance of his brother, a medical doctor who himself excavated and published papers on box graves in East Tennessee in the 1860s.

In the other two graves, which might more properly be called ossuaries, Jones found "the bones of more than one skeleton lying in disorder, and carelessly piled in without any regard to regularity." Jones includes a long discussion of the funerary customs of the indigenous people all across the Midwest and southeastern United States, but he was apparently unable to preserve any of the bones, most of which disintegrated, he wrote, "when exposed to the air." The most "remarkable" find was a copper axe some ten inches long and nearly three inches wide that had been buried with the tall man, whom Jones thought must have been a chief or other leader. The axe was finely worked, "a uniform thickness of a little less than the tenth of an inch," and weighed a little less than ten ounces. Jones noted that the copper was from the shores of Lake Superior "and that probably the implement itself was there made."[45]

Beneath the axe was a fragment of cane matting, consisting "of thin layers of split cane, about the quarter of an inch in width, interwoven at right angles with each other." Jones described how the cane was prepared for weaving and how important it was to the Cherokee. He made an interesting comparison between the fragment and Cherokee basketry, quoting James Adair's eighteenth-century description of Cherokee basketry—"the handsomest clothes baskets I ever saw"—as he noted the similar black and yellow dyes that remained in the Nacoochee fragment. Jones described "two fine specimens of the *Cassis flammea*" taken from the grave. The presence of the shells of these marine gastropods in the graves at Nacoochee is, like the copper axe, evidence of the long-distance trade common in pre-Columbian America. Jones also described in some detail a variety of other artifacts retrieved from the graves, including shell and soapstone pins, stone beads, and a variety of small copper implements. The current location of these items is not known.[46]

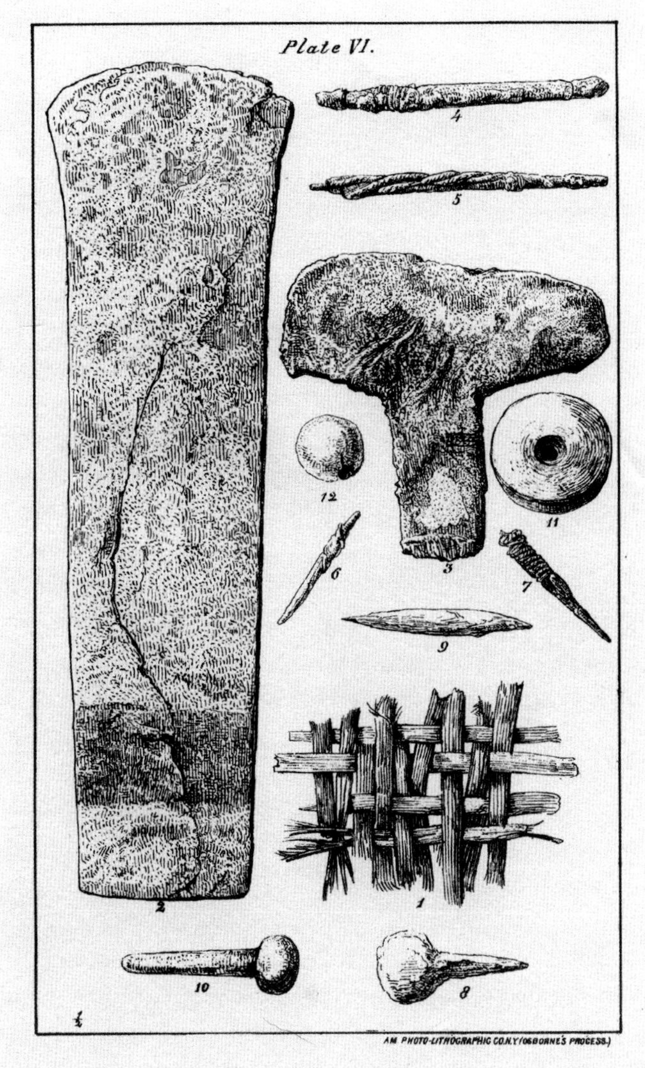

"Relics Found in Stone Graves in Nacoochee Valley," recorded by Jones in 1870, included cane matting, copper implements, shell pins, a soapstone pin, and stone beads. C. Jones, *Antiquities*, 224, plate 6.

While Jones had much to say about the graves and their contents, his description of the mound was relatively brief. He did, however, note that it rose "twenty feet or more" above the surrounding plain, three or four feet higher than it does in the twenty-first century. He also reported that "for many years, the mound's slopes and summit have been cultivated and, within the recollection of the older inhabitants, this tumulus has lost much of its original dimensions."[47]

BUREAU OF ETHNOLOGY (1891)

The U.S. Bureau of Ethnology was organized in 1879 to facilitate the transfer of records and artifacts of indigenous people from the Department of the Interior to the Smithsonian. Under the leadership of John Wesley Powell (1834–1902) from 1881 to 1894, it made enormous contributions that went far beyond records and artifacts. In 1891 Cyrus Thomas (1825–1910), a noted ethnologist and entomologist who also worked as the chief archaeologist at the bureau, published his *Catalog of Prehistoric Works East of the Rocky Mountains*. Three years later he published his landmark study on the "Mound Builders," in which he debunked the then-current notion that the mounds had been built by the Lost Tribes of Israel or some other wandering group. Many white Americans believed that the indigenous people, whose cultures they had so recently and thoroughly degraded, could not possibly have built the magnificent mounds and other earthworks in the Midwest and Southeast.

The 1891 *Catalog of Prehistoric Works* was intended to guide the production of national archaeological maps, but it was not without its errors. Thomas recognized as much in his introduction, noting that "it has been prepared to a large extent from the published notices of antiquities scattered through numerous works, periodicals, pamphlets, etc., which are often so incomplete and indefinite as to leave the precise localities uncertain." In an attempt to correct some of the more egregious errors, Thomas engaged experts to examine certain parts of the survey. For the "mountain region" of the Carolinas, East Tennessee, Georgia, and Alabama, he chose the great historian of the Cherokee, James Mooney (1861–1921).[48]

Thomas identified five sites in White County, beginning with "three mounds in Nacoochee Valley, on headwaters of Chattahoochee, on the

Nichols, Williams, and Johnson [*sic*] farms. Two of them very large and perfect."[49] He noted that they had been reported by Mooney and also gave page numbers where the features were mentioned by George White in his *Historical Collections of Georgia* and in Charles C. Jones Jr.'s *Antiquities of the Southern Indians.* He wrote of "Stone cairns on ridge, 1 mile west of Tray Mountain" and "Mound on east side of Sautee Creek, 1 mile above Chickamauga Creek," both also reported by Mooney. And he included the "Subterranean or buried village 'on Duke's Creek, 4 miles from Nacoochee Valley'" and sourced it to White and to Jones's *Antiquities.* Finally his report included, "Rock walls on Mount Yonah" that were "noticed" by Jones.[50]

The report of the "subterranean or buried village" had been reported in a local paper in the 1830s. In addition to the buried remains of log structures, the newspaper report, which White quotes at length, stated that "a great number of curious specimens of workmanship were found in situations which preclude the possibility of their having been moved for more than a thousand years." In fact, historian Dr. Thomas N. Lumsden and others believed the features date to the mid-sixteenth century, as is discussed in the following chapter.[51]

HEYE FOUNDATION (1915)

George Gustav Heye (1874–1957) was a wealthy investment banker who assembled the world's largest collection of artifacts from the indigenous people of North America. His collection included artifacts from Nacoochee and was the basis for his Museum of the American Indian, which he founded in New York in 1916 and which opened to the public in 1922. In 1989 Congress passed the National Museum of the American Indian Act, which authorized the Smithsonian to acquire Heye's New York museum—now known as the National Museum of the American Indian George Gustav Heye Center—and to build a new museum in Washington, D.C., which opened on the Mall as the National Museum of the American Indian in 2004. About a third of Heye's original collection, which included human bones and funerary objects, has since been repatriated to affiliated tribes. Although he had no professional training, his personal contact with and study of the artifacts that he collected made him an expert in his own right.[52]

George Heye at Nacoochee Mound in 1915.
Courtesy of National Museum of the American Indian George Gustav Heye Center, New York.

Heye's view of Nacoochee Valley and Mount Yonah in 1915. Heye, Hodge, and Pepper, *Nacoochee Mound*, frontispiece.

In the spring of 1915, Heye began directing the excavation of the mounds at the Garden Creek site, twenty-five miles west of Asheville, North Carolina. It must have been a hurried affair since he did not take field notes or record the provenance of items he collected. His work at Nacoochee would be more extensive and was thoroughly reported not long after the project was complete. The report also incorporated data from Jones's 1873 publication "and such knowledge respecting the earlier history of Nacoochee as has come down to us." Heye made a preliminary inspection of the Nacoochee site in May 1915, in conjunction with his work at the Garden Creek site that had begun a month earlier. As always in northern Georgia, archaeologists had to deal with local legends of gold being hidden in ancient earthworks, although none was ever found.[53]

The final project report noted that "active operations" at Nacoochee began on 19 June 1915 and continued with some interruptions until 29 October. The original intent had been to excavate the entire mound, which would have necessitated removal of the 1880s gazebo, but heavy rains made that impossible, and "a satisfactory concession for the resumption of the work in the following summer could not be obtained" from Lamartine Hardman, who then owned the site. The mound remains only partially excavated.

Heye and his team noted that "Nacoochee" is a corruption of the Cherokee word Nagu'tsĭ, the meaning of which has been lost. The report goes on to state that Nacoochee is the same site as the Guasili, or Guaxule, visited by de Soto in 1540, a notion rejected by modern historians and archaeologists. More accurately, it notes that Nacoochee was "without question" a village of the Cherokee during historic times. The introduction concludes with a diatribe against the "ridiculous yarn" of the star-crossed lovers Sautee and Nacoochee. "The Spaniards of de Soto's army are forced into the romance, and although there is not a grain of truth in the assertion that any Indians ever related such an un-Indian legend, and the entire fabric of the story is imaginary, it has received such currency that no argument can dismiss it from the minds even of those otherwise intelligent."[54]

His study confirmed Jones's observation that there was "no sign of a terrace or roadway, such as that which characterized the ancient Guaxule," visited by de Soto. The report went on to describe what they found in 1915: "Little was to have been expected in the way of remains at the surface of the summit, as it was learned from a mechanic who had been employed in building the pagoda, still preserved, twenty-seven years before [1888], that about two feet of the top had been removed at that time. Subsequently the sides had been dug away more or less, forming a steep conventional terrace which was planted in vines and shrubbery, and the mound fenced."[55]

Heye had the soil analyzed and concluded that it had come from the surrounding valley floor. He recorded the mound's height as 17 feet, 3 inches, and its oval summit as 67 feet, 4 inches, at its shortest diameter and 82 feet, 9 inches, at its longest. The perimeter of the summit he estimated at 231 feet and that of the base at about 410 feet. The number of objects retrieved from the east side of Nacoochee mound is not known, but it was substantial. The team excavated about seventy-five box graves, in which they found some objects similar to what Jones had found in 1870, including another copper axe. Of it all Heye noted that "some of the earthenware pipes bearing bird effigies as decorative

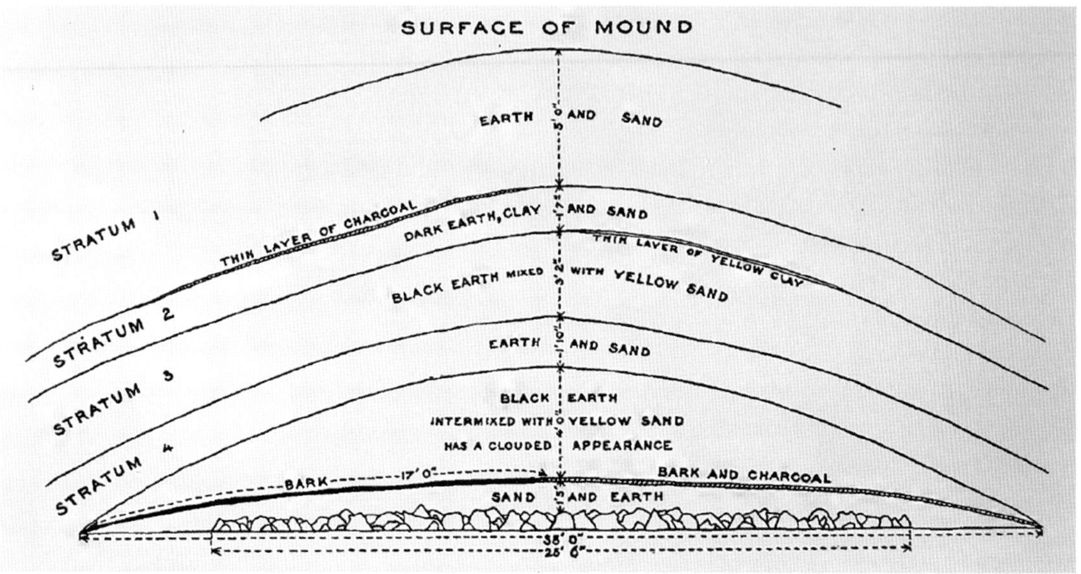

Heye's stratigraphy of Nacoochee Mound. He identified a smaller mound that had apparently existed for some time prior to the construction of the larger mound. Heye, Hodge, and Pepper, *Nacoochee Mound*, 34, fig. 4.

motives represent the height of Nacoochee art." The report is unequivocal in its conclusion: "Nacoochee may be regarded as a typical Cherokee earthwork. While not rich in artifacts, the products of material culture are such as might have been expected to be revealed by the excavation of a Cherokee site occupied from prehistoric times through a considerable part of the historic period. The culture of its inhabitants had not been greatly modified by contact with civilization, for only the upper part of the mound revealed objects of European provenience. It is not likely that more than a few feet of the height of the mound were added during later historic times."[56]

Heye believed that the mound was primarily used as a foundation for the "townhouse" and quotes Bartram's description of "a rotunda capable of accommodating several hundred people."[57] He also noted that the mounds were much older than the memory of the historic Cherokee tribe, which simply found them convenient for their own purposes. Heye's excavation at Nacoochee was considered a major achievement in the then-nascent study of native cultures in the Southeast, but it also destroyed nearly half of the mound and disturbed

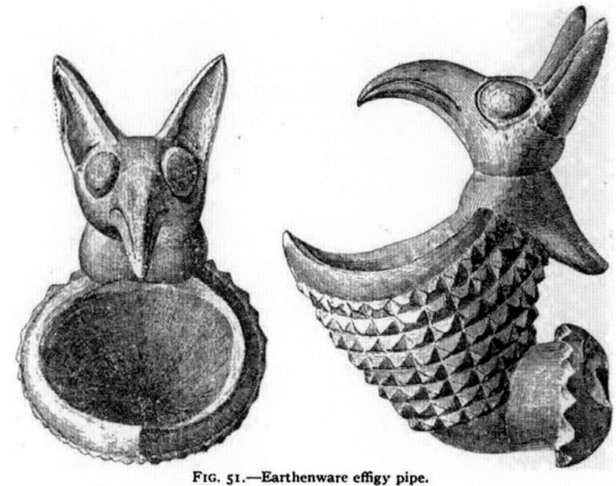

Effigy pipe retrieved from Nacoochee Mound. Heye, Hodge, and Pepper, *Nacoochee Mound*, 74, fig. 51.

thousand-year-old burials, the remains from which were only recently repatriated for tribal burial. While highly productive in terms of data retrieval perhaps, such a heavy-handed approach to archaeology would not be sanctioned today.

WAUCHOPE SURVEY (1939)

Born in Columbia, South Carolina, Robert Wauchope (1909–79) had an early interest in archaeology. He worked as an assistant to the famed archaeologist Alfred V. Kidder (1885–1963) at a site in New Mexico as a teenager and, while at the University of South Carolina in 1928–29, assisted in early excavations at the Stallings Island site on the Savannah River. He graduated from Harvard University with a PhD in anthropology and began teaching at the University of Georgia, to which he brought the school's first courses in archaeology. In 1942 he began a career at Tulane University's Laboratory of Anthropology and Archaeology and was director of the Middle American Research Institute until he retired in 1975.[58]

In 1938–40 Wauchope directed the archaeological survey of Georgia begun by the Works Progress Administration (which was renamed the Work Projects Administration in 1939) and "co-sponsored" by the

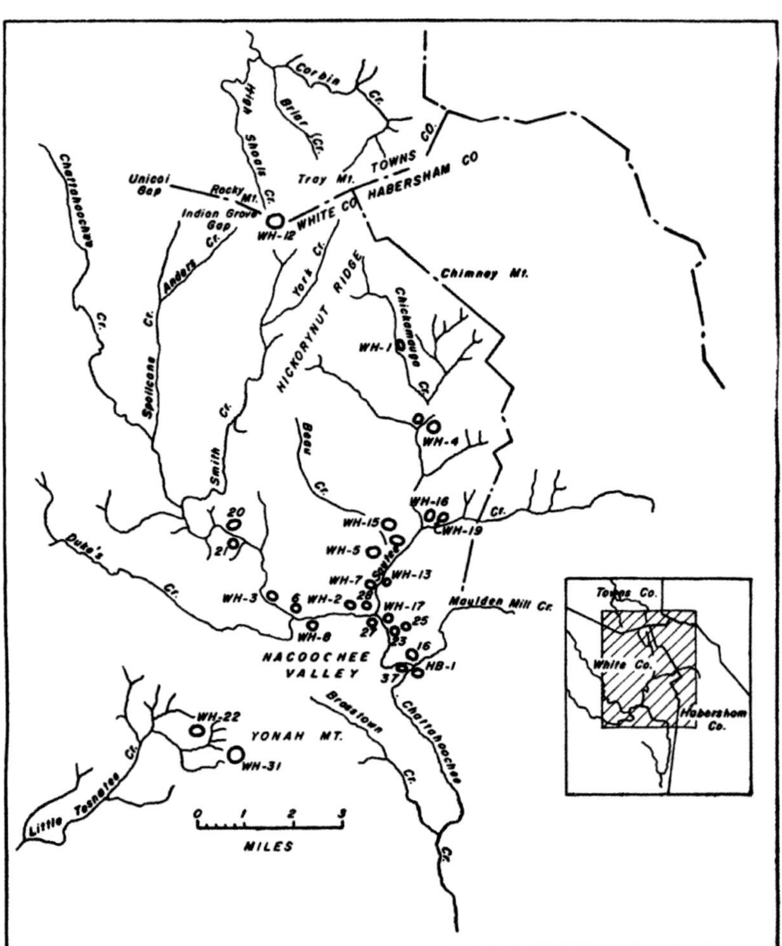

Wauchope's map of the sites he surveyed in White County and a single site across the river in Habersham County. "*Archaeological Survey*," fig. 182.

University of Georgia and the State. He surveyed two hundred sites across northern Georgia, including over thirty sites in White County, some of which he excavated. Wauchope's much-postponed report of the survey noted some of the problems he had faced, not the least of which were the time constraints and bureaucratic reporting demanded by the Works Progress Administration. Personnel, he wrote, "was invariably long on quantity and, so far as our north Georgia survey was concerned, miserably short on quality." The project's only surveying engineer could not even do long division. Nevertheless, in 1966 Wauchope looked back with pride on his work:

> We located and assessed hundreds of Indian sites in a practically unknown archaeological area. We established a ceramic sequence for northern Georgia that had been wholly lacking and that, with some refinements, is still valid. We discovered an entirely unknown period or cultural expression in north Georgia prehistory characterized ceramically by the Woodstock pottery. We identified some of the larger ceremonial sites more closely with the mature Mississippi "Etowah" ceramics than had been previously suspected. Our findings bridged many an archaeological gap between eastern Tennessee on the west, North Carolina above, and Piedmont Georgia in the south. We demonstrated unsuspected cultural continuities from Woodland to Protohistoric times.[59]

Nearly all of what Wauchope documented in White County was in and around Sautee Nacoochee. The sites ranged from Early Woodland to Protohistoric, with many sites layered through occupation and reoccupation over many centuries. Most of the sites were in the floodplain of the river and its tributaries.

In the vicinity of Sautee Nacoochee, Wauchope documented twenty-three "village" sites, a term apparently applied to all the sites that were not refuse pits or "lithic stations," where early people worked stone into spear points and other tools. The term "village" may be misleading since none of the sites contained more than a handful of houses, but

House site, typical of similar sites in and around Sautee Nacoochee. It was first occupied about 500 BCE and reoccupied after about 1350 CE. Wauchope, "*Archaeological Survey*," fig. 258A.

some of the sites appear to have been more or less continuously occupied from the Woodland through the Protohistoric Periods. Wauchope also documented three mounds in Nacoochee Valley (designated WH-2, WH-3, and WH-28). One, of course, was Nacoochee, which he did not excavate but only collected surface artifacts in the area. He made no mention of the village site that would be investigated in 2004.

The other two mounds were also on the north side of the valley and were much smaller structures. At the Eastwood site (WH-2), he described a "low mound, which could scarcely be distinguished from the natural ridge on which it stands."[60] He also investigated the hillock that stands nearly opposite the mouth of Dukes Creek. Often mistaken for a manmade mound, it was, Wauchope found, a natural feature, although he also thought the top might have been flattened by the Indians.

ARCHAEOLOGY OF THE NACOOCHEE VILLAGE (2004)

The most recent archaeological investigation in the valleys was conducted by Dr. Mark Williams, director of the Georgia Archaeological Site Files and the Laboratory of Archaeology at the University of Georgia. His work focused on the village that historical documentation suggests existed along with the mound. His goal was to determine the integrity of the village site, which many feared had been destroyed during gold-mining operations in the early nineteenth century. In addition to the reduction of the mound when the gazebo was constructed, Williams noted that its perimeter had been significantly altered by the plowing around its base, which had been ongoing since the 1820s.[61]

He created the first contour map for the mound and, most important, documented the condition of the village site just west and southwest of the mound. He concluded, "The village at Nacoochee is present and probably well preserved, primarily to the west of the mound. The gold mining activities of the early 19th century probably did not take place anywhere in the field surrounding the mound. The site has a significant early-middle Woodland Cartersville occupation and a long Mississippian period history. The most intense period of occupation was during the Lamar period (1350–1550)." Significantly, he thought, contrary to Wauchope, that evidence for "historic Cherokee occupation seems minimal based upon present analysis," but he was probably not aware of the historical documentation that suggests otherwise. He did, however, recommend additional investigation of the village site to establish a complete chronology of its development. He also saw "no good reason for archaeological excavation of the mound in the near future," if at all.[62]

3

European Contact and Tribal America

The voyage of John Cabot (ca. 1450–ca. 1500) for Henry VII in 1497 is generally credited with being the first European contact with the mainland of North America since the Vikings in the eleventh century. Even so, he landed only once, probably in Newfoundland, and then went no farther than "the shooting distance of a crossbow." They found animal traps and a wooden needle, which they took back to England to prove there were humans in this "New World."[1] In 1524 Giovanni da Verrazzano (1485–1528), in service of France's King Francis I, landed near Cape Fear in present-day North Carolina and explored north along the coast, making contact with the indigenous people and, when leaving, kidnapping a boy as an exhibit for the king. There were other voyages along the eastern coast of North America in the first decades of the sixteenth century.[2]

To the indigenous people, the fabric, beads, glass, metal implements, and other items scavenged from wreckage on the beaches of North America were wondrous in a way we can only imagine today. Objects were widely traded and given as gifts. As a result, indigenous people far into the interior knew of the strange hairy people arriving on floating islands, and they had also heard of a new world to the east long before they actually saw a European. The archaeological record suggests that Sautee Nacoochee was the site of a thriving town and the seat of a local chiefdom in the sixteenth century. Located at the crossing of two important regional paths, the people of Nacoochee would certainly have heard, though perhaps in a garbled way, of these odd newcomers. Quite possibly some of the Venetian glass beads and the iron objects retrieved in the excavations at Nacoochee were acquired through long-distance trade and not through personal contact.[3]

The Spanish *Entrada*

Spanish slaving expeditions to the Bahamas beginning in the 1490s almost certainly brought the Spanish into contact with the barrier islands off the southeastern coast of North America, long before Juan Ponce de León (1474–1521) "discovered" and named La Florida in 1513. Early Spanish attempts at colonization failed, beginning with the short-lived colony of San Miguel de Gualdape, established by Lucas Vázquez de Ayllón (ca. 1480–1526) in 1526, but European exploration and its evil companion, the slave trade, were already having a significant impact on the indigenous people. The Bahamas were depopulated within two or three decades of their discovery by the Spanish, and, as early as 1521, the slavers were raiding the coast of what is now South Carolina. The resulting chaos in the coastal tribes destabilized others far inland.[4]

THE DE SOTO EXPEDITION

In the spring of 1539, Hernando de Soto (ca. 1497–1542) began his infamous *entrada* into southeastern North America. Along with him were six hundred men, two hundred horses, several pregnant Iberian pigs, and one of their most terrible weapons, great mastiffs trained to kill. De Soto, one contemporary wrote, was "very given to hunting and killing Indians."[5] By royal decree de Soto was made *adelanto*, or "governor," with authority "to conquer, pacify, and settle whatever he wished within the ample limits of Florida." He could keep four-fifths of the plunder acquired "in the usual manner in battle" and half of what was taken from "graves, sepulchers, temples, places of religious ceremonies or buried in a public or private place."[6] The expedition landed in May 1539 on the west coast of Florida, probably on the south side of Tampa Bay, and over the next four years traversed much of southeastern North America, raping, killing, burning, and looting as they went. De Soto died somewhere in Arkansas in May 1542, but his arrogance and cruelty established a bitter precedent that warped European relations with the indigenous peoples for centuries to come.

Three firsthand accounts of the expedition survived and were eventually published, including a daily journal kept by de Soto's private secretary, Rodrigo Rangel, and an extensive narrative by the anonymous Portuguese "Gentleman from Elvas."[7] These and other contemporary sources are sometimes contradictory and vague in their descriptions, but they remain important because the de Soto expedition was the first, and last, European contact with the great Mississippian chiefdoms before they collapsed.

In 1935 Congress authorized the United States De Soto Expedition Commission to determine the precise route of de Soto's *entrada*, and an exhaustive compilation and assessment of archival sources was completed in 1939. The commission's report included a map showing the eleven routes of march that had been proposed since the French

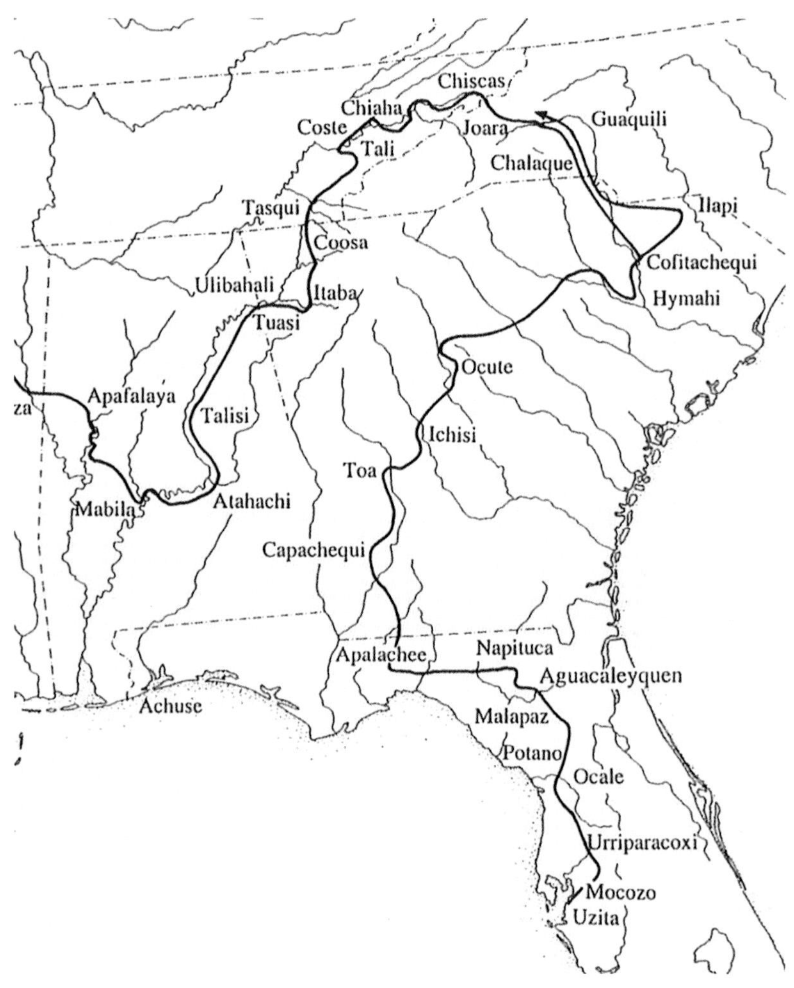

Part of Charles Hudson's map of de Soto's route through the Southeast. Juan Pardo followed a similar route through Georgia and the Carolinas. "Marker Monday."

cartographer Guillaume de l'Isle made the first attempt in 1701. Nearly all those routes showed the expedition crossing northern Georgia and passing through Nacoochee on the way. One of the routes shown was developed by James Mooney (1861–1921), the noted authority on Cherokee history and culture, and was the only one shown descending the Chattahoochee River into Alabama. The early discovery of Spanish artifacts at Nacoochee convinced historian and pioneer archaeologist Charles Colcock Jones Jr. and others that de Soto had seen the valley. Only recently has it been realized that, as noted earlier, such artifacts could have arrived without de Soto's conquistadors ever having been anywhere near Nacoochee.[8]

An enormous amount of archaeological work since World War II has added an entirely new dimension to the study of de Soto's route and to our understanding of its consequences. The noted anthropologist Charles M. Hudson (1932–2013) was at the forefront of the work to synthesize traditional historical documentation and the more recent findings of archaeology and anthropology. He was able to tie known archaeological sites to many of the locations recorded in the historical accounts of the de Soto expedition. The end result is generally considered the most accurate map of de Soto's route, which takes the entrada through the Blue Ridge along the French Broad River rather than skirting to the south along the Little Tennessee River. Most authorities now agree that de Soto never saw northeastern Georgia.[9]

LATER SPANISH EXPEDITIONS

Spain considered the de Soto expedition a failure since it had not resulted in gold, silver, or slaves; it would be nearly twenty years before there was another attempt at colonization. In 1559 Tristán de Luna y Arellano (1519–71), along with 1,500 soldiers and settlers in thirteen ships, landed at Pensacola Bay, but the venture was plagued with problems from the beginning. Short of supplies, de Luna was forced to send a delegation to Coosa in northwest Georgia for food, retracing in reverse part of de Soto's journey nearly fifteen years earlier. The colony failed; but in 1565 Spain was finally able to establish its first permanent foothold in La Florida with the founding of Saint Augustine, now recognized as the oldest colonial-era city in the United States.

Between 1566 and 1568, Juan Pardo led two exploratory expeditions from the Spanish mission at Santa Elena on what is now Parris Island, South Carolina, into the interior of the Carolinas and eastern Tennessee. Pardo's incursions were also long thought to have included passage through Sautee Nacoochee. John Swanton in the 1920s, among many others, was an early proponent of a route that took the expeditions through northeastern Georgia and down the Little Tennessee River through the mountains. As with de Soto's trail, however, most authorities now believe that Pardo, like de Soto, followed the French Broad River instead.

Some have suggested that the Spanish quest for gold in northern Georgia was in fact successful but that royal policy dictated secrecy. In his *Historical Collections of Georgia*, published in 1855, George White quoted a story that he copied from a local newspaper:

> About twenty years ago, a singular discovery was made of a subterranean village in this county. The houses were disinterred by excavating a canal for the purpose of washing gold. The depth varied from seven to nine feet. Some of the houses were embedded in the stratum, or gravel. The logs were but partially decayed, from six to ten inches in diameter, and from ten to twelve feet long. The walls were from three to six feet in height, joined together, forming a straight line upwards of three hundred feet in length, comprising thirty-four buildings, or rooms. The logs were hewn at the ends, and notched down, as in ordinary cabins of the present day. In one of the rooms were found three baskets, made of cane splits, and a number of fragments of Indian ware.
>
> From the circumstance of the land having been covered with a heavy growth of timber previous to its cultivation by the whites, twelve years before the time of its discovery, it was inferred that they were built at some remote period. The houses were situated from fifty to one hundred yards from the principal channel.[10]

Charles Colcock Jones Jr. noted these and similar features in his *Antiquities of the Southern Indians* in 1873 and included a detailed discussion of the evidence for Spanish mining in the seventeenth century. In the late nineteenth century, Mooney, who had no doubt read those earlier accounts, noted, "Long before the end of the sixteenth century . . . mines of gold and other metals in the Cherokee country was a matter of common knowledge among the Spaniards at St. Augustine and Santa Elena and more than one expedition had been fitted out to explore the

interior. Numerous traces of ancient mining operations, with remains of old shafts and fortifications, evidently of European origin, show that these discoveries were followed up, although the policy of Spain concealed the fact from the outside world."[11] Dr. Thomas Lumsden placed the Dukes Creek site in Land Lot 69 of the Third District and also reported "discovery of hewn pitch pine deep in the auriferous gravels of Bean Creek" in Land Lot 23. All of this, he thought, was evidence that the Spanish were in northeastern Georgia in the sixteenth century. Whether either of the sites, if they could be located, still retain any archaeological integrity is uncertain.[12]

During this period the prehistoric trail system in southeastern North America reoriented itself to the Spanish in Florida and along the Gulf Coast, with the primary route out of Florida running up the Flint River and across the Ocmulgee at the Fall Line, then up the Oconee to the headwaters of the Savannah River, where many trails diverged. A brisk trade developed, with the indigenous people offering

Part of Le Moyne de Morgues's map *Floridae Americae*, 1591 but commonly dated to 1565, including the earliest cartographic depiction of the Appalachian Mountains, "Montes Apalatci" (*top center*) and noting the presence of gold and silver. Item 2003623393, Geography and Map Division, Library of Congress, Washington, D.C.

deerskins, buffalo hides, and slaves in exchange for European manufactured goods, which eventually included firearms.[13]

THE AFTERMATH

When the English, who began resettlement of what they called Virginia at Jamestown in 1607, began to explore and trade with the people of the interior Southeast in the last half of the seventeenth century, they found abandoned towns and villages, overgrown orchards of peach and plum trees, and deteriorating earthen mounds. They also found far fewer people than de Soto met in the 1540s. Over the course of two or three generations in the mid- and late sixteenth century, the indigenous population of southeastern North America had undergone demographic collapse. Disease, warfare, the slave trade, and starvation reduced the population to as little as 10 percent of the estimated two million people in the region in 1540. One colonist in the 1650s wrote, "The Indians . . . affirm that before the arrival of the Christians, and before the small pox broke out amongst them, they were ten times as numerous as they are now."[14]

De Soto and other Spanish conquistadors wrought immediate havoc on the people they encountered. Armed only with bows, arrows, and spears, indigenous warriors had been slaughtered in the tens of thousands by the time the Spanish were through. That scale of loss was a tremendous blow to any community, especially when accompanied, as it typically was, by the burning of villages, destruction or theft of food crops, and the enslavement of anyone who survived.

The worst of the scourges brought on by contact with Europeans were the recurring epidemics of smallpox, measles, influenza, cholera, and other communicable diseases that swept through the indigenous people, who had no immunity to "Old World" pathogens. Often the infection preceded the arrival of the source of the contagion by weeks, months, or years. When de Soto visited the legendary chiefdom of Cofitachequi in the Carolina Piedmont in 1540, he was shown, and proceeded to plunder, the burials of "a great many" people who had been carried away in an epidemic two years before.[15]

Much as the Black Death depopulated Europe in the fourteenth century, the diseases introduced by Europeans, Africans, and the animals they brought with them decimated entire villages in the sixteenth century, sometimes leaving no one even to bury the dead, much less tend crops and do the necessary work to stay alive. The entire social structure collapsed under the weight of this loss, exacerbated by Spanish humiliation of indigenous rulers and total disregard for their culture. In a world where the favor of the gods determined all earthly outcomes, the demonstration of Indian powerlessness was devastating. Considering the terrible damage done to cultural memory by the disruption of oral traditions, one historian wrote, "It is small wonder that in later years, local Indians sometimes expressed ignorance of the builders of the ancient temple mounds. The Indians encountered by the English in the eighteenth century were merely the survivors."[16]

In dealing with this cultural trauma, many communities relocated; others consolidated or simply disappeared. The towns along the Hiwassee and Little Tennessee Rivers, the headwaters of which are just a few miles north of Sautee Nacoochee and very close to the headwaters of the Chattahoochee, were abandoned in the sixteenth century. The same was true of the settlements in northwestern Georgia, where survivors from several of the towns visited by de Soto migrated to new settlements along the lower Coosa and Tallapoosa Rivers in Alabama. Likewise, all the mound centers in the Oconee River valley visited by de Soto were abandoned before 1600. Over northern Georgia and elsewhere, the greatly reduced population dispersed to upland areas or sometimes to single isolated homesteads; no large villages are known from the seventeenth century. The Indian towns around Sautee Nacoochee may never have been completely abandoned during this period, but there is likely to have been significant disruptions in the traditional use of the mounds as far fewer people occupied the area.[17]

The Indigenous People of the Southeast

In the fourteenth century, the climate across much of the northern hemisphere began to cool with the onset of the so-called Little Ice Age, which would last until the mid-nineteenth century. Shorter growing seasons were reducing yields, and an unwieldy social structure was beginning to break down even before disease, warfare, and their enslavement by one another and the Europeans tipped the indigenous people into demographic collapse. As that happened, these Americans of the sixteenth century devolved into smaller, less centralized social groups. In less than a century, the Indian tribes of the historic period emerged, including in southeastern North America the Catawba, Cherokee, Chickasaw, Choctaw, and Muscogee. Nearly all the tribes of the historic period were, as one historian put it, "created in the melting pot set boiling" by sixteenth- and seventeenth-century epidemics.[18]

Europeans often made the mistake of seeing tribes as nation-states, but they were nothing of the sort. Although there were strong cultural and linguistic ties within a tribe, relationships were generally egalitarian and revolved around clan and family ties and not necessarily around some centralized power. There were respected elders and chiefs, to be sure, but most of them did not, and could not, rule by fiat, as European kings were wont to do. In a very real sense, a chief's rule depended entirely on the consent of the governed.

All four of the major language families of eastern North America were represented in the southern Piedmont and Blue Ridge at the time of European contact: Muskogean, Algonquian, Iroquoian, and Siouian. Each represents ancient origins in other regions, subsequent migration, and, over many centuries in southeastern North America, division into the multiplicity of dialects spoken by the indigenous people of the early historic period. Sautee Nacoochee's heritage is with the Cherokee and their ancestors. In the seventeenth and eighteenth centuries, the valleys were on the fringe of the Cherokee world and always subject to depredations from the Creek, the Catawba, and other mortal enemies.

THE MUSCOGEE

As early as 1715, "Creek" was the name English traders gave the Muscogee people, the Southeast's largest and most powerful confederation of tribes in the eighteenth century. James Adair (1709–83), the Irish-born trader who lived with the indigenous people for forty years beginning in 1735, may have taken the name too literally when he wrote, "It is called the Creek country, on account of the great number of Creeks, or small bays, rivulets and swamps, it abounds with." More likely, the name was a truncated version of "the Indians of Ochese Creek," who were among the first Muscogee to come into contact with Carolina traders in the late seventeenth century. All the tribes of the historic period have origin myths, with most of the southern tribes having their ancestors migrating from the west under a variety of circumstances. Adair wrote that the Muscogee "believe their original predecessors came from the west, and resided under ground, which seems to be a faint image of the original formation of mankind out of the earth, perverted by time, and the usual arts of priest-craft."[19]

John Swanton, C. C. Jones, James Mooney, and most other historians and ethnologists of the period—as well as the early archaeologists who believed them until their data began to accumulate—accepted these origin myths as "distorted cultural memories of actual historical events." More recently, and benefiting from the vast amount of archaeological work in the last half of the twentieth century, archaeologist Vernon James Knight stated flatly that "there is no archaeological evidence to support the Creek myth of a prehistoric Muskogean invasion of the Southeast." Moreover, he continued, "it is now recognized that these texts have little to do with 'history' in the Western sense. One of the main purposes of these migration stories was to spell out relationships of a social and ceremonial nature in symbolic language. . . . These native 'histories' were conceived not as linear narratives of past events but rather as charters establishing the status of the local group in a present and continually unfolding social environment."[20] Knight may have been too dismissive of the utility of myth and legend in reconstructing human history, but archaeology trumps them all if the goal is an approximation of real events. As a result, all the sources that did not have the benefit of the enormous amount of archaeological work in the last half of the twentieth century must be read with that in mind.

The Muscogee were organized in more or less independent towns (*talwas*), of which there were as many as fifty in the mid-eighteenth century. In the early historic period, each *talwa* was part of one of three larger polities centered on the lower Chattahoochee River, the Tallapoosa River, and the Coosa River. Each of these had a main town, a chief (*mico*), and several tributary towns, and each, Knight believed, was "in a very real sense descended from a corresponding Mississippian chiefdom extant in the sixteenth century." It was not until the early eighteenth century that pressure from the English drove the various Muscogee *talwas* to begin acting as a sort of confederacy. Even then the English failed to understand that, above the level of the town, any military or civil alliances among the Creek, and most other tribes as well, were by and large conditional and might be abandoned at any time. Only in the 1780s did the Muscogee become institutionalized as a nation.[21]

In the 1770s Adair observed that, while other tribes were declining "on account of their continual merciless wars, the immoderate use of spirituous liquors, and the infectious ravaging nature of the small pox," the Muscogee were not. Adair thought that this was because "the Muskohge have few enemies, and the traders with them have taught them to prevent the last contagion from spreading among their towns, by cutting off all communication with those who are infected, till the danger is over." In addition, Adair noted that "the men rarely go to war till they have helped the women to plant a sufficient plenty of provisions, contrary to the usual method of warring savages," and he thought this helped ensure the Muscogee prospered. Adair also admired the Muscogee's "artful policy of inviting decayed tribes to incorporate with them," which he thought had doubled their numbers in thirty years' time. In addition to Guale, Hitchiti, Timucua, and Apalachee, there were bands of Natchez, Shawnee, and others who found a home among the Muscogee.[22]

Creek towns of the early historic period resembled Mississippian-period towns and varied in size, from as few as thirty houses to as many as a hundred. "These settlements," White notes, "were strung out along the rivers, with more density around the ceremonial center."[23] William Bartram, among others, described the typical indigenous housing on the upper Oconee River as usually consisting of "two houses nearly the same size about thirty feet in length, twelve feet wide, and about the same height." The buildings had bark roofs and walls of wattle-and-daub, which were generally kept whitewashed inside and out. One of the houses was divided into two rooms, one of which was used for cooking, while the other house was used for food storage and had a porch that created "a pleasant, cool, and airy situation, and here the master or chief of the family retires to repose in the hot seasons, and receives his guests or visitors."[24]

A few Creek settlements were in the upper Georgia Piedmont, and their claim to hunting grounds stretched to the Savannah River. Legends, but few facts and no archaeology, describe great battles between Muscogee and Cherokees over these hunting grounds in northern Georgia. One of those battles was supposed to have given names to Slaughter Gap and Blood Mountain, a few miles west of Sautee Nacoochee. Precisely when that might have occurred is not clear, but, if it did, it was probably in the mid-seventeenth century. In the Piedmont of Georgia, by the early historic period, the Chattahoochee River was generally taken as a boundary between the two tribes, but raids continued on both sides of the river throughout the colonial period.[25]

SAVANNAH RIVER TRIBES

The Yuchi, or Uchee, passed through northeastern Georgia early in the historic period. Calling themselves Tsoyaha, or "Offspring of the Sun," they originated in the sixteenth century out of the Mississippian culture in eastern Tennessee but spoke a language unlike any of their neighbors. Like the Creek, they lived in independent villages or towns along the rivers and streams. In the mid-seventeenth century, some bands of Yuchi migrated, perhaps to take better advantage of their trading ties with the indigenous people at Spanish missions, perhaps to escape constant warfare with the Cherokee. As early as 1661, one band was in the Savannah River valley, which for reasons that are unclear had been more or less depopulated for most of two centuries. Later they relocated to the lower Chattahoochee River valley, where the tribe became part of the Muscogee Confederacy.[26]

Shawnee people also passed through northeastern Georgia in the seventeenth and early eighteenth centuries, but Suwanee in Gwinnett County is thought to have been their only permanent, if short-lived, settlement on the Chattahoochee River. The Shawnee were known as traders in the early historic period and spoke an Algonquian language that one contemporary said was "fine, smooth, and easy to be got."[27] As a result, it became a sort of lingua franca on the frontier. Relatively few in number, the Shawnee were caught up in the so-called Beaver Wars in the mid-seventeenth century, when the Iroquois, in their quest for hunting grounds to replace what they were exhausting in the fur trade with the French and the English, drove the Shawnee from their homeland on the Ohio River and into settlements that ranged from Pennsylvania to Illinois to Georgia.

By 1670 one of those Shawnee bands had migrated to the Savannah River, where three Shawnee towns were recorded in 1708. The Shawnees were known to the English as "Savanoes," or Savannahs, and they formed an alliance that lasted until the disaster of the Yamasee War. After that the Savannah River villages were abandoned, and the Shawnees there relocated, many to join another band in Pennsylvania but some migrating to join the Creek on the lower Chattahoochee.

The Westo were also forced to relocate by wars with the Iroquois, and they too settled in the Savannah River valley around 1660. They were among the first of the indigenous people in southeastern North America to acquire guns and used them to terrible advantage as they raided neighboring tribes for people whom they could sell into slavery in exchange for more guns and other manufactured goods. Since the Spanish attempted to keep firearms from the Indians under their control, those at the missions in what are now southeastern Georgia and northern Florida were a prime target of the Westo, but they raided other tribes as well. The Carolinians soon recognized that the constant threat of the Westo slaving raids was not conducive to trade, so in 1680 some of the colonists secretly armed the Savanoes, who proceeded to exterminate the Westo tribe more or less completely.

Finally, there were the Yamasee, who appear to have been, like the Creek, composed of remnants of earlier tribes, including the Guale of coastal Georgia. They, too, emerged as a tribe mainly in response to the European invasion but were able for a brief period to pose an existential threat to the nascent colony of Carolina when they began the so-called Yamasee War in the spring of 1715. Allied with Cherokees, Muscogees, Yuchis, and others, they attacked settlements all over South Carolina and inflicted heavy casualties before the colonial militia were able to put down the revolt.

The Cherokee

The tribal name first appears as "Chalaque" in the report by the de Soto expedition's Gentleman of Elvas, published in 1557, and the same spelling appears in some early maps. By the late seventeenth century, the English were referring to the "Charakee" or "Charikee," their corrupted pronunciation of the proper tribal name, Tsa'lagi'.[28] The Cherokee people speak an Iroquoian language, and there is archaeological evidence that they have occupied the Southern Appalachians since at least the Woodland Period one or two thousand years ago. Their heartland was the narrow mountain valleys of eastern Tennessee, the western Carolinas, and northern Georgia, but they claimed hunting grounds from the Cumberland Plateau and Kentucky to the western Carolina Piedmont.

In 1776 the famed naturalist William Bartram (1739–1823) journeyed into Cherokee country and made some of the earliest observations of Cherokee life. From foodways to architecture, his journals are one of the more reliable windows into the Cherokee world during the colonial period. In his journals he commented on the mounds he saw in the Cherokee towns, noting that they were of a "much ancienter date" than the Cherokee houses that were usually built atop them. "The Cherokees themselves," he wrote, "are as ignorant as we are by what people or for what purpose these artificial hills were raised . . . but they have a tradition common with the other nations of Indians, that they found them in much the same condition as they now appear, when their forefathers arrived from the West and possessed themselves of the country."[29]

Detail from L'Isle and Guérard's *L'Amerique septentrionale*, first published in 1700. Much of the interior Southeast remained unexplored. Item 71005434, Geography and Map Division, Library of Congress, Washington, D.C.

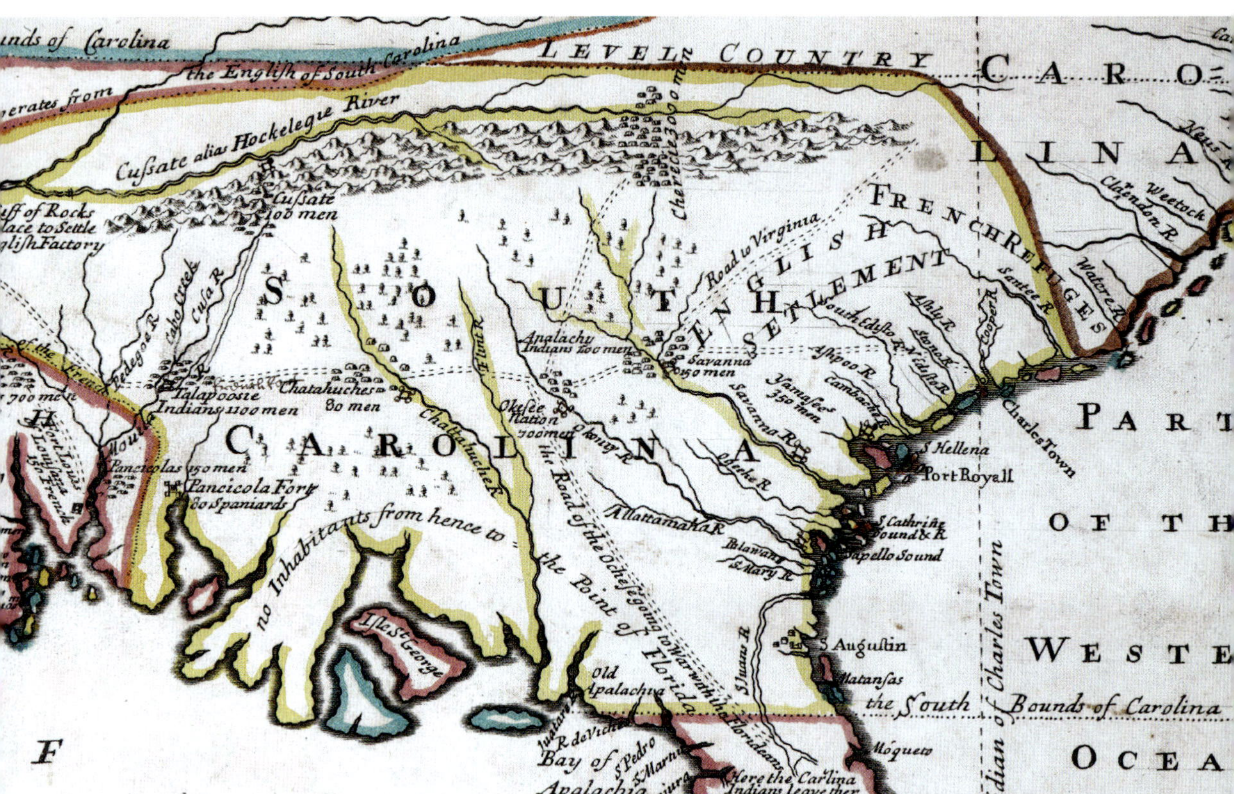

Part of Thomas Nairn's *Map of South Carolina*, published about 1711 as part of Edward Crisp and others' map, *A Compleat Description of the Province of Carolina*. Major crossroads were at the "Ochese Nation" on the Ocmulgee River and at the falls of the Savannah River (*near center*). Note the trail from Ocmulgee to what is probably meant to represent Tugaloo and the Lower Towns of the Cherokee (*top center*). Item 2004626926, Geography and Map Division, Library of Congress, Washington, D.C.

Bartram wondered if the mounds were built "as great altars and temples similar to the high places and sacred groves anciently amongst the Canaanites and other nations of Palestine and Judea." He may have read Adair, who spent a good portion of his time arguing that the indigenous people were descended from the Ten Lost Tribes of Israel.[30] In sorting out the history of the indigenous people—"the devious maze of Indian chronology," Mooney called it—the effects of demographic collapse on cultural memory should not be underestimated.[31] Bartram's comments about the antiquity of the mounds confused understanding of Cherokee history for the better part of two centuries.

Only with the benefit of twentieth-century archaeology were mistaken characterizations of the tribe as "prehistoric carpetbaggers" finally put to rest. Archaeology has demonstrated a continuity between the historic Cherokee and some of the earliest inhabitants of the Southern Appalachians.[32] Most definitively, the character of Cherokee pottery is considered to be "the end product of the complicated stamped tradition, the tradition which is the hallmark of the South Appalachian province from the beginning of the Middle Woodland period," around two thousand years ago.[33]

Yet, as C. C. Jones, Cyrus Thomas, and other linguists, historians, and ethnographers figured out early on, Cherokees spoke an Iroquoian language, a fact that in itself suggests that they did not originate in the Southern Appalachians, since the Iroquois heartland was in northeastern North America. According to Cherokee tradition, too, their original homeland had been the upper Ohio River valley. Linguistic evidence demonstrates fairly conclusively that the ancestors of the Cherokees were immigrants to the Southern Appalachians but that the migration from the original Iroquoian homeland must have occurred far earlier than previously imagined, perhaps as early as three or even four thousand years ago.[34]

Towns and Town Houses

The Cherokee the Europeans encountered in the seventeenth century appear to have organized themselves in a fashion similar to the Muscogee—that is, in more or less independent towns stitched together by a common language and culture. By the late seventeenth and early eighteenth centuries, however, the forces unleashed by European intrusions led, much as with the Muscogee, to the development of a "loose tribal state," a development made easier for the Cherokees because of their relative homogeneity as a tribe.[35]

The Cherokee generally reserved the "town" designation for those settlements with a "town house," which served as a focal point for tribal ceremonies and community life. These towns were not the compact towns of Europe but, like the Mississippian settlements had been, only loosely centered around the town house. The English arbitrarily divided the Cherokee towns into Lower, Middle, and Overhill according to geographic distribution, a useful distinction nonetheless since towns within each group did tend to act in concert.

The number and location of towns varied over time, but in the early eighteenth century there were generally around three dozen towns, more or less equally divided between the three regions. Several eighteenth-century accounts of Cherokee towns have survived, and much of what is described is reminiscent of what archaeology has revealed about the appearance of Mississippian towns prior to European contact. Of the Cherokee towns, James Adair noted, "Their towns are always close to some river, or creek; as there the land is commonly very level and fertile, on account of the frequent washings off the mountains, and the moisture it receives from the Waters, that run through their fields. And such a situation enables them to perform the ablutions, connected with their religious worship."[36]

Bartram left an evocative description of a town he saw in northwest Georgia in the 1770s: "Riding through this large town, the road carried me winding about through their little plantations of Corn, Beans, &c. up to the council-house, which was a very large dome or rotunda, situated on the top of an ancient artificial mount, and here my road terminated; all before me and on every side appeared little plantations of young Corn, Beans, &c. divided from each other by narrow strips or borders of grass, which marked the bounds of each one's property, their habitation standing in the midst: finding no common high road to lead me through the town." He also recorded a Cherokee town house he saw in southwestern North Carolina in 1776: "The council or town-house is a large rotunda, capable of accommodating several hundred people; it stands on the top of an ancient artificial mount of earth, of about twenty feet perpendicular, and the rotunda on the top of it being above thirty feet more, gives the whole fabric an elevation of about sixty feet from the common surface of the ground."[37]

An unidentified nineteenth-century source described much the same while demonstrating the evolution of the use and purpose of the town house:

> Every town has a house, or particular spot of ground, appropriated to dancing, holding council, and of late, courts. This public house, generally called Town House, is built in a circular form, with perpendicular walls six or eight feet high; from thence it ends at a point, giving the roof a conical form, which is supported by interior posts. From the floor to the highest point of the roof is fifteen to twenty feet. Puncheons are laid around on the inside to serve as seats. The house is covered with the bark of forest trees, confined on with the bark of hickory shrubs, the hickories themselves, or white oak shreds. A doorway is left in building the house, and on the outside a small shed or portico is made; and in front of this is a level yard laid off in a square, and made smooth for the purpose of dancing, on particular occasions.[38]

Heye's partial excavation of Nacoochee Mound in 1915 did not locate post holes, since the top of the mound had been reduced when the gazebo was built in the 1880s, but he did locate remains of large central firepits of the sort expected in a Cherokee town house. He provided a plausible explanation for why the town houses were almost always on a mound, besides the obvious one of emphasizing the town's main civic and religious structure. He wrote, "Since mountain valleys are often overflowed, it is quite probable that in order to place these sacred houses above the floods, they were, as stated in tradition, located on artificial mounds. The town house was the depository of numerous ceremonial objects which could not readily be removed in a sudden emergency. And, as it is said traditionally that a sacred fire was kept burning on a peculiar excavation in the center of the earthen floor, this could not be removed from the hearth-place, and hence some provision for its protection was necessary."[39]

Bartram also described what he took to have been typical housing in the Cherokee towns of the 1770s:

> The Cherokees construct their habitations on a different plan from the Creeks, that is but one oblong four square building, of one story high; the materials consisting of logs or trunks of trees, stripped of their bark,

> notched at their ends, fixed one upon another, and afterwards plaistered well, both inside and out, with clay well tempered with dry grass, and the whole covered or roofed with the bark of the Chesnut tree or long broad shingles. This building is however partitioned transversely, forming three apartments, which communicate with each other by inside doors; each house or habitation has besides a little conical house, covered with dirt, which is called the winter or hot-house; this stands a few yards distance from the mansion-house, opposite the front door.[40]

Log-walled buildings, as Bartram described them, were not typical of the earliest houses at Sautee Nacoochee. Wauchope and others have shown that pre-Columbian houses were mostly round, wattle-and-daub structures, transitioning to a rectangular plan during the Mississippian Period. Although the Cherokees used large timbers in framing their prehistoric buildings, conventional log-wall construction did not become widespread until the indigenous people acquired iron tools in the seventeenth century.

TRADERS

A century passed after de Soto's *entrada* before English traders began to venture into the southern wilderness. English efforts to colonize at Roanoke in 1584 failed, and not until 1607 were they able to establish their first colony, Jamestown. In 1663 a royal charter was granted for the Province of Carolina, and in 1670 Charles Town was laid out on the Ashley River, three hundred miles southeast of Sautee Nacoochee. The English soon realized the value of the Cherokees to them, not only in terms of resources that might be extracted but also the security against the French and Spanish that their presence provided.

The cartographer Emanuel Bowen summarized the arrangement on the map he published in 1755: "The Cherokee Indians is a Numerous & Warlike Nation & as they are in Amity & Alliance wth. the Subjects of ye King of Great Britain, they serve as a powerful Barrier to Carolina & Georgia in ye present War against France & Spain. The Emperor of ye Cherokees & the King of ye Catawbas renew'd their League of Friendship wth Govr Glenn at Charles Town in Sth. Carolina in May 1745."[41] By the time Charleston was founded, British traders were already venturing into the hinterlands and vying with the Spanish and French traders, who had been working the territory for the better part of a century. It was a brutal contest, with one Carolina trader commenting that "the french are verry buise in setting these People [the Cherokee] to knock us in ye head but as yet can't Prevail nor I hope will not."[42]

Most of these traders married or, as contemporary sources often put it, "took" Indian women. Regardless of who the father was, the children of Cherokee women were always considered Cherokee, not "half breed," as the whites called them. Many of them and their descendants were important tribal leaders in the late eighteenth and early nineteenth centuries. On the other hand, white women married to Indian men (there were seventy-three such couples in 1826) had no status in the tribe at all, and their mixed-blood children did not become citizens of the Cherokee Nation until 1825.[43] Something of the extent of the trade between the Europeans and the indigenous people is suggested by the statement on John Mitchell's map of 1755 that "the English have Factories & Settlements in all the Towns of the Creek Indians of any note, except Albamas, which was usurped by the French in 1715 but established by the English 28 years before." The map could have also stated that there were English traders in most of the Lower and Middle Towns of the Cherokees as well.[44]

The early British explorers found cultures that had already been significantly affected by European trade. By trading deerskins, which at the time seemed to be in endless supply, they could have a range of European manufactured goods, including cast-iron cookware, which they immediately recognized as far superior to their traditional pottery, at least in terms of utility and durability, if not aesthetics. As a result, it did not take long for the Cherokee pottery tradition that dated back to the Woodland Period to disappear, although their tradition of making fine baskets from river cane continued, since their baskets remained superior to anything that could be imported. Skiagunota, one of the old warrior chiefs of the Lower Towns, recognized the change this trade with the Europeans had wrought: "My people . . . cannot live independent of the English. The clothes we wear we cannot make ourselves. They are made for us. We use their ammunition with which to kill deer. We cannot make our guns. Every necessity of life we must have from the white people."[45]

The volume of deerskins being shipped out of Charleston ballooned, and by the eighteenth century the transportation of tens of thousands of hides out of the interior had become a major problem. The Cherokees and other interior tribes still depended on human "burtheners" for transporting hides and other trade goods, but, after the Virginia traders began supplying horses for that purpose, it became almost impossible to find people to carry the hides to market. In the 1720s South Carolina built a horse path to Tugaloo, and in 1740 the Cherokees themselves opened a horse path down the west side of the Savannah River to Augusta.

In the early eighteenth century, the bitter Cherokee complaints against the traders were legion, as they were routinely swindled and cheated. Those complaints played a large part in precipitating the Yamasee War, which broke out in the spring of 1715 and involved Creek, Choctaw, Catawba, and the smaller Savannah River tribes. Instigated perhaps by the Ochese Creek, Indian attacks on white settlers, as well as the murder of nearly all of the hundred or so traders spread across the region, filled Charleston with refugees and for a time seemed to threaten the existence of the colony. In July, however, Col. George Chicken (1685–1727) and his Goose Creek militia decisively defeated the Yamasee, whose attacks had hit closest to home, and put the rebellious tribes on the defensive. It is within that context that Nacoochee and most of the other Indian towns in northeastern Georgia first enter the historical record.

NACOOCHEE, CHOTA, AND THE LOWER TOWNS

Twenty-five Lower Towns have been identified as existing in the second and third quarters of the eighteenth century. One of the few towns shown on nearly every map of the period was Tugalo, or Tugaloo, forty miles southeast of Sautee Nacoochee, where major trading paths between Charleston and the Cherokees crossed the Tugaloo River near the mouth of Toccoa Creek. Now flooded by Lake Hartwell, the site had been occupied since the fifth or sixth century CE and was being visited by English traders as early as 1690.[46]

The town of Chota was almost certainly associated with the Nacoochee Mound, which Wauchope's 1937 archaeological survey designated WH-3. The neighboring town of Nacoochee was likely the "Eastwood Site" that Wauchope documented a mile and a half east of the famous mound and designated WH-2. When Land Lot 75, which encompasses that site, was sold in 1829, the deed description mentioned that it lay "on both sides of the Chatahoochee River known by the name of the Nau-cu-cha old Indian town." Both sites, Wauchope reported, were "occupied contemporaneously . . . throughout their two long histories," beginning in the Early Woodland Period two or three thousand years ago.[47]

Nacoochee and its companion town, Chota, which were among the most westerly of the early Lower Towns, first appear in the historical record in the journals that Colonel Chicken kept during the course of a diplomatic mission to the Cherokee over the winter of 1715–16. On several occasions during the course of that winter, he passed through and stayed in Nacoochee and Chota, which was often called Little Chota to distinguish it from Chota, or Echota, the Cherokee capital and one of the Overhill Towns.

In December 1715 Colonel Chicken was dispatched to Tugaloo, the closest Cherokee town to the Carolinians, in an attempt to persuade the tribe to stay out of the ongoing Yamasee War. With a contingent of his militia, he arrived at Tugaloo on 29 December 1715 and was toasted with the ceremonial "black drinke" that only "great men and captains" can drink. The next day he met with Tugaloo's "Congger," or conjurer, the term the English applied to Cherokee spiritual leaders.[48]

On New Year's Day 1716, Chicken received instructions to go to "Chottee" and that afternoon "marched west about 4 miles to a Town called Tawcoe," or Toccoa. The journey continued on 2 January, with Colonel Chicken commenting that "ye way that we came is very hilley and stoney with severall small Crickes." They apparently spent the night at "Suckee," or Soque, where the trail from Tugaloo crossed the Soque River a mile or two north of present-day Clarkesville, and arrived in "Nocouchee," which was "about 12 miles" from Soque, before dark. He waited, he wrote, "until ouer Company come up and ye Indians painted them sealves then we marcht to a Town of peace adjoin to it called Chottee, whare wee was mett by ye head men of that place and most Towns from ye other sides of ye Hilles."[49]

They were met with great ceremony by the assembled conjurers and headmen at the "Round House," or town house, at Chota, but the initial meeting was interrupted by young warriors who began dancing a war dance that went on all night long, in spite of pleas from some of the elders to stop. The next day Colonel Chicken continued hearing the tribal leaders' complaints about the traders, who the Cherokees said, "had ben verry abusefull to them of latte and not as whitte men used to be to them formerly." That night he sat again in "ye round house" and listened to the warriors continue their threat to raise the Red Stick of war with the English against the Creek. In spite of the war dances, Colonel Chicken left Chota feeling optimistic and spent a snowy evening at Soque before returning to Tugaloo, where he heard that the Yamasee had fled to Saint Augustine. "The Charrikees are our friends," he wrote in his report, "and the head men of the Creeks [*sic*] are coming here to us in order to go down wth us to ye English Governmt to sue for a Peace and Trade wth us."[50]

On 20 January 1716, Colonel Chicken left Tugaloo again, this time bound for Quanassee, one of the Middle Towns, on the upper reaches of the Hiwassee River. Again he spent nights at Soque and Chota before setting off for Unicoi Gap, on a road so steep and rocky that they were forced "to light and walke more then ride." At the gap he noted the short distance, less than a mile, between the headwaters of the Chattahoochee and Hiwassee Rivers. He no doubt drank from what was later known as Herbert's Spring, one of the origins of the Hiwassee, where "it was natural for strangers to drink thereof, to quench thirst, gratify their curiosity, and have it to say they had drank of the French waters."[51] When he arrived at Quanassee, he heard more complaints against the traders and talk of war before traveling all day and arriving back at Chota after dark.

The morning of 27 January 1716, "ye warre houpe" broke across Chota, as news arrived that the Creek emissaries expected at Tugaloo had arrived and been promptly murdered by the Cherokee, who believed them to be part of a larger Creek war party. Colonel Chicken rushed back to Tugaloo to help contain the damage, but the incident set off a war between the two tribes that would not be put to rest until the Battle of Taliwa in 1755. Even then there were Creek raids on Tugaloo as late as the 1780s. For a while the threat of a Cherokee-British alliance kept the Creek at bay for the Carolinians, but it was not long before the Creek renewed their raids against the Cherokee. One of those raids was on Nacoochee and occurred sometime in 1717. It was noted by William Hatton, South Carolina's principal factor for the Cherokee trade, in a detailed report to his superiors in Charleston concerning the ongoing problems with the Indian trade: "In the meantime

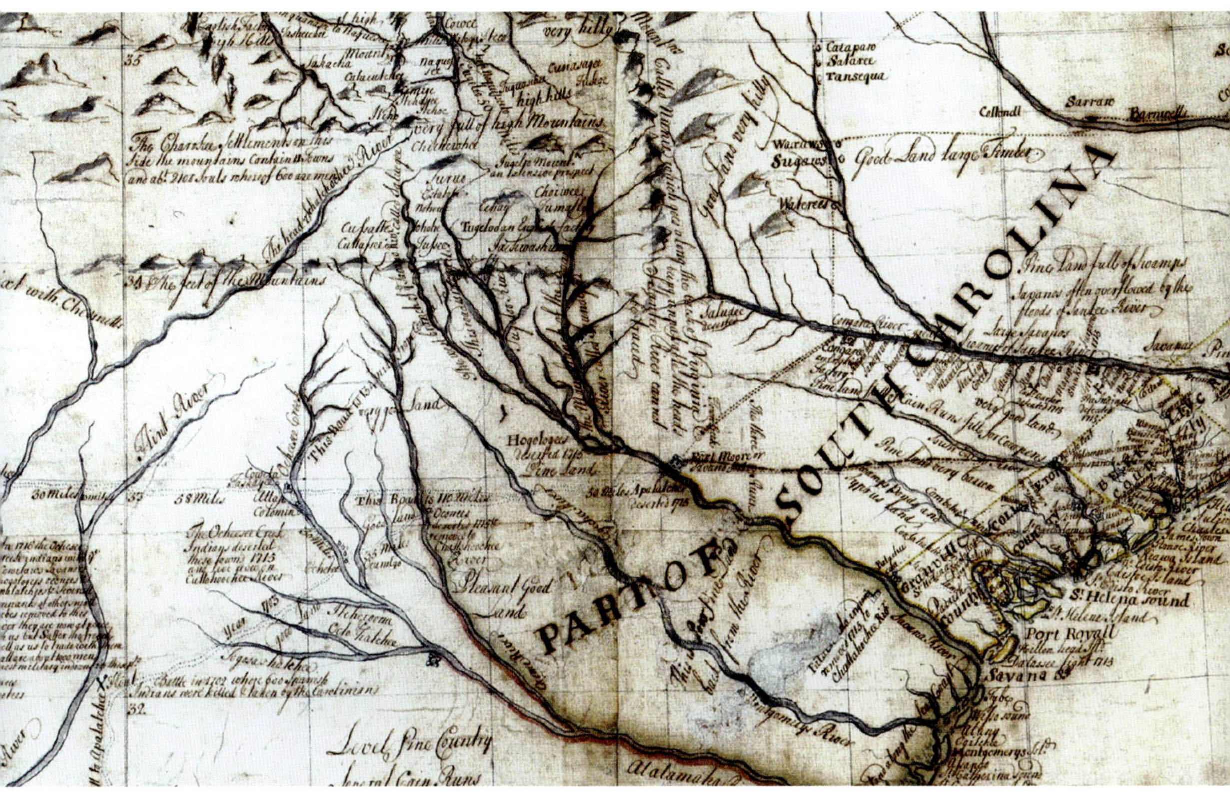

Part of Hammerton's *Map of the Southeastern Part of North America*, 1721. No towns are shown along the upper Chattahoochee, but where they would have been located (*upper left*) is the notation "the Charokee Settlements on this Side of the Mountains contain 11 towns and about 2,100 souls, whereof about 600 are fighting men." "Toogeloo an English Factory" is noted just above center here. Cummings, *Southeast in Early Maps*, plate 48.

the Creeks [*sic*] came upon a town called Nogoutchee and destroyed it, carrying off an abundance of slaves and killed most of the rest of the inhabitants. The next morning before they left the place they killed three of our factors on the path between Chottee and the aforesaid town, [the two] being about a mile asunder."[52]

The human toll was tremendous, although it is not clear if Nacoochee was burned or simply plundered by the Creek.[53] Around the same time, Hatton reported on the ongoing series of attacks on trading posts among the Cherokee:

> I went to Chottee a Town about 30 miles distant from my Residence [at Tugaloo] to Trade. At my entrance I found that Store was broken and most of the Goods and some Skins Stolen. Whilst I was there I receiv'd news that Quanissee Store was broken and likewise Tunnissee Store. So that Scarce a Store we had in the Nation escap'd; this was not all for those that did it was head Men, and was not asham'd to own it, but would often bid us begone out of the Nation. They did not want us nor our goods among 'em, and if that we Staid till such a time we Should be kil'd. All their Cry was that the Virginians was very good, but they valued us of Carolina no more then dirt.[54]

In 1721 Francis Varnod conducted a census of the Cherokee. He counted 221 people at Chota, including 59 men, 97 women, and 65 children; at Nacoochee he counted 90 people, only 11 of whom were women and only 7 children. Nacoochee continued to be shown on later maps, and, as the Lower Towns in South Carolina were abandoned, at least some of the refugees may have made new homes in the towns at the upper reaches of the Chattahoochee River. Also in 1721 the Cherokee ceded their lands along the Santee, Saluda, and Edisto Rivers in the South Carolina Piedmont, which were part of a hunting range shared with the Catawba and other tribes. It was only the first of what would be thirty-six separate land cessions forced on the Cherokee over the next 114 years.[55]

The end of the Yamasee War brought an end to the Indian tribes on Charleston's doorstep, but tensions with the Muscogee and Cherokee increased. In 1725 Colonel Chicken conducted another diplomatic mission to the Cherokee, whom the French were still trying to lure into an alliance. His journal from that trip does not mention Nacoochee or Chota, although the text of his journal has been interpreted to say that it does. Chicken spent the night of 21 July at a place he called

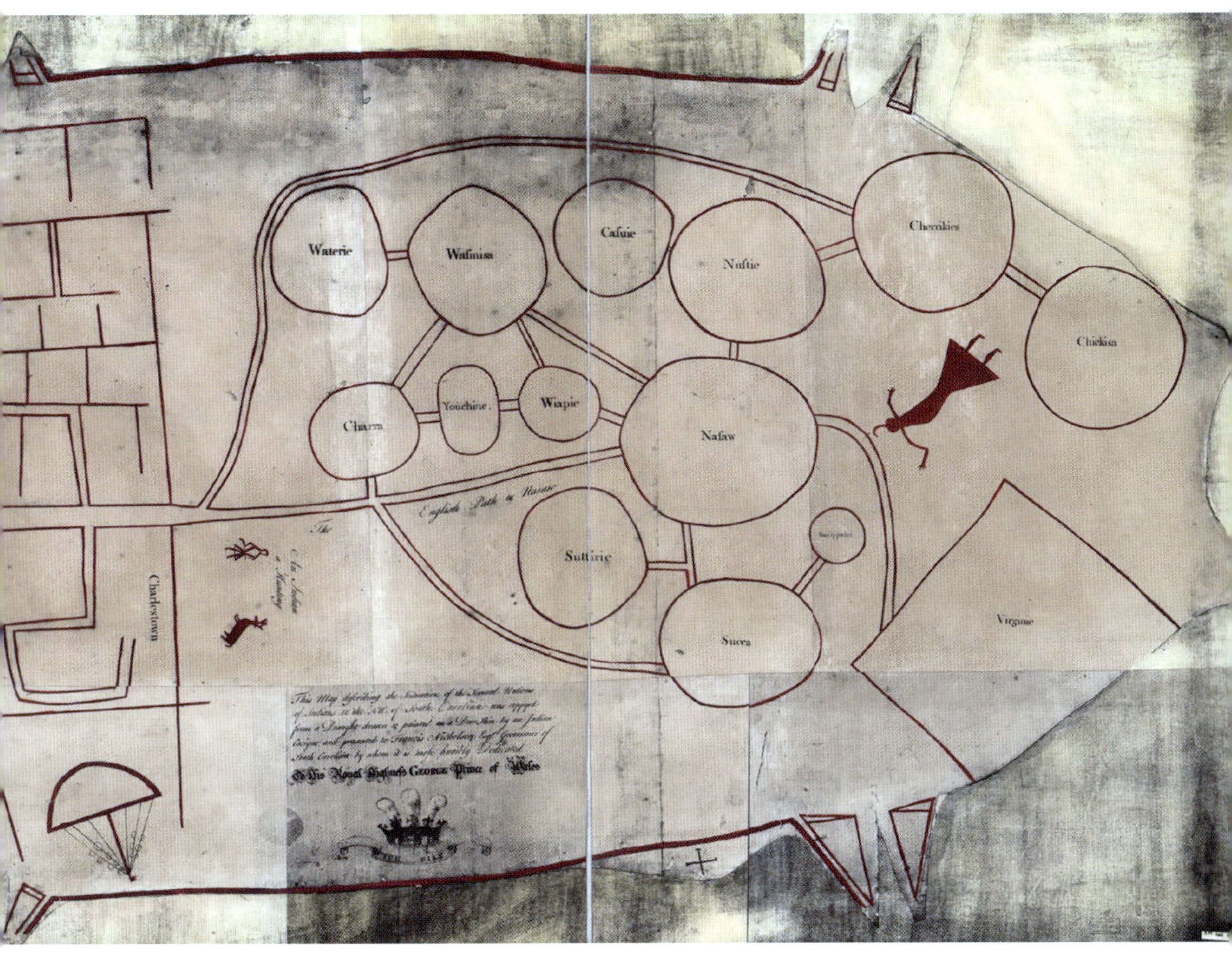

The so-called *Catawba Deerskin Map* dates to the 1720s and illustrates something of the disconnect between European and Indian world views in the eighteenth century. Item 2005625337, Geography and Map Division, Library of Congress, Washington, D.C.

Part of Herbert's *New Mapp*, dated 1725 but actually a 1744 copy of the original map, which was lost. It is the earliest map yet located that shows the towns of Chota and Nacoochee (*just below and left of center*). Map HMAP1725H, Special Collections, Hargrett Library, University of Georgia, Athens.

"Nocochee," but it is clear within the context of the journal that he was in one of the Middle Towns, most likely at Nequassee.[56]

For the remainder of the eighteenth century, Cherokee history is a terrible story, beginning with a smallpox epidemic in 1738–39 that James Adair said carried away half of the tribe. The loss of so many young adults, the disease's prime victims, was devastating. "The old magi and religious physicians, who were consulted on so alarming a crisis, reported the sickness had been sent among them on account of the adulterous intercourses of their young married people, who the past year had, in a most notorious manner, violated their ancient laws of marriage in every thicket and broke down and polluted many of the honest neighbors' bean-plots by their heinous crimes, which would cost a great deal of trouble to purify again." It was a sign of end times to some Cherokees and ultimately confounded the tribe's conjurers. When their healing rituals failed, Adair wrote "all the magi and prophetic tribe broke their old consecrated physic pots, and threw away all the other pretended holy things they had for physical use, imagining they had lost their divine power by being polluted, and shared the common fate of their country." Being "naturally proud," he continued, "a great many killed themselves" after seeing themselves disfigured by the scars that the disease invariably left, if one survived.[57]

Throughout the eighteenth century, Cherokees were involved with what Mooney called "chronic warfare with their neighbors." Enmity between the Cherokees and the Creeks was hereditary, Mooney thought, and only worsened with the founding of the Georgia colony at Savannah in 1733 and of Augusta at the falls of the Savannah in 1735. By midcentury, as many as five hundred thousand deerskins were shipped from southern ports, and, as the deer became more scarce, competition between the tribes grew apace.[58] In 1750 Creek warriors destroyed the Cherokee Lower Towns of Echy and Estatoe on the Tugaloo River and, two years later, Keowee, Tugaloo, Oconee, and Tomassee as well. Against such a blow, the Cherokee could only pull back and consolidate. Mitchell's 1755 map shows as much, with most of the Cherokee towns west of the Savannah included among "Deserted Cherokee Settlements." Chota and Nacoochee were both marked with Mitchell's symbol for abandoned towns.

Deserted Cherokee Lower Towns, including "Chotee" and "Nanquchee" (*left and below center*), 1755. South Carolina claimed what is now North Georgia until after the Revolution. Mitchell, Kitchin, and Millar. *British and French Dominions.*

FRENCH AND INDIAN WAR

The English had a difficult time maintaining an alliance with the Cherokee as the French and Indian War erupted in 1754. Henry Timberlake (1730–65), a British officer whose memoirs provide much of what is known about the Cherokee during that period, thought he knew why.

> I found the [Cherokee] nation much attached to the French, who have the prudence, by familiar politeness—which costs but little and often does a great deal—and conforming themselves to their ways and temper, to conciliate the inclinations of almost all the Indians they are acquainted with, while the pride of our officers often disgusts them. Nay, they did not scruple to own to me that it was the trade alone that induced them to make peace with us, and not any preference to the French, whom they loved a great deal better. The English are now so nigh, and encroached daily so far upon them, that they not only felt the bad effects of it in their hunting grounds, which were spoiled, but had all the reason in the world to apprehend being swallowed up by so potent neighbors or driven from the country inhabited by their fathers, in which they were born and brought up, in fine, their native soil, for which all men have a particular tenderness and affection.[59]

In 1755 the Cherokee signed a treaty with South Carolina, ceding claim to more hunting land in the Carolina Piedmont between the Wateree and the Savannah Rivers. The following year a treaty secured a Cherokee-British alliance and allowed the British to construct Fort Prince George at Keowee, the most important of the Lower Towns, and Fort Loudon near Tanasi, among the Overhill Towns. After fighting with the British in Virginia, Cherokee warriors returning home suffered a series of unprovoked attacks by backcountry settlers, who had the audacity to take the scalps of those they killed as "French Indians," for which they could collect a bounty. Even as the Cherokee were mourning their dead, a few soldiers at Fort Prince George began assaulting Cherokee women, "which set fire to the fuel, and kindled it into a raging flame," threatening to set off another war. Attempts by the Cherokee to extract justice were met with British arrogance and double-dealing, and in November 1759 the British themselves declared war. As Adair put it, "In brief, we forced the Cheerake to become our bitter enemies, by a long train of wrong measures, the consequences of which were severely felt by a number of high assessed, ruined, and bleeding innocents."[60]

In June 1760 a British force of 1,600 men, led by the Scottish general Archibald Montgomery, Eleventh Earl of Eglinton (1726–96), descended on the Cherokee Lower Towns. Mooney described the campaign, "Crossing the Indian frontier. Montgomery quickly drove the enemy from about Fort Prince George and then, rapidly advancing, surprised Little Keow'ee, killing every man of the defenders, and destroyed in succession every one of the Lower Cherokee towns, burning them to the ground, cutting down the cornfields and orchards, killing and taking more than a hundred of their men, and driving the whole population into the mountains before him."[61]

Over the Blue Ridge, he continued down the Little Tennessee River until he was met by a large Cherokee force on 27 June 1760, defeated, and forced to retreat to Fort Prince George. The following summer Col. James Grant led a second expedition of some 2,600 men, including some Chickasaw and nearly all the few surviving Catawba warriors, and proceeded to take up where Montgomery had left off. After defeating the Cherokee in a second battle, he proceeded to destroy all the Middle Towns, including crops, orchards, and granaries, and drove survivors into the mountains. Crushed, the Cherokee surrendered.

The Treaty of Paris in April 1763 ended the French and Indian War, and that fall a royal proclamation was issued prohibiting any British settlement west of the Blue Ridge. The proclamation did nothing to control the growing tide of European settlers pressing against the frontiers. Maintaining the integrity of treaty boundaries all along the southern frontier was an exercise in frustration, and the Cherokee were subjected to relentless pressure for more cessions. So, on 1 June 1773, the tribe signed the Treaty of Augusta, ceding claim to two million acres in the Broad River basin in the eastern Georgia Piedmont, encompassing most of what are now Hart, Madison, Elbert, and Oglethorpe Counties and parts of several others.

The Americans

When the Revolution broke out in 1775, the Cherokees again allied themselves with the British, whose strength they were counting on to control the increasingly lawless frontier. In the summer of 1776, the British attacked Charleston from the sea, while the Cherokee launched a series of violent raids that struck fear in the backcountry. The American response was fast and furious. In July 1776 Maj. Andrew Williamson (ca. 1730–86) collected some 1,100 backcountry militia, including 240 from Georgia under the command of Col. Samuel Jack and 230 from South Carolina under Maj. Andrew Pickens (1739–1817). By the middle of August, all the Lower Towns had been destroyed again. That there were only two towns (Tugaloo and Chota) to destroy in northeastern Georgia was a sign of the severe depopulation of the area, as the Cherokee crowded westward in their futile attempt to escape the violent frontier.[62]

Williamson then crossed Rabun Gap to join forces with the North Carolina militia under Col. Griffith Rutherford and continue its work in the Middle Towns. Rutherford's militia had assembled at Old Fort in southwestern Burke County and no doubt included many of the families that would move to Habersham County in 1822. Mooney described the result,

> As the army advanced every house in every settlement met was burned—ninety houses in one settlement alone—and detachments were sent into the fields to destroy the corn, of which the smallest town was estimated to have two hundred acres, besides potatoes, beans, and orchards of peach trees. The stores of dressed deerskins and other valuables were carried off. Everything was swept clean, and the Indians who were not killed or taken were driven, homeless refugees, into the dark recesses of Nantahala or painfully made their way across to the Overhill towns in Tennessee, which were already menaced by another invasion from the north.[63]

When the campaign was done, there was virtually nothing left of the Lower and Middle Towns, the Overhill Towns were severely damaged, and thousands of Cherokees were homeless. Crushed, they had little choice but to sign the Treaty of DeWitt's Corner in May 1777, ceding nearly all of their land in South Carolina, including several of their

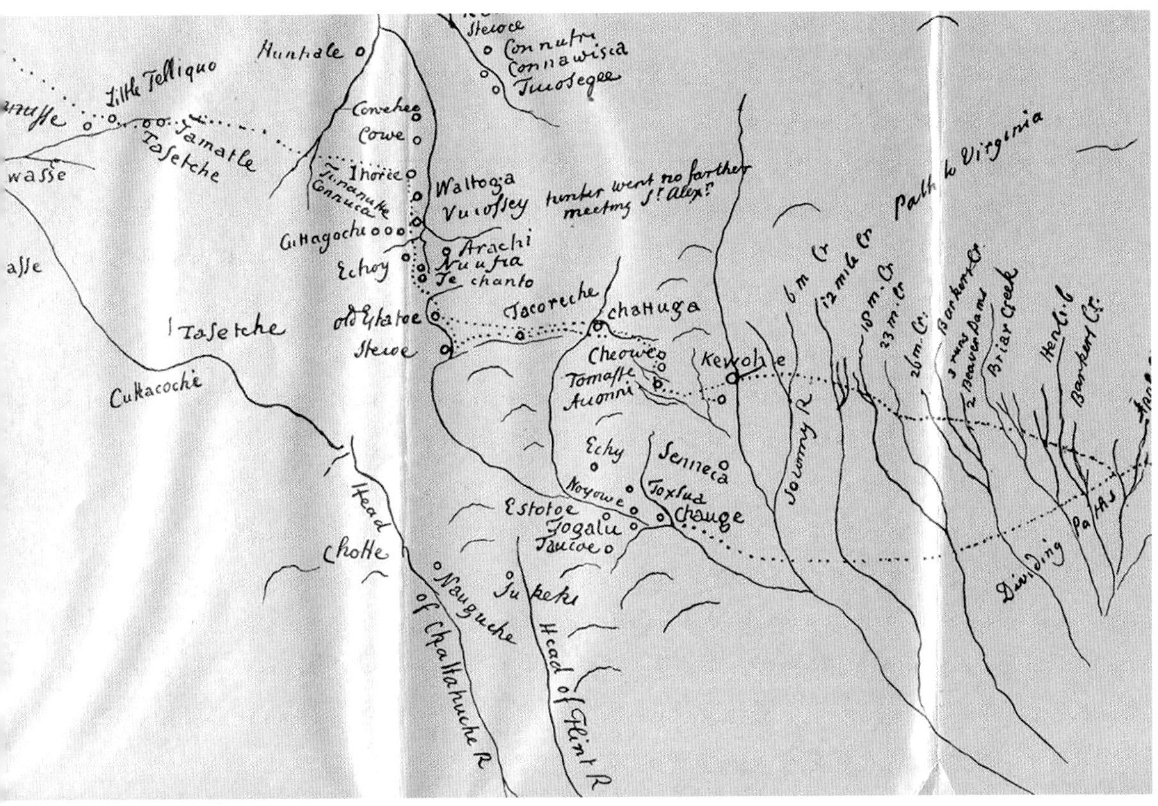

Lower Towns. The treaty ended any organized involvement by the Cherokees in the American Revolution, but the Cherokees in Georgia were apparently enough of a problem that in October 1781 Major Pickens launched a surprise attack on Sautee and the other settlements on the upper Chattahoochee River. Baker reports seventy-seven Cherokee were killed and another forty taken prisoner and likely sold as slaves.[64]

Part of George Hunter's *Map of the Cherokee Country and the Path Thereto in 1730*, a 1917 redrawing of the original map. The Lower Towns of Tugaloo, Soque, Nacoochee, and Chota are just below center in this image. South Caroliniana Library Map Collection, South Caroliniana Library, University of South Carolina, Columbia.

A preliminary draft of the Treaty of Paris was agreed on in November 1782, effectively ending the Revolution and leaving the Cherokee with little choice but to sign another treaty with Georgia, at Long Swamp Creek in May 1783. The treaty established the northern border of Georgia by an arbitrary line drawn from a point on the Tugaloo River in "a direct line to the top of the Currohee mountain," twenty-five miles southeast of Sautee Nacoochee, "thence to the head of the south fork of

Oconee river," forty miles to the southwest in what is now northeastern Gwinnett County. The boundary was untenable as soon as it was drawn, perhaps intentionally so. There were virtually no natural landmarks to mark most of its course, and it left the natural boundary, the Chattahoochee River, well within Cherokee territory. That fall another smallpox epidemic swept through the desolated tribe.[65]

TREATY OF HOPEWELL

In November 1785 the Cherokee and the United States signed the Treaty of Hopewell, which reaffirmed earlier state treaties. The treaty also established national control over relationships with the tribes, much to the relief of the Cherokee and others tired of dealing with the states' varying approaches to tribal relationships. The treaty also purported to guarantee the integrity of the Cherokee Nation's borders in perpetuity, even as several hundred white families were already squatting on the Cherokee side of the boundary. The Hopewell treaty was significant in its recognition of the sovereignty of the Cherokee, but the tribe was bitterly divided. The national councils were dominated by older chiefs, whom a vocal group of mostly younger chiefs accused of being too accommodating to the white man's demands.

Dragging Canoe (1738–92) advocated using any means necessary, including violence, to protect the tribe's lands. In 1776 he led a few hundred Cherokee to withdraw from the Overhill Towns on the Little Tennessee River to a new settlement on the Tennessee River near present-day Chattanooga, not far from the mouth of Chickamauga Creek. There would eventually be eleven so-called Chickamauga towns, all in northwest Georgia and southeast Tennessee. There were no more Cherokee land cessions in Georgia in the eighteenth century, but Dragging Canoe and his followers continued to wage a grinding guerrilla war across the frontier for most of the next decade.

The present study has been unable to document why Chickamauga Creek, which drains the northern end of Sautee Valley, is so named. The field notebook from the 1820 survey of the Third District names the Chota and Chattahoochee Rivers and Dukes, Smiths, Brasstown, Shoal, Amies, and Sauty Creeks. Among the numerous unnamed branches and creeks recorded in that survey, a later hand has added "Chickamaugy" to one of the streams. However, "Chickamaugee Creek," flowing south into the Third District, was shown on the original plat of survey for the Sixth District. The name may have meant the presence of Cherokees sympathetic to Dragging Canoe's cause, or it may reflect the fact that John Martin's two wives were raised in northwestern Georgia around Chickamauga, and so they gave a familiar name to the creek running through their plantation. The new Chickamauga towns were part of a major shift in the Cherokee Nation between 1775 and 1794. Relentless pressure from Virginia, the Carolinas, and Tennessee forced the Cherokee to relocate over and over, shifting the demographic center of the nation some hundred miles to the southwest.[66]

Dragging Canoe died in 1792, but fighting continued even after his followers suffered decisive military defeat in the fall of 1794, and the war did not end until the signing of the Treaty of Tellico Blockhouse in 1798. The crisis of defeat forced the tribe at last to establish a formal national council, empowered to speak for all the towns. Little Turkey (1758–1801), First Beloved Man, was made the tribe's first national chief. Over the next twenty years, the traditional ethnic state of the Cherokee steadily evolved into the Cherokee Nation of the early nineteenth century.[67]

THE NEW CHEROKEE NATION

One historian of the Cherokee has noted that the tribe's experience of "culture shock reached its most desolating stage" in the last quarter of the eighteenth century. Old institutions and authorities failed and new ones had not yet matured, leading to a chronic breakdown in law and order, especially in outlying areas. A few Cherokee simply gave up and surrendered themselves to the oblivion of alcohol. Adair noted as much when he reported that some of the Cherokee were excessively immoderate in drinking—"They often transform themselves by liquor into the likeness of mad foaming bears."[68]

Most of the Cherokee, of course, remained determined that they would survive as a people, but, as early as 1782, a few families abandoned hope of peaceful coexistence with the Americans and began migrating to Arkansas. By 1795 there was a "steady stream" of westward emigrants. The majority of the Cherokees remained in the East and resented those who abandoned the efforts to preserve their homeland, since any outmigration automatically gave the federal government

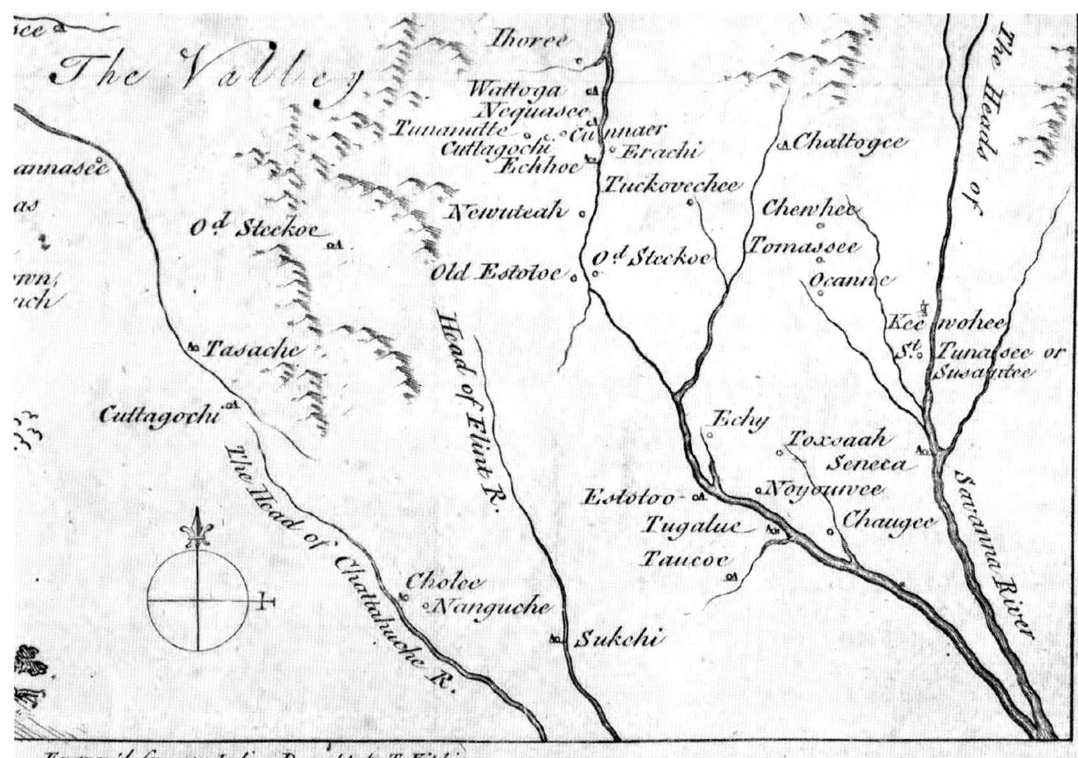

Part of Kitchin's *New Map of the Cherokee Nation*, showing the Lower Towns of the Cherokee, published in 1760. Chotee, Nanguche, Sukehi, and Tugalue are in the lower portion of this image. This is one of several maps that mistakenly place the head of Flint River in northeast Georgia. Item 2021586055, Geography and Map Division, Library of Congress, Washington, D.C.

another excuse to seize Cherokee territory. As the national council gained power in the first decade of the nineteenth century, the very definition of Cherokee changed. No longer were ethnicity, language, and culture enough to define the Cherokee Nation. Faced with treaties built on European concepts of private property and the nation-state, the tribe was forced to demarcate and attempt to harden a border around their homeland as they had never done before. After 1809 actual residence inside those tribal boundaries was required for citizenship; Cherokee became not just a people but a place.[69]

The men had the most difficult time in adjusting, as steady encroachment on their hunting grounds made a traditional way of life increasingly untenable. The sale of the last of the millions of acres of tribal territory in Tennessee in 1805–6 sparked a rebellion among a faction of mostly younger chiefs and led to the assassination of Doublehead (1744–1807), the principal chief who had signed the treaty and been one of the most vigorous supporters of acculturation. Efforts by federal agents, missionaries, and some tribal leaders to turn the Cherokee to farming were difficult, since there was a strong cultural bias against agricultural work by Cherokee men, who traditionally left such work to women and the elderly. Nevertheless, the Cherokee Nation transformed itself into a nation of farmers in the early nineteenth century and, with the Choctaw, the Cherokee were at the forefront of efforts to acculturate and preserve their homeland.

Ultimately the tribe was led politically by two or three hundred mostly mixed-race families, descended from white traders who had married Cherokee women and the many fewer Cherokee men who had married white women. These families profited most from accommodating white Americans, while the vast majority of the Cherokee struggled to subsist, always one crop failure away from disaster. By the first decade of the nineteenth century, Cherokee towns were fast being drained of life as families moved out to widely dispersed farmsteads and embraced an agricultural lifestyle. The town house remained important to the community, but it was no longer at the center of a dense settlement. The adjustment to a sedentary, agricultural way of life was not easy, and a cycle of "starving times" (1804, 1807, and 1811) and smallpox epidemics (1806, 1817, and 1824) plagued the Cherokee Nation as it struggled to maintain its integrity and rebuild its institutions.[70]

The tribe steadily refused to break up communal ownership of property into severalty and resisted federal efforts to eliminate the Cherokee tradition of matrilineal inheritance. The emptying towns were contributing to deterioration of the tribe's clan system, especially the ancient law of clan revenge, known as "the law of blood" or "blood revenge." Fundamental to traditional Cherokee concepts of order and justice, revenge for the killing of a clan member by someone from another clan too often resulted in a series of retributive killings between clans that spiraled out of control.[71] In 1808 the council established "regulating parties" consisting of six men each, organized as a militia, "whose duties it shall be to suppress horse stealing and the robbery of other property within their respective bounds." Crucially, the law also exempted militia members from the law of clan revenge if they accidentally killed someone in the line of duty. In 1810 the council, with the consent of the seven

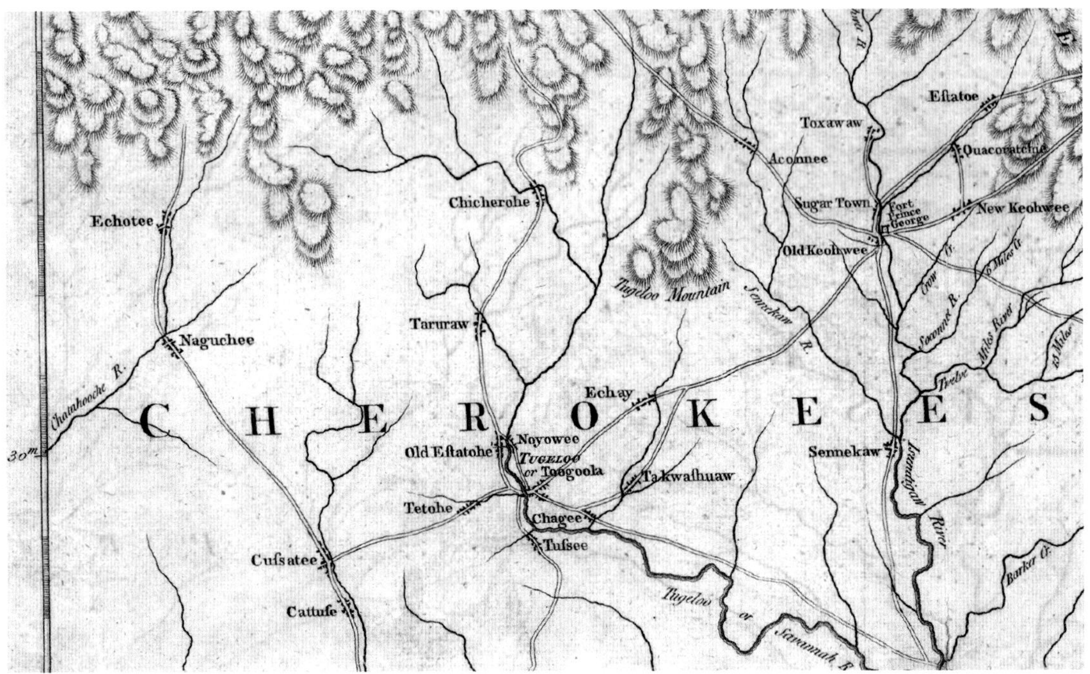

Detail from Mouzon, Laurie, and Whittle's *Accurate Map*, originally drawn in 1775 and published in 1794, which includes part of what is now Georgia. "Naguchee" and "Echotee" are shown at left, but, as with many early maps, the relative location of towns and streams is sometimes problematic. Item 2006626006, Geography and Map Division, Library of Congress, Washington, D.C.

clans, went further and passed "an act of oblivion for all lives for which they may have been indebted, one to the other."[72]

While absolving the clans of the necessity of revenge for past murders, the 1810 law did not abolish clan revenge entirely but rather established tighter parameters within which the clans might exact revenge. These changes in the years before the War of 1812 reflected the gradual transfer of power from the clans to the national council and marked a major step in establishment of a national rule of law.[73] The Cherokee were the exemplar for tribal acculturation, leading the way among the so-called Five Civilized Tribes—Creek, Seminole, Choctaw, Chickasaw, and Cherokee. Moravian, Congregationalist, Baptist, and Methodist teachers established schools in the Cherokee Nation as early as the 1790s, but, except for the Moravians, all the Christian schools tended to proselytize rather than educate.

In the 1820s the tribe established a new capital, named New Echota, in northwest Georgia and more or less completed their nation building with the adoption of a constitution in 1827. Meanwhile, in the early

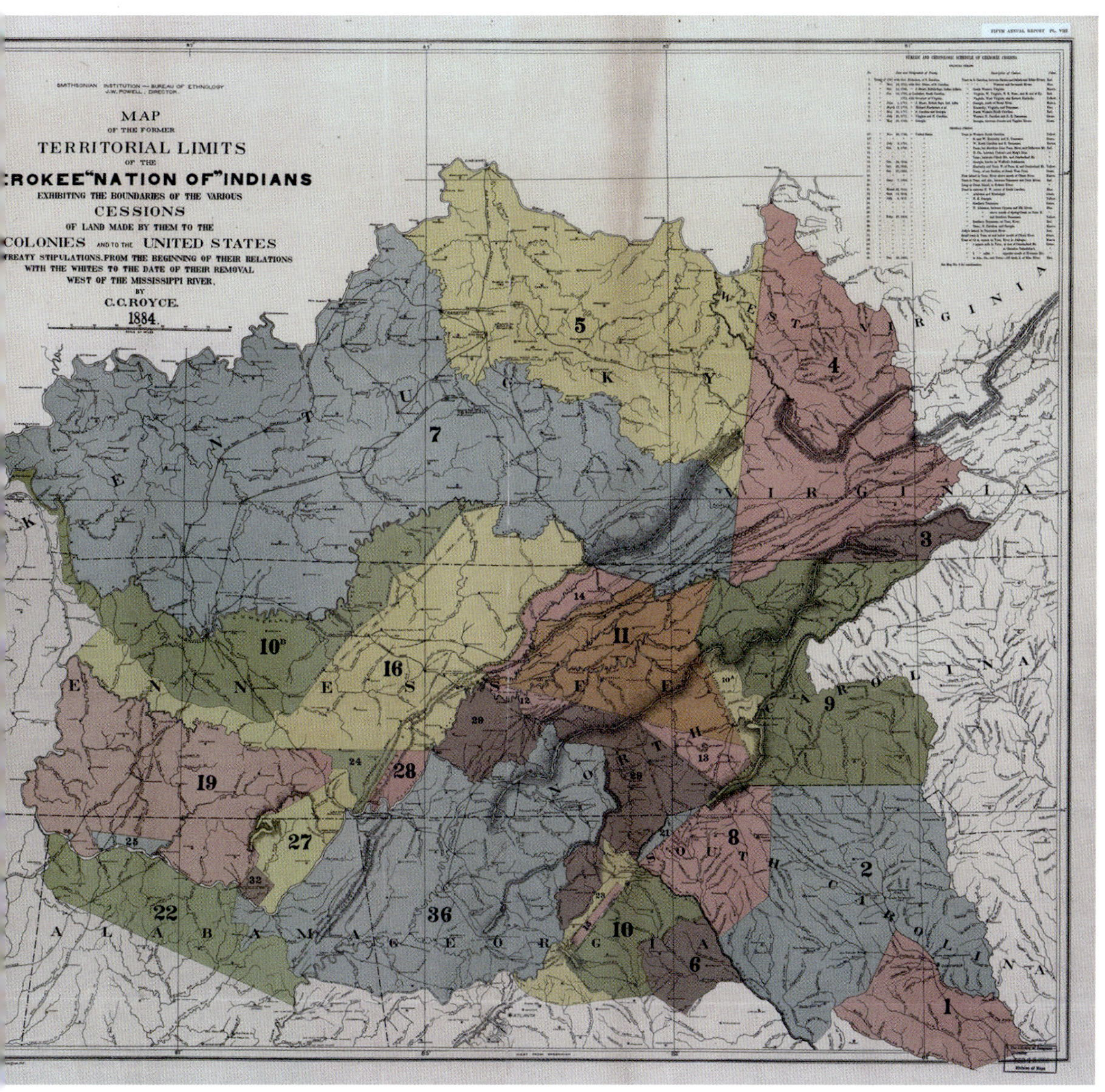

Royce's *Map of the Former Territorial Limits*, 1887, depicting the extent of the thirty-six land cessions forced on the Cherokee, 1721–1835. The numbers represent the order in which the treaties were signed, ending with number 36, the Treaty of New Echota. *Fifth Annual Report*, plate 8.

1820s, the great Sequoyah (ca. 1770–1843) completed development of the first Cherokee syllabus, which was adopted by the tribe and used to publish the first tribal newspaper in 1828. Literacy helped tie the tribe together and gave them a powerful weapon in their ultimately doomed battle to preserve their homeland against encroachment by white settlers. The Baptist preacher Evan Jones lived among the Cherokee for many years and, with other sympathetic whites, did what he could to support the tribe's efforts to maintain their homeland. It was a depressing fight. "Oh! Christians, pray for the Cherokees," he wrote. "Oppression scowls about their borders."[74]

GHOST TOWNS

At the turn of the nineteenth century, northern Georgia remained a part of the Cherokee Nation, but its people were in turmoil. According to one historian of the period, the mountains and valleys between the Chattahoochee and Conesauga Rivers were filled with "confused, disorganized, and discouraged" refugees from the destroyed Lower Towns. It is no wonder that there was "a persistent breakdown in law and order" as well.[75]

William Bartram missed Sautee Nacoochee in his famous travels in the 1770s, although he saw many of the other Lower Towns, and his journals remain an invaluable resource for understanding the Cherokee during that period. Benjamin Hawkins (1754–1816), general superintendent of Indian Affairs from 1796 to 1816, saw the valleys in 1796 when he made a journey from upstate South Carolina across northern Georgia into Alabama. In his journal he left one of the earliest descriptions of Sautee Nacoochee, which was then still part of the Cherokee Nation. Unlike Colonel Chicken, who arrived from Tugaloo on a route generally following today's State Route (SR) 17, Hawkins arrived from the Dividings, now Clayton, on more or less the route of today's SR 255.

On 27 November 1796, he entered Sautee Nacoochee and left the only eighteenth-century description of the Sautee village. Located on the north side of the junction of Chickamauga and Sautee Creeks and documented by Wauchope in 1939, the site was called "Santa" by Hawkins, who noted that it was "formerly an Indian town" but now presented a forlorn appearance:

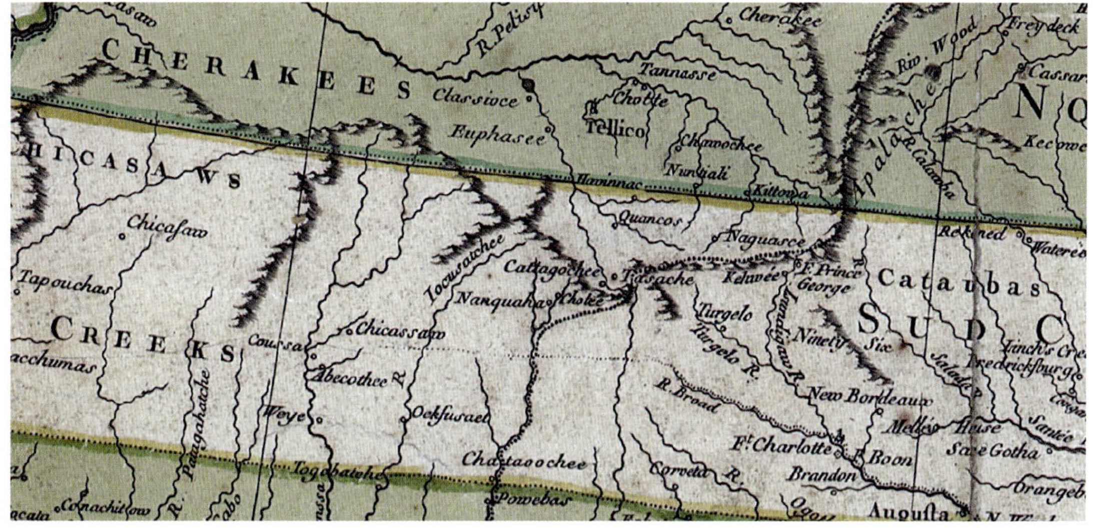

Part of Lattré's *Carte des Etats-Unis de l'Amerique*, 1784. "Nanquaha" and "Chotee" are at center and must represent Nacoochee and Chota. South Carolina claimed what is now northern Georgia until after the Revolution. Item 73691628, Geography and Map Division, Library of Congress, Washington, D.C.

> There was one hut, some peach trees and the posts of the town house . . . the lands about the town poor, gravelly, covered with dwarf trees, a mountain to our left, very barron. 4 miles farther south we passed through Little Chota, a creek 35 feet wide run through the town, there remains one small field of corn, some peach, plumb and locust trees, the border of the creek covered with small cane. The lands in the neighbourhood poor. As I approached this town I had a fine view of a mountain to the S. The water which oozed from near its summit formed a sheet of ice of.[76]

Hawkins' measurements of distance are considered reliable, so it seems the reference to four miles of travel is for that entire segment, it being a little over four miles from the site of Sautee to the crossing of Dukes Creek, which would have been a few hundred yards west of where SR 75 crosses today.

He failed to note the site of the town of Nacoochee, perhaps because it had disappeared entirely or perhaps because his route took him to the west, along what is now Rabun Road. He might also have forded the river just upstream from Sautee Creek and continued on the road that paralleled the south side of the river. In either case it seems clear that he was referring to a view of Mount Yonah as he traveled in a generally west-southwesterly direction along the valley. The "creek, 35 feet wide" flowing through the village must refer to the Chattahoochee

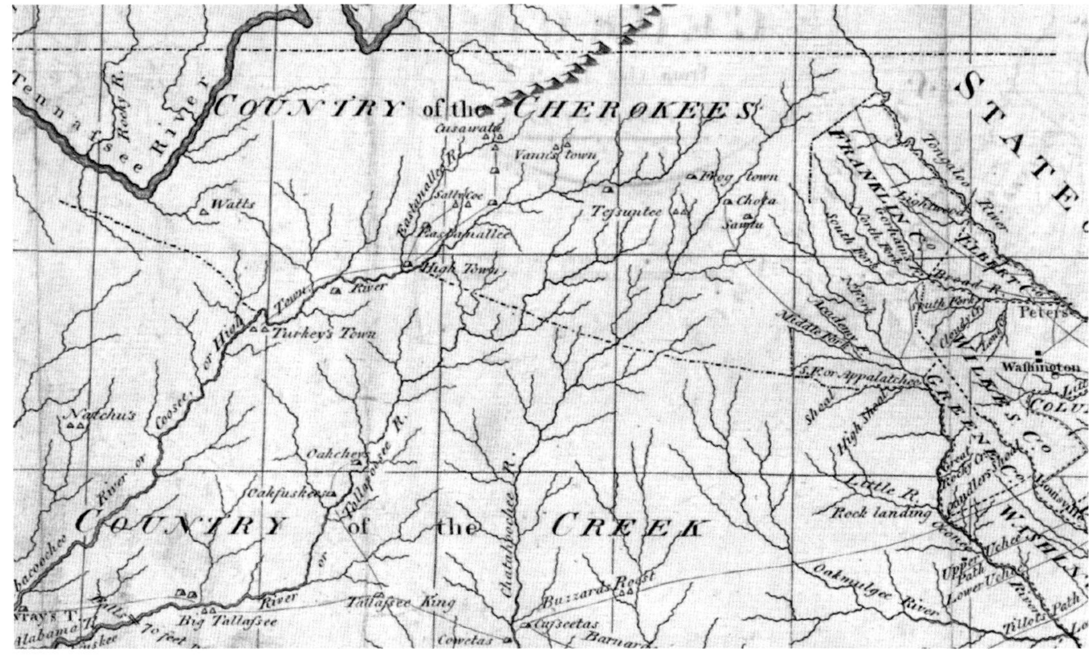

Part of Barker and Carey's *Georgia, from the Latest Authorities*, published in Philadelphia in 1795. This is the earliest map to more or less correctly depict the locations of "Sawtu" and "Chota" (*upper right*) and the first to depict "Frogtown" in what is now northern Lumpkin County, Georgia (*right of center*). Item 71000975, Geography and Map Division, Library of Congress, Washington, D.C.

River; in the eighteenth century the part of the river upstream from the Soque River was called Chota River by the Cherokees and was not considered to be part of the Chattahoochee River. Hawkins saw the ruins of the Cherokee at Sautee Nacoochee, even if the valleys might not yet have been totally depopulated and abandoned. They remained part of the Cherokee Nation, if only nominally, and would for a few more unsettled years.

4

Resettlement of the Valleys

The Treaty of Hopewell in 1785 confirmed Georgia's frontier with the Cherokee, including the cession in 1783, out of which the state had organized two large counties: Franklin and Washington. Sautee Nacoochee and much of northeastern Georgia remained well within Cherokee territory, but a huge wave of white immigration was beginning to crash against the Cherokee Nation. In 1790, when the first federal census was taken, Franklin County, which encompassed all of what are now Jackson, Clarke, and Oconee Counties, as well as most of Madison, Banks, Hart, and Stephens Counties, had a population of around a thousand people.[1] Ten years later population growth was such that Jackson County was created out of the southwestern half of the original Franklin County, and the two counties had a combined population of 14,595. The state's population as a whole only doubled during that period, suggesting something of the huge influx of new settlers into the upper Georgia frontier.

In January 1795 Georgia governor George Mathews signed the Yazoo Land Act, transferring thirty-five million acres, encompassing what are now Alabama and Mississippi, to four speculative land companies for $500,000. The scandal that erupted when the corruption behind the sales came to light remade Georgia politics. Part of the resolution of the fraud came in the so-called Compact of 1802, by which Georgia relinquished to the federal government any claim to those western lands. In exchange Georgia received $1.25 million and a promise that the claims to land inside the state by the Creek and the Cherokee would be extinguished "as soon as such purchase could be made upon reasonable terms." In the end that would be a long slog.[2]

Northeast Georgia

The egalitarian ideals of the Enlightenment that guided much of the early federal policy toward the indigenous peoples foundered against the raw violence and animus on the southern frontier. Not a few of the Cherokee's adversaries were the Scots-Irish, many of whom immigrated from Ulster and the Scottish Borders between 1717 and 1776 or were descendants of those who did. No other group of immigrants had a more profound impact on the development of the southern Piedmont and highlands. Far afield from the Enlightenment's understanding of the basic equality of humankind, many of them saw the indigenous people as little more than vermin to be exterminated. Long before the discovery of gold in northeastern Georgia, Hugh Montgomery, a federal agent with the Cherokee, noted "the prevailing idea in Georgia (especially among the lower class) . . . [is] that they are the rightful owners of the soil, and that the Indians are mere tenants at will; and indeed, Sir, there is only one point on which all parties both high and low in Georgia agree, and that is, that they all want the Indian lands."[3]

THE WOFFORD SETTLEMENT

Col. William Wofford (1728–1823) was one of those Scots-Irish. Born in Maryland, he moved to South Carolina in the 1760s and established an ironworks near Spartanburg. He was a lieutenant colonel during the Revolution and participated in Colonel Williamson's dreadful campaign against the Cherokees in 1776. In 1792 he bought a tract of four hundred acres near Toccoa, some twenty-five miles southeast of Sautee Nacoochee. With his son Nathaniel and about fifty relatives and friends, Wofford settled on what was then the frontier with the Cherokee Nation. Unfortunately, not until February 1798 did Benjamin Hawkins complete a survey of the actual boundary line established in the 1785 treaty, and only then was it realized that "Wofford's Settlement" was on the wrong side of the line.[4]

Wofford had no use for the Cherokees, and they in turn had "great resentment of his conduct." He and other intruders on Georgia's northern frontier were a constant aggravation, and tribal leaders demanded they be removed. The state, however, was loathe to remove these settlers who were there in part because of Georgia's own failure

Carey and Gridley's map *The State of Georgia*, in *Carey's General Atlas*, 1818, does not show the Wofford Settlement, but it does show "Chuta T[own]" and "Santee T[own]" (*just left of center*). Item 2006635240, Geography and Map Division, Library of Congress, Washington, D.C.

to properly survey its boundaries.[5] In the summer of 1801, a delegation of Cherokee chiefs traveled to Washington, D.C., to meet with the secretary of war and demand that Wofford and the other intruders be removed. A promise by the secretary that they would be removed by Christmas 1801 was not kept, and political pressure mounted to force a cession of the land. Twenty years later President Thomas Jefferson recalled, "No case of intruders ever occurred which excited more anxiety of commiseration with us than that of Wofford's settlement."[6]

In February 1803 negotiations were renewed, but it took until the fall of 1804 for a treaty to be signed. Negotiated by Sen. Daniel Smith (1748–1818) of Tennessee and federal agent Return J. Meigs (1734–1823), it was signed by ten Cherokee chiefs, one of whom was James Vann (1766–1809). Along with the original instructions to Smith and Meigs was the authorization to pay Vann up to $300 to secure his help in accomplishing the negotiations, although Meigs thought Vann too rich to bribe.

By the treaty's terms, the Cherokee gave up a strip of land four miles wide and over twenty-three miles long, much larger than the three-by-twelve mile tract that Wofford had originally claimed. In exchange the tribe would receive $5,000, an annuity of $1,000, and, it was hoped, peace along the frontier. Even then the treaty document was misplaced and never ratified. The mistake was at last discovered after repeated inquiries from the Cherokee, and it was finally ratified in 1824, by which time the Wofford Settlement tract had been extended, somewhat arbitrarily, ten miles to the southwest.[7]

The treaty did not stop other intruders, all well armed and "some of them shrewd and desperate characters, having nothing to lose and hold[ing] barbarous sentiments towards the Indians." In February 1808 federal agent Meigs reported to his superior, "The Cherokee complain that intrusions on their land on the frontier of Georgia by the white people are still continued and are increasing." To Georgia governor Jared Irwin, he wrote on the same day, "It is out of all dispute that . . . all of those on the waters of the Chattahoochee are really aggressors . . . in open violation of the laws, which justifies the complaints of the Indians and disturbs their quiet."[8]

Meigs knew from experience that attempting to remove the squatters was an exercise in frustration. Troops might be sent in to burn log cabins and destroy crops, but no sooner than the troops were gone, the cabins were rebuilt and the crops replanted. One Cherokee chief wrote Meigs that the intruders return "as thick as ever, the same as crows that are startled from their food again." The land was tax free, and all the squatters had to do was to wait for the inevitable land cession, when they could acquire the property cheaply and legally from the state. By 1808 Meigs had given up on the old policy of assimilating the indigenous people and expressed an increasingly common view when he wrote that "it is my opinion that there never will be quietness on any of these frontiers until the Indians are removed over the Mississippi."[9]

The Cherokee, like other tribes of the southeastern United States, were often bitterly divided among themselves over tribal support for acculturation. Opposition reached a new high in the years before the War of 1812, as Tecumseh (1768–1813), the great Shawnee chief, attempted to build a pan-Indian alliance that would permanently roll back the tide of white settlement. The Cherokee never adopted the angry and vindictive approach advocated by some of the tribes, but in May 1811 the Cherokee National Council instructed Meigs to remove all white people from the nation. Meigs thought that was "madness" but tried, unsuccessfully, to carry it out. The order was later modified to allow whites married to Cherokees to remain, along with a few "useful" whites such as teachers and gristmill operators.[10]

In late spring 1811, the Great Comet appeared, "enormous and brilliant" and portending "all kinds of woes and the end of the world," as Leo Tolstoy described it in *War and Peace*.[11] What it meant to the Cherokee is not clear, but Tecumseh and the Shawnee took the comet as an omen of good luck for their pan-Indian alliance. It was still brilliant on 16 December 1811, when the entire Southeast was rattled by the first of three massive earthquakes centered on New Madrid, in southeastern Missouri. The first is thought to have measured 8.1 on the Richter scale, the second 7.8, and the last as high as 8.8. Church bells rang in Boston, and, according to one report, "houses in the Cherokee Nation were knocked off their foundations and large sinkholes, thirty yards wide, suddenly appeared and then filled slowly with murky, green water." The Moravians at Spring Place in northwest Georgia recorded twelve significant tremors between December 1811 and April 1812.[12]

THE CREEK AND THE WAR OF 1812

In June 1812 war broke out between the United States and Great Britain, with the early campaigns all fought in the north, where the British had made Tecumseh an officer in their army. In June 1813 civil war broke out among the Muscogee, and, with some of them allied with the British, the Muscogee Confederacy posed a formidable threat to the United States. In August 1813 the Creek war faction, the Red Sticks, descended on Fort Mims on the Tallapoosa River in southern Alabama, where they massacred at least 250 men, women, and children and took many more prisoners.

News of the massacre terrorized the entire southern frontier, and there was an uproar demanding action against the Muscogee and their allies. Tennessee, Georgia, and Alabama militias were called out, and previous offers from the Cherokees to ally themselves against the British and the Muscogee were finally accepted by the Americans. When Gen. Andrew Jackson marched his army against the Muscogee in October 1813, there were as many as 600 Cherokees in his ranks. The fighting was vicious, culminating in the dreadful Battle of Horseshoe Bend in March 1814. General Jackson led a combined force of 6,600 against 1,000 Muscogee fortified in a bend in the Tallapoosa River in eastern Alabama. In the end only 200 Muscogee managed to escape to Florida, as Jackson's men cemented a terrible reputation by their desecration of the dead. The Creek War was over, and their nation was crushed. In the Treaty of Fort Jackson, signed 4 August 1814, they were forced to cede twenty-three million acres in what are now southern Georgia and Alabama.

Success in the Creek War and at the Battle of New Orleans in January 1815 made Jackson a national hero. It also unleashed a wave of violence against the Cherokee that began with the Tennessee militia's indiscriminate trail of destruction as they passed through the Cherokee Nation on their way home from Horseshoe Bend. Terrorizing the Cherokees, the militiamen shot livestock for sport, looted and burned houses, and destroyed crops in a large swath through northern Alabama. It was a sorry repeat of the backwoods attacks suffered by returning Cherokees after fighting as allies in the French and Indian War.

Meigs meticulously documented the destruction, much of it through testimony of white officers. His official estimate of property damage was $22,863.65, but privately he thought that it was actually twice that much. Jackson dismissed the charges out of hand and did nothing to fulfill the promises to pay Cherokee volunteers in the war. Even widows and orphans of the Cherokees who had been killed fighting alongside the Americans were ignored. There were also unresolved issues surrounding the border between the Cherokee Nation and the land that the Muscogee had been forced to cede in northern Alabama. In spite of Jackson and an army of speculators eyeing the area's rich cotton land, the secretary of war signed a treaty with the Cherokee on 22 March 1816 that resolved the boundary dispute in favor of the Cherokee. The treaty also gave the United States "the right to lay off, open, and have the free use of, such road or roads, through any part of the Cherokee nation . . . as may be deemed necessary for the free intercourse between the States of Tennessee and Georgia and the Mississippi Territory." Free passage for U.S. citizens was guaranteed, while the Cherokee agreed "to establish and keep up, on the roads to be opened under the sanction of this article, such ferries and public houses as may be necessary for the accommodation of the citizens of the United States."[13]

A second treaty was signed in September 1816, by which the Cherokee ceded the last of their land in South Carolina for $5,000.

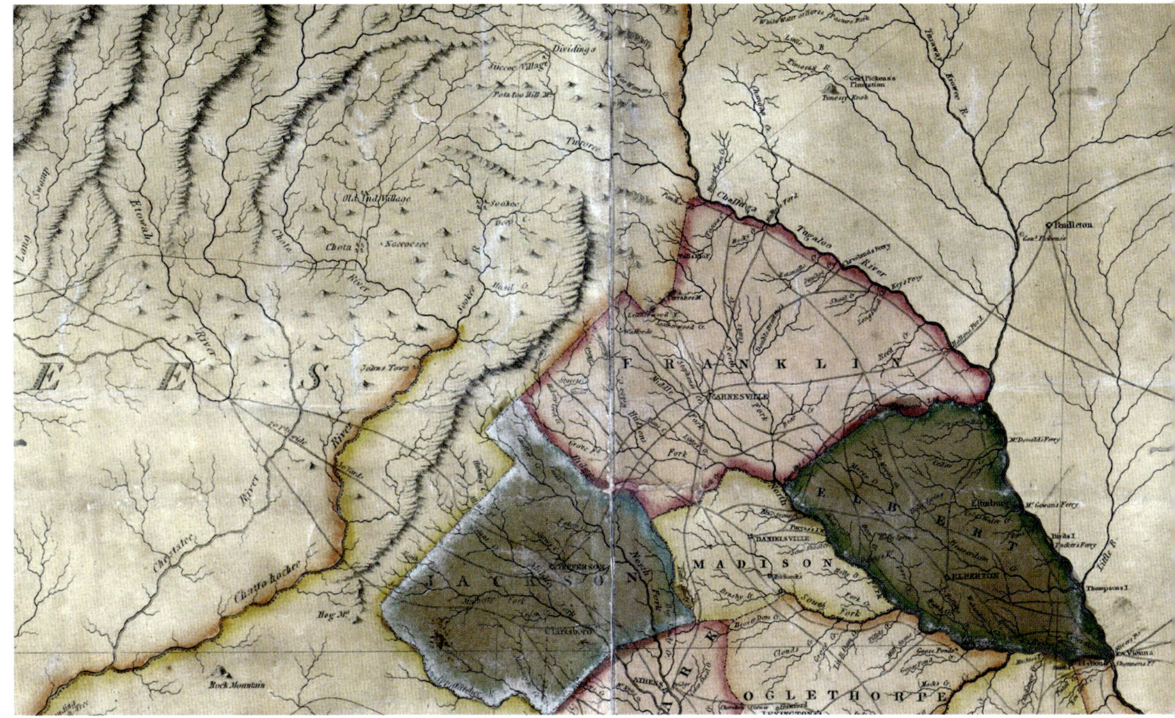

Part of Early's *Map of the State of Georgia*, published in Savannah in 1818. Note the road from Dividings, today's Clayton (*top center*), to "Old Indian Village," which was Sautee, and on to "Chota" and "Naccocsee" (*center left*). David Rumsey Map Center, Stanford Libraries, Stanford, Calif.

Most of the Cherokees had already left that narrow strip of land, which encompassed several thousand acres on the east bank of the Chattooga River, but the 1816 cession may have sent a few refugees into northeastern Georgia, which still remained a part of the Cherokee Nation. The failure to seize the Cherokee land in Alabama proved to be a political disaster for the Madison administration. By late summer the government was forced to go back to the Cherokee and negotiate a treaty extinguishing all of their claims in Alabama and some in Tennessee as well. The treaty was signed in September 1816 and promised "to perpetuate peace and friendship between the United States and Cherokee tribe, or nation, of Indians, and to remove all future causes of dissension which may arise from indefinite territorial boundaries."[14]

Peace and friendship were not to be. Jackson and his allies were already pushing an old proposal by Jefferson that ceded land in the East be exchanged for equivalent acreage west of the Mississippi River. Several thousand Cherokees had already given up and migrated to lands along the lower Arkansas River, but, in addition to fighting with the Osage and other tribes that resented the intrusion onto their own lands, the Cherokees had yet to receive title to any of that land from the federal government. And all the while the Americans continued to push their way into the Cherokee Nation and, under the guise of providing post roads, were actually creating military roads to facilitate the slow-motion invasion of the Indian nation.

THE UNICOI TURNPIKE

The first attempts by the federal government to push roads through Creek and Cherokee territories were in the 1790s. In 1801 Doublehead (1744–1807), chief of the Lower Towns, voiced the opposition of many Cherokees: "When you first made these settlements there were paths which answered for them. . . . We do not wish to have them [wagon roads] made through our country. Our objections to these roads are these: a great many people of all descriptions would pass [along] them . . . and you would labor under the same difficulties you do now. We hope you will have no more to say on this subject; if you do, it would seem as if we had no right to refuse."[15]

Many of those in the Upper Towns, however, especially some of the rich mixed-blood Cherokees such as James Vann, saw money to be made and supported the roadbuilding, and even Doublehead eventually changed his mind. In 1803, after much acrimonious negotiation, the Cherokee finally agreed to a federal road through their nation. Within a few years, Doublehead and Vann would both be assassinated, in part for their support of that road treaty.

The Creek, too, had given the United States "a right to a horse path, through the Creek country, from the Ocmulgee to the Mobile," ostensibly built for mail delivery. They were irate to discover that simple foot logs over the creeks, as stipulated in the treaty, had morphed into bridges, and the horse path itself into a wagon road, a sequence of events that were part and parcel of Gen. Andrew Jackson's strategy to destroy the Creek Confederacy. With these roads, one historian noted, "the palimpsest was scored again as the trading paths, post roads, and traveler's routes were rutted with commissary wagons and heavy weaponry."[16]

In 1813 the Cherokee National Council took a new tack and met with Nicholas Byers and David Russell, who were agents for Tennessee and

Georgia, respectively, and authorized them to establish a "Turnpike Company." The council resolution was adopted on 8 March 1813 and stated,

> We, the undersigned Chiefs and Councillors of the Cherokees in full council assembled, do hereby give, grant, and make over unto Nicholas Byers and David Russell, who are agents in behalf of the States of Tennessee and Georgia, full power and authority to establish a Turnpike Company, to be composed of them, the said Nicholas and David, Arthur Henly, John Lowry, Atto[rney], and one other person, by them to be hereafter named, in behalf of the state of Georgia; and the above named persons are authorized to nominate five proper and fit persons, natives of the Cherokees, who, together with the white men aforesaid, are to constitute the company; which said company, when thus established, are hereby fully authorized by us, to lay out and open a road from the Tennessee River, to be directed the nearest and best way to the highest point of navigation on the Tugolo River.

The company was authorized "to erect their publick stands, or houses of entertainment" on what was to be a "free and publick road" and agreed to pay $160 annually for twenty years, after which the road would revert to Cherokee ownership. By the time Tennessee and Georgia got around to passing charters for the road in 1815 and 1816, respectively, the particulars of the enterprise were significantly different.[17]

The "Unaca or Unacoi Turnpike Company" [*sic*] cited in the Georgia charter envisioned a road "from the head of boatable water on the Tugalo river the most direct route to fall into the road at Samuel Thompson's on Nine Mile creek, in East Tennessee," which was essentially the same route described in the 1813 resolution from the Cherokee. Russell Goodrich, Nicholas Byers, David Russell, Arthur H. Hanley, and John Lowry again made up the company, but there was no mention of Cherokee representation. The state evaded accepting any responsibility by also noting that the road should be "agreeable" to the company's "treaty with the Cherokee Indians on that subject." In addition, the road was no longer to be a "free and publick" road but rather a toll road, with profits to be retained by the company. Tolls were set at 6½ cents for someone on horseback, 6½ cents for pedestrians, $1.00 for a wagon team, and $1.25 for a coach or carriage.[18]

James Rutherford Wyly (1782–1855), who had recently bought a plantation at Tugaloo, forty or so miles southeast of Sautee Nacoochee,

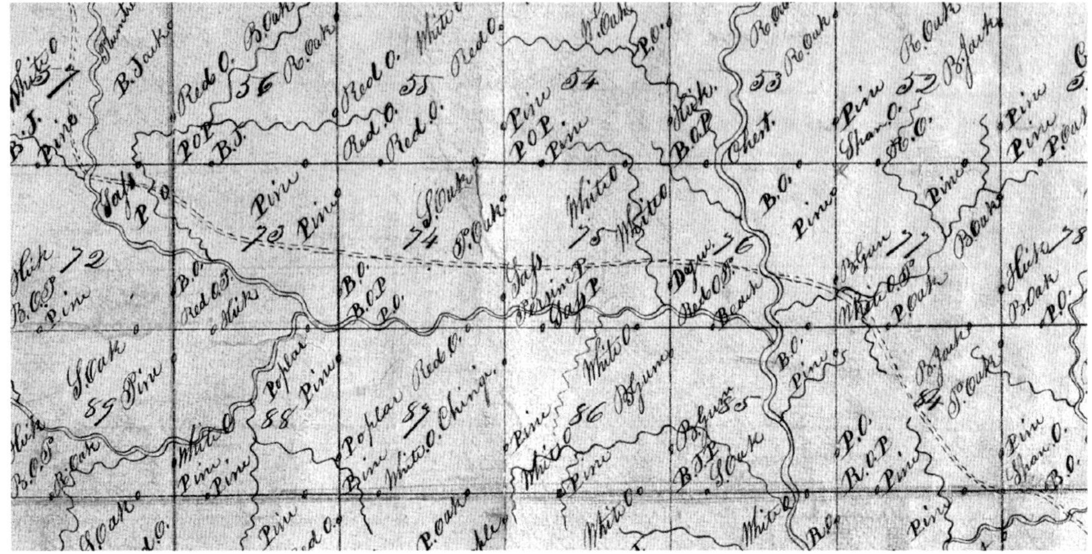

Part of the plat of survey for District 3 of Habersham County, 1820, encompassing what is now Sautee Nacoochee. The double-dashed line marks the route of Unicoi Road through the district, from lower right to upper left in this image. Habersham County, District 3 Survey, Records of the Surveyor General, Georgia Archives, Morrow.

was appointed superintendent of construction. The road was to be cleared of rocks and trees for a width of twenty feet, unless digging was required, in which case the width could be reduced to twelve feet. It was possible to widen the ancient trail into a road in many places, but in others a different route had to be chosen. Even so, the road crossed the Chattahoochee River twenty-eight times in eight miles between Nacoochee and Unicoi Gap.[19]

Construction may actually have started in 1814, although that does not seem likely. Roadbuilding remained a contentious issue with the Cherokee, and on 22 March 1816 they only reluctantly signed the now-revised treaty, which included a provision that essentially stripped the Cherokee of any control over travel through their territory: "It is expressly agreed on the part of the Cherokee nation that the United States shall have the right to lay off, open, and have the free use of, such road or roads, through any part of the Cherokee nation, lying north of the boundary line now established, as may be deemed necessary for the free intercourse between the States of Tennessee and Georgia and the Mississippi Territory. And the citizens of the United States shall freely navigate and use as a highway, all the rivers and waters within the Cherokee nation."[20]

The original deadline for completion stated in the Georgia charter had to be extended to 1 November 1818, but it is assumed that the road opened shortly after that, "with as much convenience as any other road through the Cherokee country." The *Knoxville Register*, in announcing the road's opening in April 1819, also reassured travelers that there had been "established a number of houses for the entertainment of travellers, (which are and will be kept by white men and their families) so as not at any part of the road to have a distance of more than 20 miles without a house of accommodation of travellers."[21]

Wyly was able to profit from the new road by operating at least three of those entertainment houses, beginning with Traveler's Rest at Walton's Ford near Tugaloo, where the road originated. About twenty miles to the west, he established an inn in Nacoochee Valley, possibly on Land Lot 74, Third District of Habersham County (now White County), which he bought in November 1820. It is not clear if there was already a residence or other structures when Wyly acquired the land

Part of Royce's map of the thirty-six land cessions by the Cherokee, 1721–1835. The yellow area, marked "23" (*left and below center*), represents the cession of 1817. The irregular brown arc to the northwest of that represents the cession of 1819. *Fifth Annual Report*, plate 8.

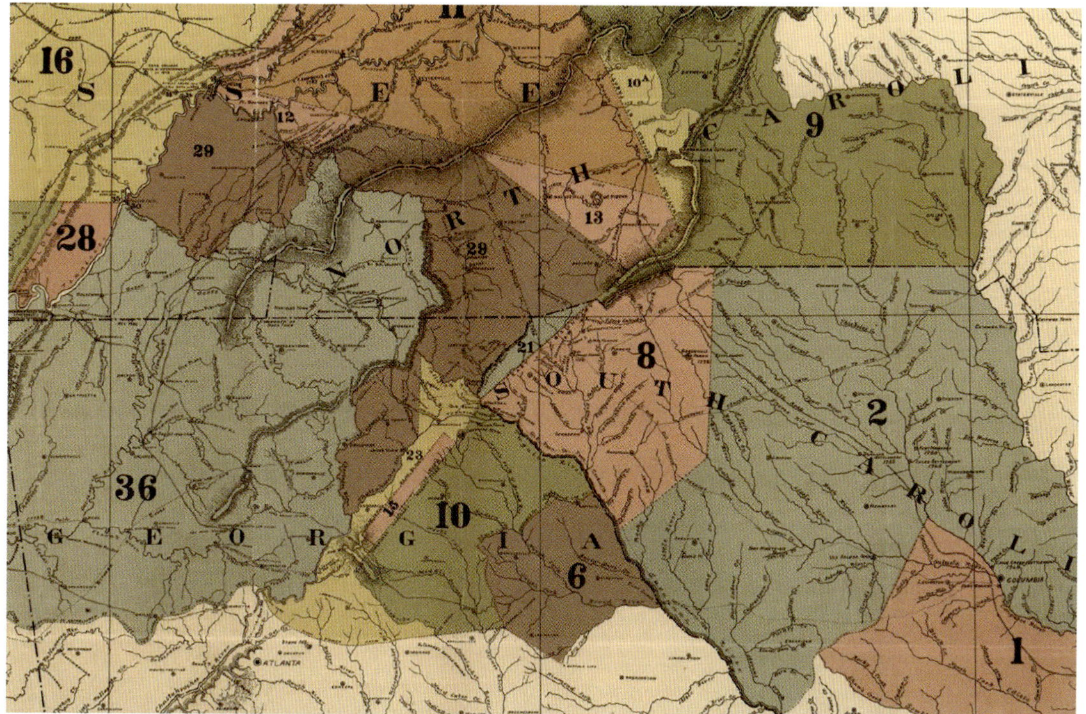

lot, which encompassed the intersection of the Unicoi Turnpike and the ancient road from the Cherokee towns in western North Carolina now known as Rabun Road; if not, there soon would be. Wyly is also believed to have operated a hotel on Land Lot 75 in the 1830s and possibly another on Lot 76 in the 1840s. Although the inns and taverns along the turnpike were probably profitable, the company itself had to borrow $3,000 from the state in 1821, which was not repaid on time, leaving the Cherokee to complain in 1826 and 1828 that they had not gotten their annual payments for the road.[22]

The route of Unicoi Turnpike was one of the very few roads depicted on the original plat of survey for Habersham County in 1820. Comparison with modern maps shows that some of its route can still be traced in modern thoroughfares, including parts of SR 17 in Habersham and White Counties. There have been significant changes, of course, particularly around Clarkesville, which did not exist when the turnpike was first built. The turnpike originally crossed the Soque River about two miles north of where Clarkesville was laid out in 1821. Most of the original route of the turnpike has been lost in that vicinity but can be picked up to the west in Alec Mountain and Amy's Creek Roads to SR 17. With the growth of Clarkesville, a new, more direct, route developed to the south of the original turnpike, and out of it developed the modern route of SR 17.

1817 and 1819 Treaties

Early in 1817 negotiations resumed to resolve new boundary disputes and to formalize the terms by which tribal land could be traded for new land in the West, and a treaty was signed on 8 July 1817. Its first provision addressed the cession of land in Georgia in exchange for land on the White River in Arkansas:

> The chiefs, head men, and warriors, of the whole Cherokee nation, cede to the United States all the lands lying north and east of the following boundaries, viz: Beginning at the high shoals of the Appalachy river, and running thence, along the boundary line between the Creek and Cherokee nations westwardly to the Chatahouchy river; thence, up the Chatahouchy river, to the mouth of Souque creek; thence, continuing with the general course of the river until it reaches the Indian boundary

> line, and, should it strike the Turrurar river, thence, with its meanders, down said river to its mouth, in part of the proportion of land in the Cherokee nation east of the Mississippi, to which those now on the Arkansas and those about to remove there are justly entitled.[23]

The Cherokee would be giving up some 337,000 acres in Georgia, and another treaty provision cost them 278,000 acres in Tennessee. Although the treaty language is confusing, the 1817 treaty described the western boundary of the Georgia cession as following the course of the Chattahoochee River to the mouth of Soque River, where it ran in a straight line due north, two or three miles east of Sautee Nacoochee. From the Cherokee perspective, there were more insidious provisions in the treaty, beginning with its implicit recognition of tribal divisions that the Cherokees themselves did not recognize. In particular, by including the western band of Cherokees in the negotiations, the United States purported to deal with the entire Cherokee Nation, when in fact the Cherokee had redefined citizenship to include only those living within the nation's boundary, which did not include the Cherokees in Arkansas.

The treaty negotiators also played on the division between Upper Towns, who opposed removal, and Lower Towns, who supported migration. The Cherokee Nation strenuously resisted the federal government's efforts at detribalization too, especially those designed to break up communal ownership of property and a matrilineal society. The treaty's terms were such that male heads of Cherokee families "who may wish to become citizens" had the option of claiming a reserve in the East and being granted citizenship, albeit the second-class citizenship of free people of color. Alternatively, one might opt to move to a proposed reservation in Arkansas, which of course was the preferred alternative, as far as the Monroe administration was concerned. There was simply no option to stay and carry on as a sovereign nation.

The prospects for assimilation were dim too, given the commonly held attitudes expressed by Georgia governor George Troup, when asked about the possibility of assimilating "civilized" Cherokee: "If such a scheme were practicable at all, the utmost of the rights and privileges which public opinion would concede to Indians would be to fix them in a middle station between the negro and the white man; and that, as long as they survived this degradation, without the possibility of attaining the elevation of the latter, they would gradually sink to the condition of the former."[24]

By this time Meigs supported Cherokee removal, but he also recognized the real cost to their people:

> Some of the Cherokees may be destitute of local attachment, but these are few. The greater number have strong local attachments—the rivers, the springs, and the mountains, their old hunting grounds, and more than all, the bones of several generations of their ancestors, lie buried in their plantations and on the battlegrounds in their wars with the northern tribes. . . . Although told that they are going to a country with the same mild climate and temperature as this, their philosophy is not sufficient to silence local recollection.[25]

In 1818 there were the inevitable negotiations for more land, with the resulting treaty signed on 27 February 1819. Between that treaty and the 1817 treaty, the Cherokee lost four million of their remaining fourteen million acres, most of it in North Carolina and Tennessee, but also another two hundred thousand acres in Georgia, where the boundary was pushed to the upper branches of the Chestatee River. With that Sautee Nacoochee passed out of the Cherokee Nation and became part of the State of Georgia.

THE CHEROKEE RESERVES

Article 8 of the 1817 treaty offered citizenship to any Cherokee family head that wished to remain in their homeland east of the Mississippi if they, in effect, renounced tribal ownership and accepted "a reservation of six hundred and forty acres of land in a square to include their improvements which are to be as near the centre thereof as practicable, in which they will have a life estate with a reversion in fee simple to their children."[26] The register of Cherokees applying for reserves in the two years after February 1817 included over three hundred names, but only thirty-one reserves were actually named in the treaty of 1819, for those "believed to be persons of industry, and capable of managing their property with discretion, and have, with few exceptions, made considerable improvements on the tracts reserved."[27]

The locational references in the register of the reserves are very general and often ambiguous, since they were not surveyed until after the applications had been accepted. Nevertheless, the register indicates that at least fifteen of the several hundred reserves that were claimed were located in the upper Chattahoochee River watershed. Among the five reserves listed in Georgia in the 1819 treaty were those of John Martin, Jeter Lynch (mistakenly named "Peter" in the treaty), and Daniel Davis, all brothers-in-law.[28] The census roll notes which of the reserves were being claimed by a "native" and which by right of an indigenous spouse or child, but many, if not all, of the families were related by blood or marriage. One contemporary observed of the Martins, the Adairs, and some of their in-laws, "I know all these people and were it not that I did know that they had Indian blood in them—I should never have detected by looking at them."[29]

Several individuals with Cherokee connections can be located in the 1820 Habersham County census, including members of the Adair, Martin, and Ward families. In December 1828 Georgia extended the borders of Habersham, Hall, Gwinnett, DeKalb, and Carroll Counties to the northern and western borders of the state, setting the stage for the final assault on Cherokee sovereignty. As a result, when the federal census was taken in 1830, the Habersham schedules included residents living west of the Chattahoochee and Chestatee Rivers, as well as a few living between the two rivers. Several of the Cherokee-related names appear in both censuses, but Don Shadburn's analysis of the census schedules suggests that the census taker did not always include a white man's Cherokee wife.[30]

THE MARTINS

The first residents of Sautee Nacoochee about whom much is known were two prosperous mixed-race families, the Lynches and the Martins.[31] John Martin Jr. (ca. 1784–1840), later known as Judge Martin, was the son of John "Jack" Martin Sr. (1745–1803) and his wife, Susannah Rebecca Emory (ca. 1745–ca. 1802), also known as Ani'Gilâ'h of the Long Hair Clan of the Cherokee. Her father was a Scottish trader, while her mother was the daughter of the Cherokee chief Old Tassell and his Creek wife. In addition Jack Martin's brother was the noted Revolutionary War general Joseph Martin (1740–1808), who is credited with maintaining the Cherokee alliance during the Revolution and whose third wife was Elizabeth "Betsy" Ward (ca. 1758–1803), daughter of the trader Bryant Ward and Nancy, the Cherokee widow of Kingfisher.[32]

John Martin was only one-eighth Cherokee, but, in keeping with tribal tradition, he was probably raised by his mother and her brothers. Although she died about 1802 and his father a year later, Martin was well educated nevertheless. While in his twenties, he married sisters Lucy and Eleanor McDaniel, and eventually had eight children. Although he had to learn to speak the Cherokee language, by 1818 Martin had achieved such status that he was named to the delegation sent to Washington to negotiate the treaty of 27 February 1819. At an unknown date, he established at least one plantation on Sautee Creek near his sister Nancy Martin Lynch and her husband, Jeter. In March 1819 Martin accepted a 640-acre reservation on Sautee Creek, along with U.S. citizenship, as described in both the treaties. As early as May 1819, however, Martin was complaining to Meigs about rude treatment from his white neighbors.[33]

On 9 October 1823, Martin relinquished his reserve for $2,000 and moved sixty miles west to the area where his wives had lived before their marriage. It is believed that his polygamy played a role in his decision to remain with the Cherokee Nation. He established plantations for his two wives about fifteen miles apart, one on the Salacoa River and the other on the Coosawattee River. He apparently had a major change in attitude toward relationships with the United States and, after his move, continued his involvement in tribal politics. He was one of the authors of the first Cherokee constitution and joined its signing in July 1827. He was a judge in the first Cherokee Supreme Court when it was established that same year and, after removal to Oklahoma, treasurer of the Cherokee Nation.[34]

His plantations in northwest Georgia were confiscated after the state's land lottery in 1832, and he moved to the Red Hill Valley of Tennessee. He had just completed a dogtrot log cabin there when the Treaty of New Echota was signed on 29 December 1835, setting the Trail of Tears in motion. Judge Martin died at Fort Gibson in eastern Oklahoma in 1840. The tribe erected a monument in his honor at the Cherokee courthouse in Tahlequah, Oklahoma, in 2013.[35]

Lynch Mountain, which flanks the eastern side of Sautee Valley, gets its name from Jeter Lynch (1775–1819) and his wife, Nancy Martin Lynch (ca. 1778–ca. 1823), who was Judge Martin's sister. Living in the mountain's shadow, the Lynches were also a mixed-blood family who ultimately failed to assimilate. Jeter Lynch was born in Pittsylvania County, Virginia, the son of Capt. William Lynch (1742–1820) and his wife, Elizabeth. The captain is believed to be the origin of "Lynch's law" by virtue of the rough justice he administered as a backcountry judge in Virginia. In 1798 Captain Lynch and his family moved to Pendleton District, South Carolina.[36]

In 1799 Jeter Lynch and Nancy Martin were married, and over the next twenty years they had at least eight children. When they moved to Sautee Nacoochee is not clear, but it was perhaps shortly after their marriage. The exact location of their farm is uncertain, but it seems to have encompassed all or part of Land Lot 46, which is traversed for a mile or so by Lynch Mountain Road. When that land lot was sold in 1823, the deed included a caveat about the uncertainty of title due to "the Indian reservation."[37]

Jeter Lynch, too, registered for a reservation under the treaty terms, but he died before it could be surveyed. His widow accepted the terms of the reservation and remained there even after she married Jonathan McWhorter in 1821. They had a son before both she and McWhorter died in 1823, perhaps still living on the Lynch reserve on Sautee Creek. There is no record of where Jeter Lynch or Nancy Martin Lynch McWhorter are buried.[38] Their untimely deaths clouded title to the property, which had already been included in the 1820 land lottery. In a letter summarizing the fate of the Cherokee reserves, federal agent Hugh Montgomery noted, "Mr [unclear] Linch died shortly after the Treaty [of 1819] and a Notification was filed by his widow who died shortly thereafter. The heirs did not get the Reservation but were paid $3000 by the Government for it."[39]

In the spring of 1826, the Habersham County Inferior Court ordered the auction of "the whole of that Tract or parcel of Land containing six hundred & forty acres more or less lying and being in said county on Sota [*sic*] Creek and known as Geter Lynch's reservation." On 2 May 1826 Lynch's reservation was "sold at Public Outcry to the highest bidder," for $2,000. No legal description of the property has been located, and details on the purchaser, John Williams, are scant. He may have been the John Williams enumerated next door to Moses Harshaw in the 1830 census of Habersham County, or the John D. Williams enumerated near Daniel Brown on another page in that same census.[40]

THE ADAIRS

Adair is a name often noted in Cherokee history, beginning with the legendary James Adair, so frequently quoted in the previous chapter. In 1770 brothers John and Edward Adair, who may have been relatives of James, immigrated from England and became traders with the Indians. Both married Cherokee women. In 1779 John Adair (1756–1815) married Ga-hoga, of the Deer Clan, and they settled in one of the Lower Towns in upstate South Carolina. They had several children, one of them being Walter Adair (1783–1835), often known as "Black Watt" to distinguish him from a cousin, also named Walter, who was known as "Red Watt" and who married the daughter of Judge Martin's half sister Lucy Fields.[41]

In 1804 Black Watt Adair married Rachel Thompson (1786–1876), one of three or four unions of the Adair and Thompson families in that generation. When they settled near the site of the ancient town of Soque, nine miles southeast of Sautee Nacoochee, is not clear, but one of the border commissioners attempting to establish the boundary between Georgia and Tennessee in 1818 spent the night with Adair and wrote that "it is one of the handsomest places I ever saw." In the summer of 1817, Walter Adair was among the first to claim a reserve. Numbered 22, it included 100 acres and a lime kiln "east of Chattahoochee," as well as 540 acres around his house at Soque.[42]

Two of Walter Adair's brothers also claimed reservations. His oldest brother, Samuel Adair (ca. 1780–before 1838), made claim #97 at some unspecified location on the Chestatee River. Their younger brother, Edward Adair (1789–1864), also claimed a reservation at Soque. Like most other Cherokees, the Adairs soon decided that second-class citizenship on the unruly Georgia frontier was insufferable. They sold their reservations to the state in October 1823 and by 1824 were in northwest Georgia, where they temporarily reestablished their fortunes. Walter Adair died there in 1835, and Samuel was dead by the time the Cherokees were forced to Oklahoma three years later. The town of Adairsville was named to remember the family, which had settled about two miles north of the present town.[43]

THE ENGLANDS

As Matt Gedney has noted, there were a number of men named England in and around what is now Helen, "some of whom had taken Cherokee wives." The family's ties of kinship are difficult to understand, but David and William England also claimed reserves under the 1817 treaty, both "near Chattahoochee." William England applied for a reserve in July 1817, and, although the register does not note the basis for his application, it appears to be because his wife was a daughter of Caty McDaniel and John Ward, noted earlier.

In May 1818 Caty Ward also made an application "in proxy for William England," apparently because he was unable to make the application in person. The applications were not granted, and in February 1821 William England's son Elijah paid the fortunate drawer in the 1820 land lottery $1,000 for Land Lot 39 in today's Helen Valley. The deed noted it as the place "where said Elijah England now lives."[44] The relatively high price suggests that the property had already been improved, perhaps with a house and outbuildings.[45]

David England (1790–1845), who may have been William England's son or nephew, also applied for a reserve. He was, according to descendants, a "farmer, trader, miner, land speculator, and entrepreneur." He married Susannah Fields, who was at least part Cherokee, and through her claimed a 640-acre reserve in May 1818. Like the Adairs, the Englands ultimately rejected the reserve arrangement, and in October 1823 David England was paid $4,000 to extinguish his claim.[46] No fewer than three heads of household named "William England" can be found in the 1820 federal census of Burke County, North Carolina, and the name appears in the legendary list of presumably white families who first settled in and around Nacoochee in 1822. Whether any of these were related to the Cherokee reserve applicants is not certain but seems probable.

THE MCDANIEL-WARDS

Catherine "Caty" McDaniel Ward (1763–1823), William England's mother-in-law, was the daughter of the Scottish trader William David McDaniel Sr. (1716–52) and Sookie (Su-Gi) "Granny Hopper" McDaniel (1739–1820). Sookie was herself daughter of the great Cherokee chief Kanagatoga "Old Hop" Moytoy (1690–1761), First Beloved Man of the Cherokee, and Su-Gi Standing Turkey (ca. 1690–1735).[47]

Around 1778 Caty, or Katy, McDaniel married John Jackson Ward (1750–1815), the son of Bryant Ward, an Irish trader, who had married a Cherokee woman in the 1740s. Caty and John Jackson Ward had at least fifteen children, and in May 1818 the widowed Caty and five men who were probably her sons claimed reservations. Three of them were located simply on "Deep Creek," which might have been the same Deep Creek that empties into the Soque River in Habersham County but could have been over the mountains in North Carolina or Tennessee, where there are creeks with the same name; the others were "on Chattahoochee," which could have been almost anywhere north of what is now Gainesville. Caty died on her reserve on the Chattahoochee in 1823 or 1824. Although she had already signed away her reserve, her children continued to challenge its ownership as late as 1846.[48]

THE KEYSES

In July 1817 three brothers claimed separate reserves "on waters of Mud Creek": Samuel Keys (1787–1855), Isaac Keys (b. ca. 1790), and William Keys (ca. 1794–before 1871). All three married daughters of the Virginia trader Samuel Riley (1774–1825) and his wife, Gu-Su-Sti-Yu (1756–1802) of the Long Hair Clan, a daughter of the great Cherokee chief Doublehead (1744–1807), who was assassinated, in part, for his complicity in ceding the Cherokee hunting grounds in Tennessee and Kentucky in 1805. It is not certain the Keyses' claims were on the same Mud Creek that empties into the Chattahoochee west of Cornelia, but if those claims had been granted—and not all were—the Keyses would have enjoyed use of nearly two thousand acres in what is now northern Hall County.

In addition to those relatively prosperous Cherokee families, there were surely many less prosperous farming families, some of the thousands of families in the Cherokee Nation who had been leaving the towns for dispersed farmsteads for a generation or more, but none of them have been identified by name. Most of them were full-blooded Cherokee whose ancestors had occupied the hills and valleys of Sautee Nacoochee for hundreds, if not thousands, of years. And they had no title to their land, since it was not owned in fee simple but was part of the tribe's communal ownership of the land. The "Legend of Nacoochee" is a poor substitute for the stories they might have told.

Local Government

The 1817 treaty with the Cherokee was signed in July, but it was nearly eighteen months before the state authorized organization of the cession into counties and the mapping of district surveys. The reason for the delay is not clear, although it appears that the treaty negotiators' success in wrangling more land from the tribes got ahead of the state's ability to organize the cessions. The uncertainty generated by unorganized Cherokee cessions must have added tremendously to the turmoil on the Georgia frontier, which, as noted earlier, contributed to the failure of the system of 640-acre reservations for individual Cherokee families. It should be noted, too, that Georgia's new territory was platted for the 1820 lottery without any regard for those reservations.[49]

HABERSHAM COUNTY

In December 1818 Georgia governor William Rabun signed legislation creating seven large counties out of the recent Creek and Cherokee cessions—Early, Irwin, and Appling in southern Georgia and Gwinnett, Walton, Habersham, and Hall in northern Georgia. Habersham County included all the 1817 Cherokee cession northeast "of a line to begin at a place where captain John Miller now lives, on the line of Franklin County, and running north, thirty degrees west, to the Chattahoochee river." That line still marks the southwest boundary of Habersham County. In addition the act took part of the Wofford

Settlement, which had been added to Franklin County in 1812, and included it in the new Habersham County.[50]

The new county was named for Joseph Habersham (1751–1815), son of James Habersham (1712–75), a wealthy merchant and planter in Savannah. The Habersham family, which included brothers James Jr. (1745–99) and John (1754–99), were a major force in the economic and political life of colonial-, revolutionary-, and federal-era Georgia. Joseph Habersham was a leader of revolutionaries in Savannah and colonel in the Continental army, even as his father remained loyal to the king.[51] He served as the first U.S. postmaster general under Presidents George Washington and John Adams. He also built a summer house that still stands a few miles north of Clarkesville.

Also in December 1818 Governor Rabun signed legislation organizing county government in each of the new counties and authorizing elections. Until a county seat could be established, the legislative act specified where the elections were to be held, locating those for Habersham County "at the schoolhouse near Walter Adairs [*sic*]." Although that section was repealed in December 1819 and more latitude granted in polling places, the county's first elections must have occurred at that schoolhouse, which was located near the intersection of what are now New Liberty and Low Gap Roads in northern Habersham County.[52]

Those first elections were held on 10 April 1819 and resulted in the election of William Benton Wofford (1791–1858) as sheriff, the most powerful position in the county. Somewhat ironically, William Wofford Sr., founder of the ill-fated Wofford Settlement in the 1780s, was elected county surveyor. Other first officials were Miles Davis, clerk of superior court; Joseph Dobson (1787–1871), clerk of inferior court; William Steedly (or Steedley), coroner; and Absalom Holcombe and James R. Wyly, inferior court judges. At a second election that fall, Benjamin Cleveland (1784–1858), who was already speculating in real estate in the county, especially in and around Sautee Nacoochee, was elected the county's first state senator and James Blair its first state representative. By the time the legislature met again in December 1819, another land cession had been forced on the Cherokee, which necessitated another legislative act that expanded some of the earlier counties westward and added a new county, Rabun, that would be included in the land lottery. As a result, the expanded incarnation of Habersham County stretched from Tallulah Falls to the Chestatee River and included all of present-day Habersham and White Counties, as well as portions of Stephens, Banks, and Lumpkin Counties.[53]

The state gave the Habersham County Inferior Court the authority to establish a county seat and acquire land "for the public buildings . . . which shall be near the center thereof as conveniency will admit, at which place the court and general elections shall be held."[54] Perhaps as early as the summer of 1819, the justices chose a site just down the Soque River from the old Cherokee town of Soque and near Walter Adair's plantation and schoolhouse. It is not clear where court was held in the early years. Not until after the land lottery in the fall of 1820 could a site for the county seat be formally established.

Benjamin Cleveland and Benjamin Chastain were prominent in the early history of the county and of Nacoochee, especially as real-estate speculators but also as early inferior court judges. In February 1823 Chastain conveyed thirty-two acres in Land Lot 2 of the Twelfth District, and Cleveland conveyed ten acres in neighboring Land Lot 19 of the Tenth District, a total of forty-two acres on high ground just south of a bend in the Soque River, to the inferior court judges for a county seat. The town of Clarkesville was not incorporated by the legislature until 26 November 1823, but by then most of the town lots had already been sold. When Adiel Sherwood described Clarkesville in his *Gazetteer of the State of Georgia* in 1829, he found "twenty-three houses, four stores, six mechanic and two doctor shops, four law offices, C[ourt] H[ouse], Jail, and a neat two-story Academy." It was, he thought, "as healthy a spot as is in these United States."[55]

There appears to have been too much already invested in the site to consider relocating when the 1819 Cherokee cession led to a large expansion of Habersham County to include what is now White County, leaving the county seat far from the geographic center of the expanded county. The distance to the county seat from the western part of the county was a major reason for the creation of White County in 1857.

WHITE COUNTY

For myriad reasons new Georgia counties were constantly being hived off of old ones throughout the nineteenth and early twentieth centuries, until the last one was created in 1924. Three counties were consolidated in 1931, and the state's 1945 constitution capped the number of counties at the present 159, the most of any state except Texas. Population growth was the main reason for creating a new county, but a new county automatically meant a new court, county commissioners, state senator, and employment for more people, while perhaps making a bustling town out of a sleepy country crossroad community. Convenience was also a factor, and that certainly was the major reason for the creation of White County. There had been talk of creating a new county seat that was more centrally located than Clarkesville, but there were too many vested interests in Clarkesville to allow that to happen. In the campaign for elections in 1856, William B. Shelton campaigned for a seat in the state house on the need for a new county and introduced legislation to that effect when the legislature convened in 1857.

Shelton was born in Virginia around 1801 but came to Habersham County in the early 1830s. A merchant, he lived at Mount Yonah, the small crossroad community nine or ten miles southwest of Nacoochee, and was joined there by his sister and other relatives. Shelton's bill reportedly proposed naming the new county Wofford, in honor of his old friend and Habersham County's first sheriff, William B. Wofford. With eleven other bills for new counties in the hopper in 1857, some members were wary of creating more, but on the last day of the session in December 1857, a bill passed creating White County.[56]

Baker states that the county was named for Rep. David T. White of Putnam County, who helped Shelton get the bill passed.[57] However, the fourth edition of Adiel Sherwood's *Gazetteer of the State of Georgia*, published in 1860, indicates that the county was named for Col. John White, "a brave and favored soldier in the Revolutionary war."[58] White, who had been born in England, was a colonel in the Fourth Georgia Regiment, mortally wounded at the Battle of Savannah in October 1779 and buried at Savannah Battlefield Memorial Park.[59]

The first elections in White County were held in February 1858. Of the five inferior court justices, two were from Nacoochee: Jehu Trammell and Edwin Poore Williams. The original line between Habersham County and the new White County was Sautee Creek, which effectively divided Sautee Nacoochee between two counties, but in 1858 the legislature shifted the line to the east of Lynch Mountain, putting most of the Sautee Nacoochee community inside of White County.[60]

Mount Yonah, the village at the intersection of the Gainesville-Nacoochee road and the Clarkesville-Dahlonega road, was named temporary county seat in the legislative act that created the county. A special election in June 1858 confirmed it as the county seat but rechristened it Cleveland, in honor of Benjamin Cleveland, who had been representing Habersham County in the legislature since the 1820s. The courthouse square and town lots were surveyed and auctioned that fall.[61] In addition Edwin Williams was hired to design and build the new courthouse. Changes to his initial design, which he said was inspired by Independence Hall in Philadelphia and probably included a columned portico, were forced by some of his more parsimonious colleagues, resulting in the sparse, utilitarian building that Williams actually completed in 1859. Much of the brick was made by Williams's slaves.[62]

County Courts

Establishing the rule of law through a functioning court system was always a priority, not only to try criminal matters but also to ensure clear titles to property. Georgia's 1798 constitution vested judicial power in a state superior court, which dealt with all criminal matters as well as issues related to property. Organized into circuits, the superior court was held in each county in the circuit once or twice a year.

Initially, Habersham County was assigned to the western circuit, but that changed as the number of counties increased and more circuits were added. In Georgia county sheriffs, superior court judges, and superior court clerks, who are responsible for maintaining the official record of deeds and mortgages, are still elected officials. Not until 1845 was there a state supreme court to which lower-court rulings could be appealed. The first term of the superior court in Habersham County convened 2 September 1819, and some sources say court was held outside on log benches. The location is uncertain, but it was probably at Walter Adair's schoolhouse. Deeds were already being recorded in the county's first deed book, which was begun on 10 July 1819, but only for transactions in the part of Habersham County that had formerly been in Franklin County.

The state's constitution vested judicial power for all civil matters, except those relating to land titles, in the county inferior court, each with five elected justices. This court was tasked with appointing census

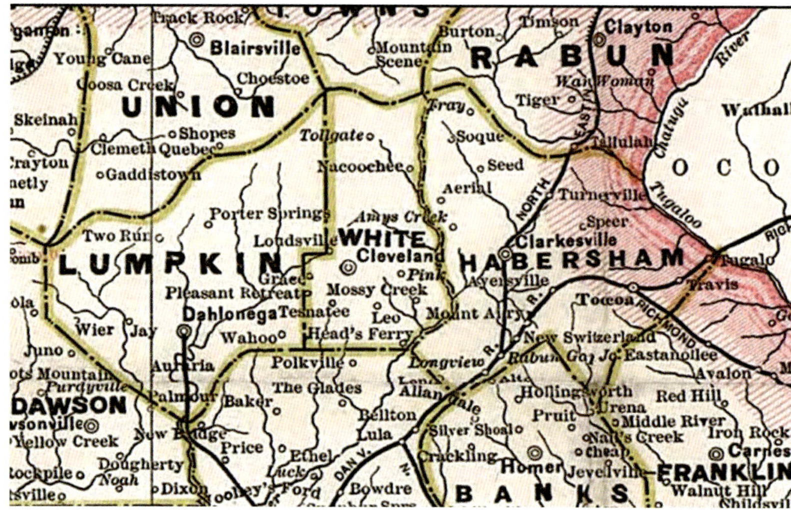

Part of Cram's *Railroad and County Map of Georgia*, 1883, showing White County, Nacoochee, and the tollgate on the Unicoi Turnpike. This map repeats the mistake of earlier maps in not showing the eastern boundary of the county as it was adjusted in 1858. Geography and Map Division, Library of Congress, Washington, D.C.

takers for the federal census and convening grand juries to assess county affairs and other issues. It had jurisdiction over county business matters, such as caring for the poor, maintaining jails, authorizing county roads and bridges, and keeping a register of wills. The first minute book for the Habersham County court was not started until 1824, although there are a variety of court records prior to that date that have never been indexed.

In 1851 the Georgia constitution was amended to provide for an elected official in each county to be known as the ordinary. In 1868 the legislature abolished the inferior court system entirely in favor of a court of ordinary, which became today's probate court in 1974. Judge Cicero H. Sutton (1822–1901) was elected as the first ordinary for Habersham County and served for many years in that capacity. At the same time, counties were authorized to elect county commissioners of roads and revenues, which most counties did and continue to do.

The lowest courts were those administered by justices of the peace, who were also elected officials. A justice of the peace was not necessarily a trained lawyer but could perform civil marriage and decide minor civil disputes at a monthly court held at a county district courthouse, or "law house" as they were often known. In 1798 districts for these justices of the peace courts were made to conform to the boundaries of the general militia districts, with each district electing two justices.

The White County courthouse, built for the county by Edwin Poore Williams in 1859. Courtesy of White County Historical Society, Cleveland, Ga.

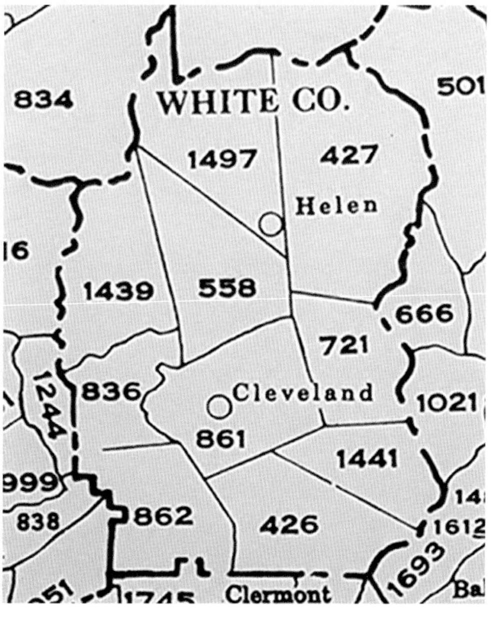

Detail from a map of Georgia's General Militia Districts. Sautee Nacoochee is in GMD 427. Georgia Archives, Morrow.

Nacoochee General Militia District

General militia districts were established during the colonial period and continued by the State of Georgia. In each district a company of all able-bodied men who could be called up for the state's defense was organized, with the captain of each company keeping rolls of militia members and often giving his name to the district. Changing captains resulted in a new district name, which was often confusing, and after 1807 the state required that all districts be numbered. The Nacoochee

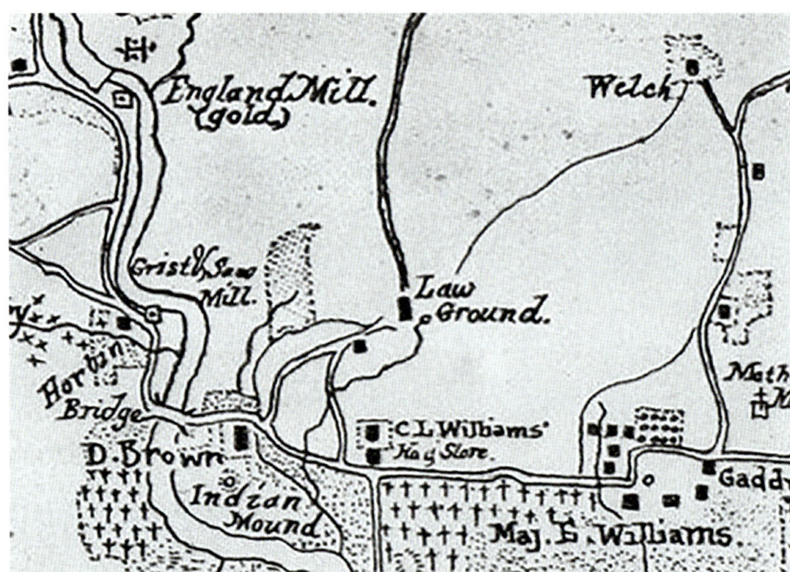

Detail from Moffat's reconstruction of Nacoochee as it existed in 1837, showing the "Law Ground" that would have been the location of the Nacoochee General Militia District courthouse. *Nacoochee Valley, 1837*.

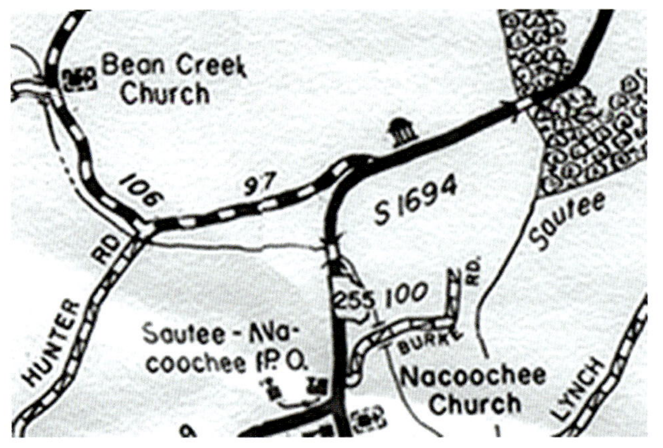

Detail from 1985 DOT map of White County, with the General Militia District courthouse symbol on the north side of SR 255, just east of Bean Creek Road.

District, which is numbered 427, appears never to have been referenced by the name of its captain. Although the military aspect of these districts disappeared after the Civil War, the general militia districts have remained a significant civil administrative unit. Among other things, election district boundaries, property tax digests, and census tracts are often organized at the militia-district level.[63]

Moffat's 1837 map of Nacoochee and local tradition locate the Nacoochee District's first "Law Ground" north of Charles Lathrop Williams's store. As discussed in the following chapter, the roads depicted on the Moffat map can be shown to correlate with roads mapped by the state in the twentieth century. The present analysis suggests that this justice of the peace court may have been located near the intersection of Bean Creek Road and the now defunct Hardman Road, both of which were most likely well traveled long before the 1820s, since both follow natural corridors of travel.

There would have been a structure in which to hold district court, and it is quite possible that the Cherokee-built log structure reported to have been located across the road to the west of the present Bean Creek church was used by the congregation as their first place of worship in the 1860s. It was also the first militia district courthouse. How long the justice of the peace court might have remained there is uncertain, and the location may have changed more than once. It was last noted in the 1985 Department of Transportation highway map for White County, when it was located on SR 255, just east of Bean Creek Road.[64]

1820 Land Lottery

Georgia's land-lottery system was created as a result of the corruption revealed by the Yazoo land fraud in the 1790s. The first lottery was held in 1805 and a second in 1807, both covering Creek cessions in south Georgia. The 1820 lottery was the third of what would be seven lotteries, the last two in 1832, dividing the lands of the ruined Cherokee Nation. Through these lotteries the state distributed 129,961 lots encompassing 23,992,058 acres. Lot sizes ranged from 40-acre "gold lots" and 160-acre farm lots in northwest Georgia to 490 acres in the mountains and in parts of south Georgia.

The original act required potential lottery participants to swear that they had not "resided" on any Cherokee or Creek land during their years of state residency. When the act was revised in December 1819 to include the land between the Chattahoochee and Chestatee Rivers, the legislature also included a provision that nullified any earlier surveys or purported land grants in the area. The 1818 act called for the land cession in North Georgia to be divided into thirteen districts, "as equally as conveniently can be," to which the 1819 act added three districts in Hall County, six in Habersham, and five in Rabun.[65]

DISTRICT PLATS OF SURVEY

District surveys authorized in the 1818 act were completed in the spring of 1819, with the act specifying that survey lines be parallel or perpendicular to the boundary between Walton and Gwinnett Counties, which ran north sixty degrees east from a point on the Yellow River at the "temporary Creek boundary."[66] The skewed platting made surveys more difficult, and the 1819 act specified survey lines oriented to the cardinal points, a practice followed in subsequent lottery surveys. The district surveys of the lands ceded in 1819 began in April 1820, and were completed in July.

Surveyors were hired by the state, after posting a $10,000 bond, to run the county boundary and district lines. Each district then had a single surveyor, who had five months after the district boundaries were set to complete his work. Each district in the counties of northern Georgia was to be subdivided into lots that were fifty chains, or 3,300 feet, on each side and contained 250 acres. The exceptions were the mountainous Districts 5 and 6 in what is now northern White County and all but one of the five districts in Rabun County. In those districts land lots were to encompass 490 acres. In all there were 18,600 lots of 250 acres each and a number of fractional lots platted for the 1820 lottery. If the fractions amounted to more than 160 acres, they were included in the lottery drawing; otherwise, the fractions were sold and, along with lots numbered 10 and 100 in each district, the proceeds set aside "for the education of poor children."[67]

The 1817 and 1819 acts also established payment schedules for the surveyors. The county and district boundary surveyors were paid $5 for each mile of survey line that they ran, with a down payment of

$150. The district surveyors were to be paid $4 a mile for each line run and receive a $400 down payment. Out of that amount, however, the surveyor had to pay all of his expenses, including "chain-men, axe-men, and every other expense incidental to the said business." The legislature specified that the surveyors mark trees "if practicable" and note watercourses. Roads, which were few and far between, were also to be delineated on the surveys. Surveyors were required to keep field notebooks, which were then copied and filed with the state surveyor general.[68]

DISTRICT 3 SURVEY

Sautee Nacoochee is in District 3 in the original 1821 survey of Habersham County. The district was surveyed as a rectangle, seven and a half miles north to south and ten miles east to west. The surveyor for the 208 individual land lots in the district was Archibald Carlisle McKinley (1779–1861). He was born in Abbeville County, South Carolina, but moved to Georgia as a young man. Listed as a farmer in the 1850 and 1860 census, he is buried at the Lexington Presbyterian church cemetery in Oglethorpe County.[69]

The surveys required the use of long chains for taking boundary line measurements, which were carefully regulated in the 1817 act. Surveyors were "to cause all such lines to be measured with all possible exactness, with a half chain, containing two perches of sixteen feet and one half each, consisting of fifty equal links, which shall be adjusted by a standard, to be kept for that purpose, in the surveyor general's office."[70] McKinley had two chain bearers, which was typical: Adam W. Rankin and Alfred Livingston. No details about Rankin have been located, but Livingston (1803–1901) grew up in Georgia, was later a farmer, and is buried at the Bethany Presbyterian church cemetery in Newton County.[71]

The 1817 act also included a requirement for the surveyors "to mark or cause to be marked, plainly and distinctly, upon trees, if practicable, otherwise stakes may suffice, all lines which may be required to be run."[72] For that McKinley had an "axe man," John R. Jeter (1791–1852). He was born in Virginia and must have moved with his family to Greene County, Georgia, shortly after the War of 1812. He was also a farmer and is buried near White Plains in Greene County.

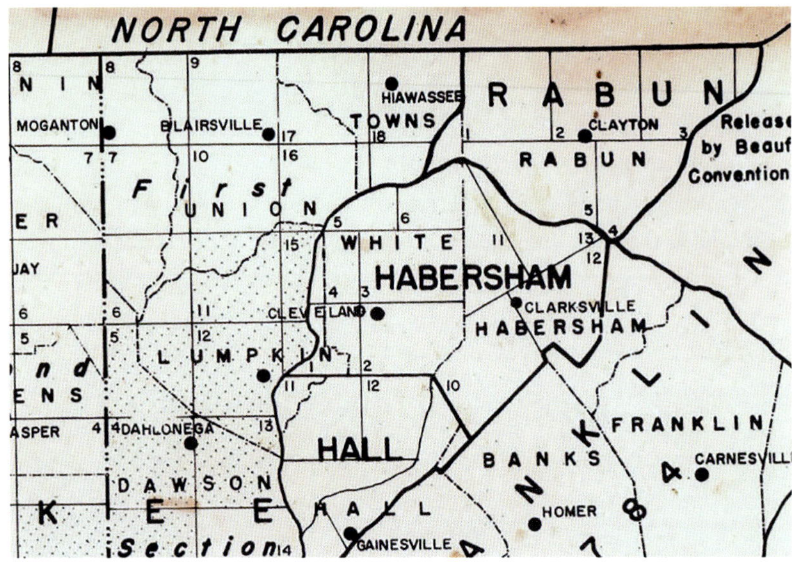

Detail from Hall Brothers' *Original County Map of Georgia*, showing original county lines and land districts overlaid on a map of counties that existed in 1895, when the map was published. Sautee Nacoochee is in District 3 of Habersham County.

McKinley began his survey on 10 April 1820 at the southeast corner of District 3, which is now in southwest Habersham County, about three miles south of SR 17 and about a mile east of the Chattahoochee River. The field notebook records the surveyor's progress, moving north along the line between Districts 3 and 11 and marking the corners of twelve lots at intervals of fifty chains or 3,300 feet. Reaching the northeast corner of the district, he turned west and marked a lot corner at fifty chains before turning south, marking corners as he went. He repeated the operation as he moved west, reversing north-south direction each time until all the lot corners had been marked.

In late May, at the district's northwest corner, not far from Dukes Creek Falls, McKinley began running the east-west survey lines that created the total of 192 land lots in the final survey, which was certified complete on 10 July 1820. The completed district plat of survey shows, in addition to land-lot lines, the course of the Chattahoochee River (or Chota River, as it was then called), Sautee Creek, Dukes Creek, and other smaller streams. Numerous native species are noted as corners to the land lots, including persimmon, sassafras, chestnut, dogwood, poplar, pine, black gum, maple, cherry, and a variety of oaks. The only road delineated is the "Unicoy Road," although, as discussed in the next chapter, the historical record suggests there were many trails and other roads in and around the valleys.

LOTTERY ELIGIBILITY

The legislation that authorized the 1820 land lottery also established a process for determining eligibility to participate in the lottery. The inferior court in each county was given three months to compile a list of those in their county eligible to draw in the lottery and send the list to the governor's office in Milledgeville. Requirements for eligibility were stated as well, but the lottery was open to "every male white person of eighteen years of age and upwards," if they were U.S. citizens and if they had lived in the state for three years prior. Veterans of "the late Indian war" were also entitled to a draw, but without having to prove residency. In addition "indigent or invalid" veterans of the Revolutionary War and the War of 1812 were entitled to two draws, as were widows of soldiers killed in those wars or the Creek and Seminole wars. Married men with a minor son or unmarried daughter were also eligible for two draws. Widows, with or without children, were eligible for one draw, if they could prove three-year residence in Georgia, as were families of one or two orphans. A family with three or more orphans would get two draws.[73]

Excluded from the lottery was "any fortunate drawer" in a previous lottery, unless they were a war veteran or "an orphan or family of orphans." Also excluded were those who refused the draft or did not hire a substitute in the War of 1812 or the Indian wars. An oath was also taken that the eligible drawer had not resided in the lottery territory prior to the extinguishing of Cherokee claims. It is likely that more than a few perjured themselves in taking that oath.[74]

"FORTUNATE DRAWERS"

The lottery was held at the state capitol in Milledgeville and began on 1 September 1820. The names of those eligible for the draw were written on individual pieces of paper and placed into one "wheel"—another wheel contained each of the numbered land lots. Since there were many more eligible people than there were land lots, blank pieces of paper were used to approximate the volume in the other wheel. The drawing went on until 2 December 1820, with the name of each "fortunate drawer" and the land lot they had drawn recorded in a register. Participants did not have to be present to win, but they were required to pay an eighteen-dollar fee within two years to receive their land grant in fee simple. The deadline could be extended, if necessitated by illness or other valid reasons, but even so a few of the lottery winners never claimed their land, and title reverted to the state. None of the fortunate drawers for land around Nacoochee appear to have occupied their property before selling it; quite likely many of them never even saw the land. The most valuable land was located along the Unicoi Road, and several of those lots were the object of the real-estate speculators who were a fixture after every lottery.[75]

Land Lot 73 encompassed what was arguably the best land in the valley and included much of the site of Chota. On 9 April 1821, the fortunate drawer, Benony Ryland of Washington County, claimed the lot and, ten days later, sold it to Henry Kendall, a merchant from Henry County, for only $100. Three months later Russell Goodrich, one of the commissioners for the Unicoi Turnpike, clearly saw the value of the property when he paid $750 for the land lot. He sold it two years later to Benjamin Cleveland for $2,425, suggesting that significant improvements had been made to the property. Not until November 1825 did Maj. Edward Williams gain title to Land Lot 73, along with Land Lot 56 to the north, both of which he purchased from Benjamin Cleveland. Unfortunately, *Deed Book C*, in which the transaction is recorded, is badly water damaged, and the sale price is illegible.[76]

On 22 November 1820, before the lottery was completed, Thomas Davis of Warren County was the first winner to claim land in the vicinity of Sautee Nacoochee, when he received a grant for Land Lot 74, which adjoins Land Lot 73 on the east and encompasses the intersection of Rabun Road and Unicoi Road in the heart of Nacoochee Valley. A few days later, Davis sold the property to Henry and Thomas Kendall for $400, significantly more than was paid for a typical land lot, which might be sold for as little as $100 or $150. The Kendalls, who bought and sold several land lots in Habersham County in the 1820s, quickly resold the property to James R. Wyly for $500. As noted earlier, Wyly was the contractor for construction of Unicoi Road and may have established one of his three "houses of entertainment" along the road, on Land Lot 74, perhaps even before the lottery.

Lachlan H. McIntosh, namesake of the famous Revolutionary War hero general Lachlan McIntosh, drew Land Lot 75, located on Unicoi Road and encompassing some of the rich bottomland along

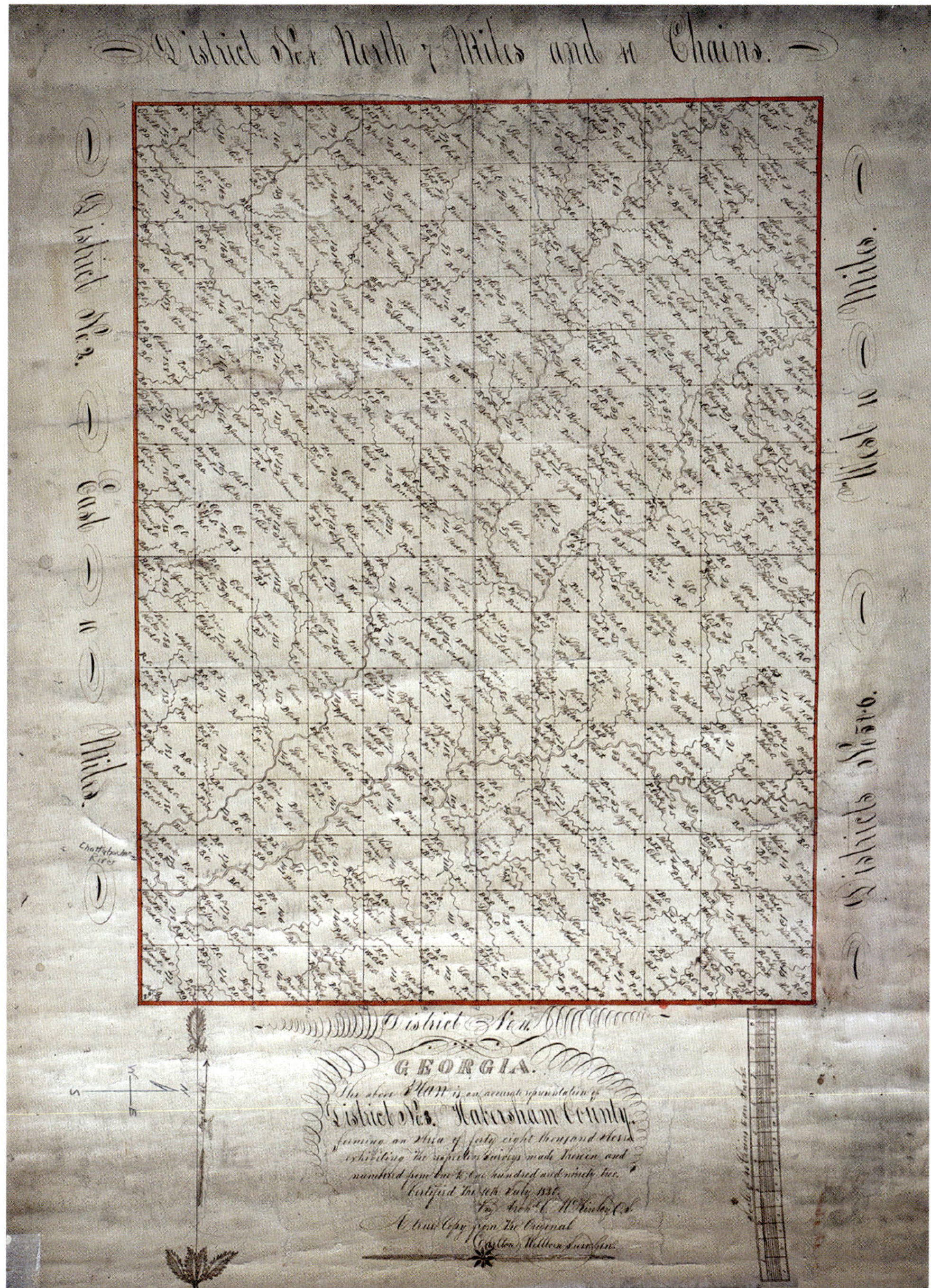

The district plat of survey for District 3 of Habersham County, certified 10 July 1820. The double-dashed lines mark the route of the Unicoi Road through the district. Secretary of State, *District Plats of Survey*.

the Chattahoochee River. It was potentially valuable real estate, but McIntosh, who lived in Savannah, delayed claiming title to the land until February 1825. He did not sell the land lot to Benjamin Cleveland until March 1826. In 1829 Cleveland sold one thousand acres, including Land Lot 75, to James R. Wyly, encompassing land "on both sides of the Chatahoochee River known by the name of the Nau-cu-cha old Indian town."[77] In January 1821 Jeremiah Lamar of Jones County claimed Land Lot 76, which surrounded the point where Unicoi Road crossed Sautee Creek. He too may have already been approached by a potential buyer, since the land lot had good potential as a site for a water-powered mill, and in August 1822 he sold the property to Michael Howell for $650.

Land Lot 57 was another valuable lot, through which ran Unicoi Road and the Chattahoochee River. Now the site of Nora Mill, the lot was granted to Peter Taylor of Walton County in February 1822. In June of that year, William Ritchie paid Taylor $300 for the lot; a year later Ritchie sold the property to Herbert Harwell for $900. Then in June 1826, Robert Westmoreland, one of the legendary sixty-one families discussed in the next chapter, acquired the lot for $1,400. Westmoreland may have built a mill before he moved away, but by 1828 it is believed that Daniel Brown was operating a sash sawmill on the river in Land Lot 57.

Land Lots 88 and 89 on the south side of the river were acquired by Jesse Richardson and his son John L. Richardson early in 1823. Land Lot 87 was granted to Nathan Talley of Greene County in September 1821, but in December he apparently sold the lot to Evin Pierson for only $20. Pierson in turn sold the lot to James R. Wyly for $150. In December 1823 John L. Richardson bought it for $500. John Richardson's father, Jesse, was one of the few early white settlers to buy property from the original grantee. Land Lot 88, adjoining 87 on the west, was on the road running to the southwest out of Sautee Nacoochee and encompassed a large tract of fertile bottomland along Dukes Creek. Granted to Henry Willet in February 1822, the lot was sold to Jesse Richardson on New Year's Day 1823 for $500.

Land Lot 44, which is crossed by Rabun Road and encompasses the intersection of Bean and Sautee Creeks, was originally drawn by William Royal of Burke County, Georgia, who had died in 1818. It was deeded by his heirs to Joseph Dobson of Habersham County on 2 March 1822, and Dobson conveyed it to William B. Wofford the following May. Three years later, on 13 May 1825, Moses Harshaw recorded his deed from Wofford, indicating a purchase price of $1,500.

A complete chain of title to the reserves of the Lynches was not established during the course of the present study. It is not clear how title to those reserves was conveyed to the State of Georgia and how the state distributed the land after that. To the north and east of Land Lot 44 lay Jeter Lynch's 640-acre reserve. Because of the Lynches' early deaths, title to the lots within that reservation was not settled until 1829. Most of the land lots in and around Sautee Nacoochee were claimed before 1825, but some were not claimed until the late 1820s. Land Lot 77, which had just a few dozen yards of road frontage and encompassed the steep hillsides along Carr Creek, was drawn by Martin White of Franklin County. He did not pay the fee and take title to the land lot until 9 September 1830.

How many of the new land lots encompassed Cherokee farms is not known, but some certainly did. And a few truly fortunate drawers were thus able to pay eighteen dollars and gain fee-simple titles to improved farmland, houses, and outbuildings left by the unfortunate Cherokee. The true value of such land lots will never be known, but some have argued persuasively that the impetus for what happened to the Cherokee in Georgia was as much a land grab as it was a gold rush.[78]

5

The Pioneers

Northeastern Georgia was being repopulated even before the Cherokee had been completely dispossessed of their land. In August 1820 the federal census enumerated 3,145 residents in newly organized Habersham County. There was as yet no county seat, and local government barely functioned. Among those enumerated in 1820 were 277 enslaved African Americans. The 1830 federal census enumerated 10,671 residents in Habersham County, 909 of them enslaved. The numbers suggest a population that had apparently exploded by an astounding 263 percent since 1820. The gold rush that swept across North Georgia after 1828 certainly accounted for part of that growth, but it was also because Georgia had extended the jurisdiction of counties along the Chattahoochee River westward to the state's border in December 1828. As a result, the actual population east of the Chestatee River was substantially less than the number stated in the census.[1]

By 1840 the Cherokees were gone from Georgia, and the census of Habersham County, which no longer included Lumpkin and Cherokee Counties, recorded a population of 7,961, including 942 slaves. The county grew slowly after that, reaching 8,895 in 1850, with 1,218, or 14 percent, of them enslaved. Census data indicates that the majority of those slaves were not in White County after it was created in 1857. The 1860 census of the newly reduced Habersham County enumerated nearly 15 percent of the population as slaves, while in White County, the slave population was less than 8 percent of the total of 3,315.[2]

Ethnicity

There were four waves of British emigration to America in the seventeenth and eighteenth centuries, each with its own folkways that echo through American culture in the twenty-first century. The first was the emigration of English Puritans, primarily from East Anglia, to New England in the second quarter of the seventeenth century. They were followed by Cavaliers escaping Oliver Cromwell's England and indentured servants, mainly from the south of England, who arrived in Virginia in the mid-seventeenth century. In the fifty or so years after 1675, a migration of Friends, or Quakers, mainly from the North Midlands of England, led the settlement of the mid-Atlantic region. Finally, a fourth great wave of immigrants from the Borders and Ulster filled up the backcountry from Pennsylvania to the Carolinas in the eighteenth century and moved into northern Georgia in the early nineteenth.[3]

The tide of settlers that washed over northeastern Georgia in the first quarter of the nineteenth century included people from a variety of backgrounds, most of them of British descent, with a few from France, Germany, and other European countries. More than a few were of African descent. Not until 1850 did the federal census begin to compile social statistics, among them numbers showing one in eight Georgians being foreign born and nearly a quarter born in other states. In the 1820s and 1830s, there would have been more foreign-born Georgians and many more born in other states, primarily the Carolinas, with a few from Virginia, Maryland, and Pennsylvania.

ENGLISH AND WELSH

Some of the settlers in and around Sautee Nacoochee were descended from early English and Welsh colonists who settled the eastern seaboard from New England to the Carolinas in the seventeenth and early eighteenth centuries. Depleted farmland coupled with a growing population sent more than a few of their descendants down the southern Piedmont as early as the second quarter of the eighteenth century. Numerous merchants and entrepreneurs from New England, too, saw opportunity on the southern frontier in the early nineteenth century.

Generally prosperous, these English and Welsh Americans had advantages of language, education, and resources that many of their neighbors did not enjoy. Typically, they were Congregational in New England, Quaker or Catholic in the mid-Atlantic, and Anglican in Virginia and coastal Carolina. On the southern frontier there were simply not enough of any of those denominations to support churches, and many became Presbyterian, Methodist, or Baptist.

SCOTS-IRISH

Of the four great sources of British immigration in the eighteenth century, those from Ulster and the Borders were the largest and longest. And, unlike the earlier immigrant groups, they were not immigrating for religious or political reasons; the Scots-Irish were primarily driven by the potential to improve their generally dreadful economic condition.[4] Many of the early pioneers in northeastern Georgia—including the Bells, Englands, and Sosebees—were part of the flood of Scots-Irish, or Ulster Scots, since as many as 60 percent of the quarter million of them who immigrated in the eighteenth century were from Ulster. Many came from the Borders of southern Scotland and northern England and were generally Presbyterian, but a few English Puritans and merchants also had emigrated to Ulster in the seventeenth century. Even Huguenots, also Calvinist, immigrated to Ulster after 1685 and were often absorbed into the Presbyterian Church.[5]

Most of these immigrants arrived at one of the Delaware River ports—Philadelphia, Chester, or New Castle. From there they and the German Protestants, who were immigrating from Holland around the same time, quickly settled Pennsylvania's rich Susquehanna River valley. The Scots-Irish, however, were "not succeeding so well . . . as the more frugal and industrious Germans," noted one contemporary observer. As a result, the Scots-Irish frequently "sell their lands in that province to the [Germans], and take up new ground in the remote counties in Virginia, Maryland, and North Carolina."[6]

These people were almost always part of a family group and were often accompanied by neighbors as they immigrated and made their way down the Great Philadelphia Wagon Road before the Revolution. Those who encountered them on the southern frontier found them often poor but always proud and belligerent—a people born of centuries of warfare with one another and with the English. Their sense of religion, too, was what David Hackett Fischer calls a "militant Christianity."[7]

Most references to "Irish" in eighteenth-century America referred to these Protestant immigrants from Ulster. By 1753 there were perhaps fifteen thousand settlers in the western Carolina Piedmont, "for the most part Irish Protestants and Germans, and daily increasing," according to Matthew Rowan, president of the North Carolina State Council at the time. By 1776 perhaps one American in ten was of Scots-Irish descent, with a much higher percentage in the southern backcountry.[8]

During the Revolution the Scots-Irish formed the backbone of the colonial forces in the South. They led the Regulator movement in North Carolina in the 1760s, culminating in one of the first skirmishes of the war when they battled Royal forces at Alamance in May 1771. And, in May 1775, it was the Scots-Irish of Mecklenburg County, North Carolina, who made the first colonial declaration of independence from the Crown. In them were found few of the conflicting interests that made Tories out of some of their more comfortable Anglican neighbors. By the mid-nineteenth century, when famine drove hundreds of thousands of Catholic Irish to America, the descendants of the Ulster Protestants who immigrated before the Revolution were fully assimilated and beginning to call themselves "Scotch-Irish," or "Scots-Irish," as some prefer today.

AFRICAN AMERICAN

A small minority of the new settlers in the Georgia Piedmont brought with them enslaved African Americans, and records of their sale are among the earliest public records in Habersham County. They were never numerous in the hills and valleys of North Georgia, where the topography limited the use of gang labor on large cotton plantations. In areas where there was exceptionally good agricultural land, however, such as in the valleys of northeastern White County, the antebellum population of African Americans rose to as much as 15 percent of the total population. In 1850 all but two of the African Americans in Habersham County were enslaved.

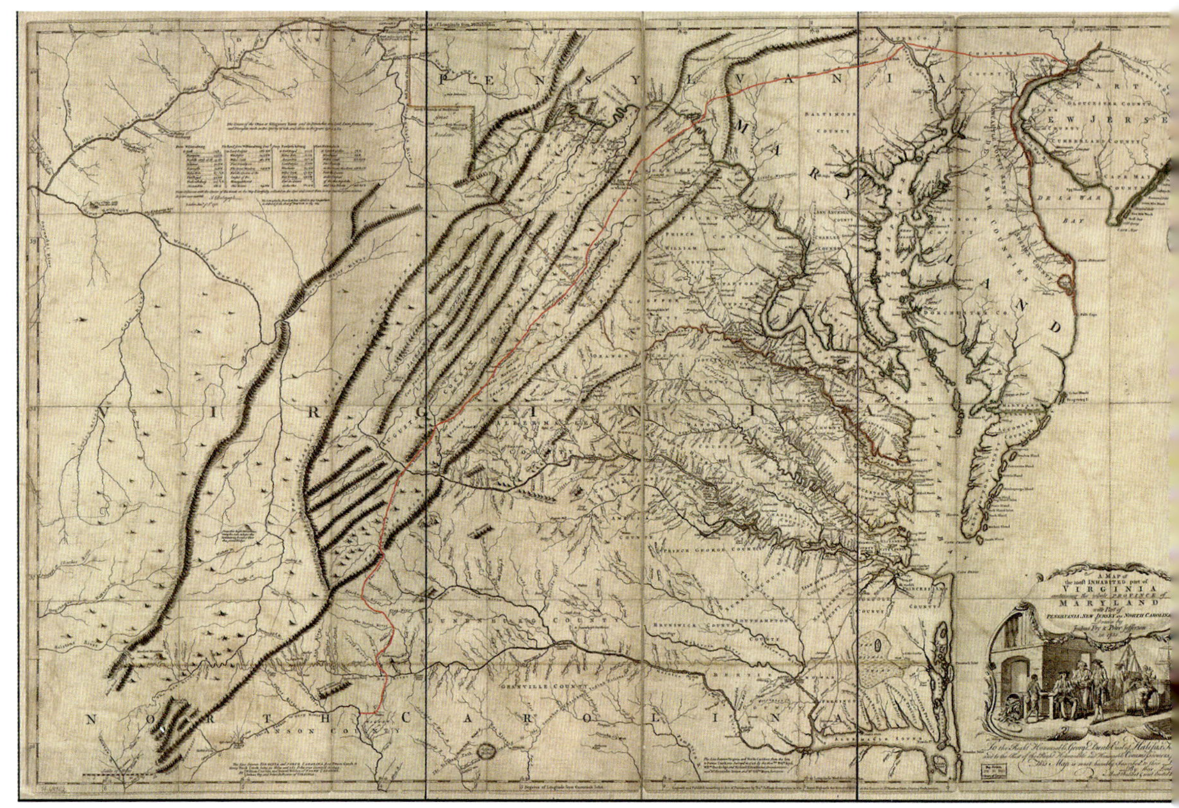

Frye and Jefferson's map of Virginia in the 1750s, annotated in bright red to highlight the route of the Great Wagon Road. Item 74693166, Geography and Map Division, Library of Congress, Washington, D.C.

The United States abolished the importation of slaves in 1807, although slaves were still being smuggled into the country in the 1850s. The internal slave trade continued, however, feeding off the natural (in some cases unnatural and forced) population increase among African Americans. Typically, their ancestors, or they themselves, were brought directly from West Africa, the Congo, or Angola, although a few came via the Caribbean sugar plantations. The 1870 federal census is the first to record the birthplace of African Americans. Among the oldest of the freed slaves, more than a few in Georgia and elsewhere recorded Africa as their birthplace. Many more, including the African American community at Nacoochee, were, like the people who had claimed to own them, born in Maryland, Virginia, and the Carolinas.

Local Tradition

Documenting local history and people depends on a variety of resources. The decennial federal censuses, 1790 through 1950 (the most recent set of population schedules now open for research), provide a wealth of personal and community information. County records of deeds and mortgages, taxes, marriages, wills and probate, and judicial proceedings, as well as newspapers and other periodicals help fill out the picture. And sometimes there are private letters, journals, and memoirs or other such personal documents that can provide texture and depth to bare statistics and legal records. In the last quarter of the nineteenth century, antiquarians and amateur historians, many of them inspired by the nation's centennial in 1876, began to search out and record details of local history.

Documentation for the people at Sautee Nacoochee is further enriched by some unique resources, including three maps of Nacoochee, two depicting the valley as it existed in 1837 and a third as it existed in 1948. Famously, there are several early twentieth-century compilations of the saga of the sixty-one families thought to have originally settled the valleys in the 1820s. For most of the past century, these documents have been important, if flawed, components of local tradition surrounding the history of white resettlement of the area in the early nineteenth century.

THE MAPS

The first and oldest of these maps is titled *Nacoochee, 1837*. Signed "R. C. Moffat, Fecit, 1891," it depicts the closely knit community that developed along the Chattahoochee River in the 1820s and 1830s.[9] Reuben Curtis Moffat (1818–94) was born in Ithaca, New York, the son of John Little Moffat (1788–1865) and Hannah Curtis (1792–1859). The elder Moffat, who is believed to have bought gold mines in North Carolina and Georgia in the 1830s, brought the family to Nacoochee, where the youngest of the Moffats' fourteen children was born in March 1837. The Moffats' daughter Adeline Margaret Moffat (1815–80) taught school at Nacoochee before her marriage to Dr. Joseph T. Curtis in New York City in June 1841. Reuben Moffat graduated from medical college in 1846 and became a renowned homeopathic physician.[10]

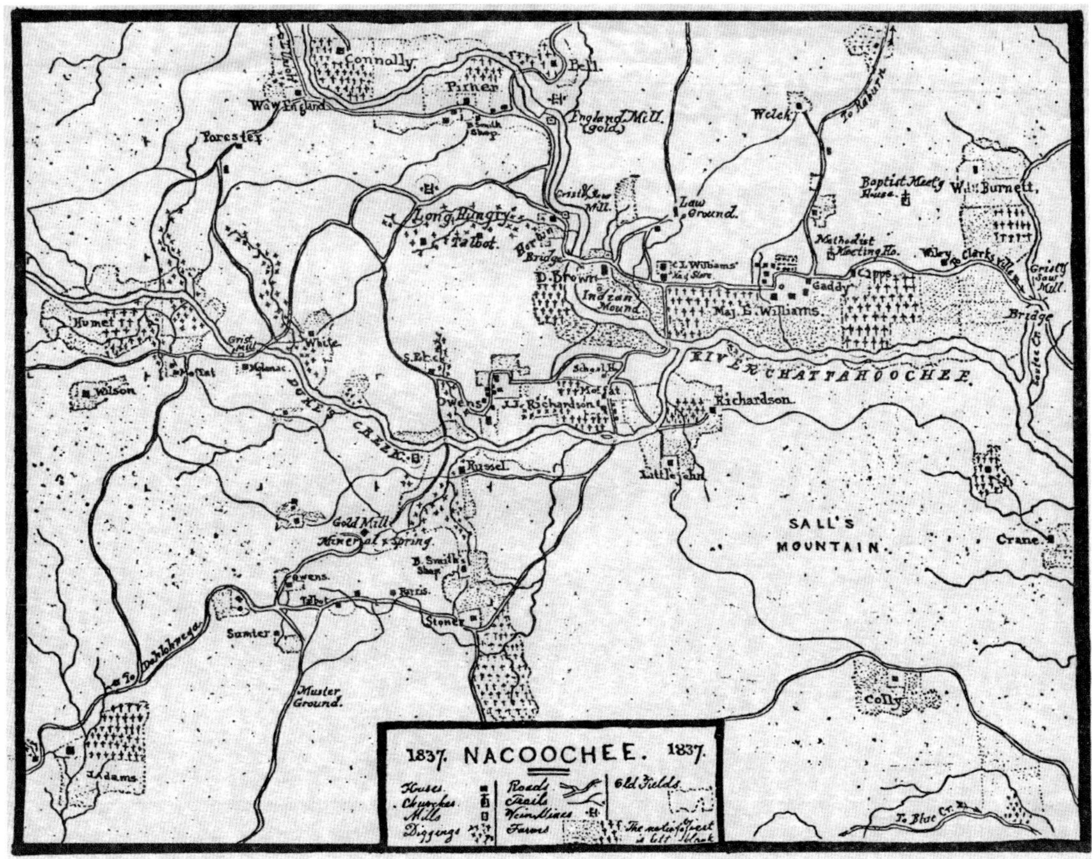

Reuben C. Moffat's 1891 map of Nacoochee in 1837. Courtesy of Sautee Nacoochee Community Association.

Moffat may have drawn the map from memory, but it is possible that he had earlier maps or notes to guide his work. Modern maps and historical documentation support the assumption that his map is a mostly accurate representation of the community's roads, residences, churches, mills, and other landmarks, although relative distances and locations are sometimes distorted. In 1922 the map was redrawn by Victor R. Hollis (1883–1963), who, with his wife, Marnie, ran the George W. Williams Orphanage at Nacoochee in the 1920s and 1930s. He added topographical features and a few names not shown on Moffat's map, while otherwise more or less replicating its information. In many cases, however, Hollis's lettering clarifies some of Moffat's.[11]

Victor R. Hollis's 1922 copy and annotation of the Moffat map of Nacoochee. Courtesy of Sautee Nacoochee Community Association.

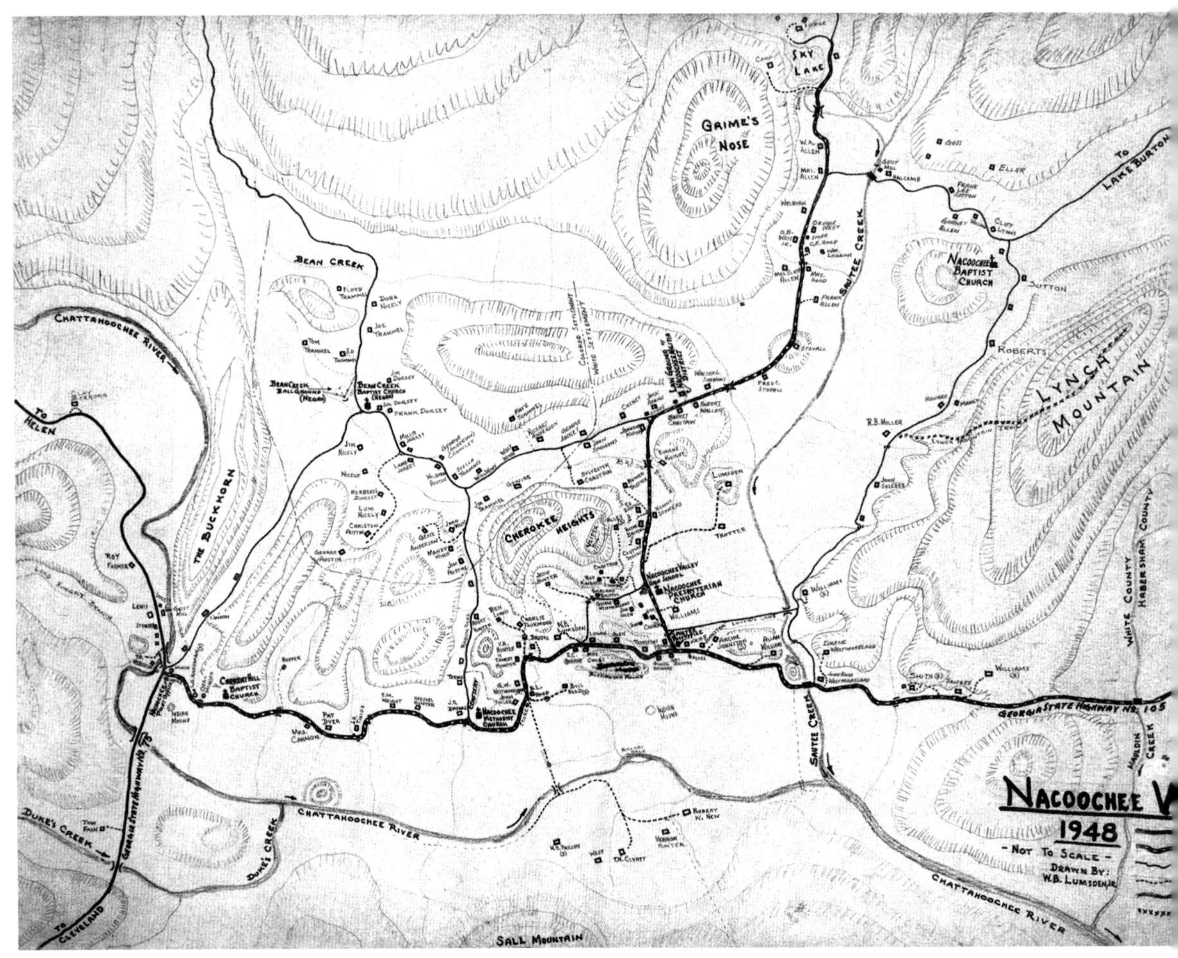

A map of Sautee Nacoochee in 1948, drawn by Walter Branham Lumsden Jr. (1919–2011). Courtesy of Sautee Nacoochee Community Association.

In 1948 Walter Branham Lumsden Jr. (1919–2011), a great-grandson of John L. Richardson, continued the cartographic tradition and drew his own map of Nacoochee. His father was the rural mail carrier for northeastern White County through most of the first half of the twentieth century. His map included most things depicted in the earlier maps, but with the addition of a boundary line between the "Colored Settlement" and the "White Settlement."

LISTS OF EARLY SETTLERS

One of the foundation legends from the early history of Habersham County is the tale of the "Sixty-one Families."[12] There are at least four versions of the story, which claims to name "the different families who came to Nacoochee Valley," with the number of families ranging from fifty-five to sixty-two and with some variation in surnames. The move is said to have occurred in early March 1822 or 1823.[13] Documentary research suggests that the story compresses multiple arrivals at different times from different places into a single grand exodus from North Carolina to Georgia. "There were sixty-one different families that came in two parties, one just one day behind the other, using the same camp fires, one party from Burke County, North Carolina, and one from Rutherford County, North Carolina. The first party led by Daniel Brown and Edward Williams left Burke County on March 1st, and the second led by Rev. Jesse Richardson, Abraham Littlejohn and Adam Pitner left Rutherford County about the same time."[14]

There is much that is clearly apocryphal, including the statement that "they were the first white people in this county and bought the land from the Cherokee at one cent an acre," but there is useful historical data as well. Whatever its faults in terms of historical accuracy, this traditional story captures a bit of Nacoochee's early history, especially relationships between families, that might otherwise have been lost or overlooked. As with any documentary source, it requires critical analysis.[15]

The tale of the sixty-one families no doubt took shape as the last of the original pioneers from the 1820s were dying in the late nineteenth and early twentieth century; Mrs. John Evans, a daughter of John Gibbs, died at the age of 107 in 1917, the last of those first settlers. Garbled stories from aging relatives were sometimes written down, such as the undated, handwritten memoir about the Lawrence Cemetery in the upper reaches of Dukes Creek. The unidentified author wrote, "In the year 1823 a crowd of immigrants moved from North Carolina and settled in Nacoochee Valley and on Dukes Creek and surrounding country. They consisted of Richardsons, Allisons, Ledfords, Haswells, Allens, Crumleys, Birds and others."[16]

These bits and pieces of information were only later compiled into a single document, perhaps in the early 1920s, as the Williams family held a reunion to celebrate a century of their living at Nacoochee. The numbered lists are especially confusing. In one version they range from number 1 to 62, but 28 is omitted. Five "families" are identified by given name only, and at least four of the numbered entries contain two separate families, while another appears to represent three. Edward Williams is listed as a leader of the party from Burke County, but he is not shown in any of the listed families. Even so, sixty-one families with surnames are in fact noted in one way or another in the documents. The story includes some biographical and locational information, but the latter is often frustratingly vague. Several families were listed as having "settled at the foot of the mountain," leaving it unclear as to which mountain; Currahee, Yonah, Tray, and Lynch Mountains are all mentioned elsewhere in the lists. None of the entries is less helpful in determining location than number 57: "The fifty-seventh family was Jim Monroe, and they settled on what is known as the Monroe place."[17]

In addition the title of some versions of the legend suggests that all the families settled in Nacoochee Valley, but the individual entries also show a number of other locations. Most were around Sautee Nacoochee, but Clarkesville, Soque, and Tray Mountain, all in Habersham County, as well as Mount Yonah, Blue Creek, and Mossy Creek in White County, are also mentioned. As Matt Gedney and others have pointed out, however, until the twentieth century, the term "Nacoochee," or "the Nacoochee," could refer to a much larger area, stretching from the Blue Ridge to Yonah.[18]

Some sources have credited the tale of the sixty-one families to Victor Goodman Bristol (1882–1969), whose paternal ancestors had settled in Burke County, North Carolina, in the eighteenth century. Around 1870 his father, George Gailer Bristol (1841–85), married Sallie J. Williams (1849–1925), daughter of Edwin Williams and granddaughter of Edward Williams. They settled near Hayesville, North Carolina, where they had five children, Victor being the youngest. In 1885 George Bristol died unexpectedly, forcing his widow and the five children to return to her father's home at Nacoochee. She and several family members are buried at Nacoochee Methodist church cemetery. Victor Bristol never married, but he was intensely interested in local history. In addition to the list of the sixty-one families, Bristol also annotated Moffat's 1837 map of Nacoochee Valley. He grew up in the valley among Williams relatives, including his grandfather Edwin, who had come to Nacoochee as a child in the 1820s. His sources were not historical documents but rather the irreplaceable, if sometimes fallible, oral traditions, in which he was a primary link.[19]

North Carolina

According to tradition, all of the sixty or so families migrated from Burke and Rutherford Counties, but historical documentation suggests that several families were from other counties in western North Carolina and even two or three from upstate South Carolina. Burke and Rutherford Counties are located adjacent to each other in the shadow of the Blue Ridge in western North Carolina.[20] Settled in the last quarter of the eighteenth century, the topography and environment of those counties are quite similar to those of White and Habersham Counties in northeastern Georgia.

The families that can be documented from these counties were mostly from what is now McDowell County, which was formed out of southwestern Burke County and northwestern Rutherford County in 1852. Over half of the family names can be found in the 1820 federal census for Burke and Rutherford Counties, including Jesse Richardson, John Wilson, John Edwards, Abraham Littlejohn, Maj. Frank Logan, James Black, Tom Edwards, William Bell, James Evans, John Adams, Benjamin Allison, John Gibbs, William England, and Daniel Brown. There were likely others who are counted in the census but are not named, since only the heads of households were listed in the censuses prior to 1850. Many of these families were descendants of those who settled western North Carolina before the Revolution. The westernmost branches of the Great Wagon Road from Philadelphia ran generally southwestward from Salisbury, North Carolina. By the 1750s they provided the white settlers a ready route to the foothills of the Blue Ridge in the western Carolinas.

The first land grants west of the Catawba River were made in 1754, but, with the turmoil of the French and Indian War, real settlement was delayed until the mid-1760s. When Mecklenburg County, North Carolina, was organized in 1762, it encompassed an area from Charlotte to the Blue Ridge as well as a large territory that was disputed with South Carolina. By 1769 the population in the area had grown to the point that the royal governor of North Carolina, William Tryon, organized a large area of what is now western North Carolina and parts of upstate South Carolina as Tryon County.

The disputed state boundary west of the Catawba was settled in 1772, and in 1779 Lincoln County was organized out of the eastern part of colonial Tryon County, Rutherford County was organized out of the western part, and the name of the royal governor was eliminated entirely. The 1780s were a period of tremendous growth in Rutherford County, especially after the end of the Revolution. The first permanent white settlement west of the Blue Ridge was established in 1785, and in 1791 a new county, Buncombe, was created there, with Asheville its county seat.

The narrative of the sixty-one families states that the 135-mile journey from North Carolina to Nacoochee took ten days. As one contributor to the traditional narrative explained, "Travel was made more difficult by reason of the fact that there were no roads a good part of the way and they had to cut their way through forests." In reality there had been passable roads out of North Carolina and through upstate South Carolina into Georgia since the 1790s. In the early 1820s, regular stagecoaches were running through Clarkesville from Franklin, North Carolina, and from Carnesville in Franklin County, Georgia, with connections to upstate South Carolina. It is unlikely that any roads required clearing for travelers to get where they were going in the 1820s; moving livestock would have been more than enough to account for the slow pace.[21]

There would have been several possible routes through South Carolina to Walton's Ford on the Tugaloo River, the eastern terminus of the Unicoi Turnpike. The most direct route would have carried them south out of Rutherford and Burke Counties by way of the Rutherfordtown Post Road, through Cowpens, site of one of the great Revolutionary battles on the southern frontier, and on to Spartanburg, Greenville, and Pendleton in upstate South Carolina. The parties would likely have crossed the Tugaloo River at Walton's Ford, although they could have decided to avoid the tolls by going northwest out of Pendleton to Clayton, Georgia, and entering Nacoochee from the northeast via what is now SR 255 and Rabun Road.

Detail from Brazier's *A New Map of the State of North Carolina*, published in 1833, showing the part of western North Carolina where many of the pioneers in the resettlement of what is now Sautee Nacoochee originated. The North Cove, home to the Browns and Williamses prior to 1822, is just above center in this image. Item 2006459002, Geography and Map Division, Library of Congress, Washington, D.C.

Early Arrivals

Several of the legendary families appear in the 1820 Habersham County census and other documentary sources years before their supposed arrival in Nacoochee. Some of them had a significant impact on the history of Habersham and White Counties. Most prominent perhaps were the Holcombes, Vandivers, and Benjamin Cleveland.

HOLCOMBE

One of the early arrivals was Jesse Holcombe (1795–1862), "who settled near Mauldin's Creek," according to the traditional list of sixty-one families, two or three miles southeast of Sautee Nacoochee and now known as Mauldin Mill Creek. He is one of no less than nine Holcombe (or Holcomb, as it is sometimes spelled) heads of household in Habersham County in 1820. A veteran of the War of 1812, Holcombe married Susannah Rusk in Franklin County, Georgia, in 1816 and apparently brought the family into Habersham County as early as 1818.

In November 1822 Jesse Holcombe bought Land Lot 151, on Brasstown Creek just east of Mount Yonah. At least two of his brothers, Hampton and John Holcombe, were in Habersham County by 1820, as was their father, Sherwood Holcombe (1763–1844), who purchased Land Lot 108, Third District, in February 1822. That land lot straddles the Chattahoochee River just upstream from Mauldin Mill Creek. Jesse and Susannah Holcombe are believed to be buried in unmarked graves at Old Blue Creek Cemetery in southern White County.[22]

VANDIVER

Adam Poole Vandiver (1787–1876) was number 19 on the list, which notes that he "settled near the foot of the mountain." Vandiver was born in South Carolina, probably the Pendleton District, the son of Rev. George Vandiver (1764–1833), a Baptist minister, and Mary Ann Poole. Adam fought in the Creek War before moving into northeastern Georgia, settling near Tallulah Falls, and appearing in the 1820 Habersham County census. He is thought to have married three times, the first time to a Cherokee woman, Gulle (The Dove) Whiting, sister to Chief Gray Eagle. Both generations of the family are thought to have been buried near their home at Tallulah Falls.[23]

CLEVELAND

Benjamin Cleveland (1784–1858) and his family were eighth on the list, but they were already in Habersham County before 1820. He was born in South Carolina and was the namesake of the noted Col. Benjamin Cleveland (1738–1806), who had made a name for himself at the Battle of Kings Mountain during the Revolutionary War. After the war the family moved to Franklin County, Georgia, where Benjamin married Anglin Blair in 1802 and where he was elected to the state legislature. He fought in the Creek War, and, when Habersham County was created in 1819, he was the county's first state representative and went on to serve six terms. Cleveland was also one of the first justices of the Habersham County Inferior Court and commanded a regiment that assisted in pushing the Cherokee people onto the Trail of Tears in 1838. He is thought to have been a carpenter and a builder, but he was also an enslaver of numerous African Americans and a prosperous merchant in Clarkesville, which had been laid out on thirty-two acres that he and Benjamin Chastain donated to the county. Cleveland served eight terms in the state senate in the 1830s and 1840s.

Habersham County land records show that Cleveland began buying property in the Third, Fifth, and Sixth Districts of western Habersham County (now White County) as early as 1821 and that many of the transactions were speculative. The author of one version of the traditional list of settlers noted that Cleveland "owned quite a lot of land, but sold it and left very suddenly in the night and no one knew where he had gone or why. He afterwards was heard of in Texas where he died." Cleveland in fact is never known to have disappeared but rather died and is buried in Clarkesville. Some of his children, however, did move to Nacogdoches, in East Texas, which may be part of the basis for the list's garbled tale. The list is correct in noting that Cleveland, the White County seat of government, was named in his honor.

FULLER

Family number 36 was John Fuller, who, according to the list of families, "settled down near Cool Springs [*sic*] Church," the Methodist church located along Unicoi Road, five or six miles east of Sautee Nacoochee. There are no Fullers in the 1820 census of Burke or Rutherford Counties, and no John Fuller appears in the antebellum censuses of Habersham County. However, Ezekial Fuller (born about 1791) and his wife, Eleanor, are documented in the federal censuses of 1820–60, and both are buried at Old Clarkesville Cemetery. The population schedules for the 1870 census show at least one of their sons living at Cool Springs. The same census also shows one John Wesley Fuller, who was born in North Carolina about 1837, working as a blacksmith in Clarkesville, but any relationship to the Fullers at Cool Springs has not been established.

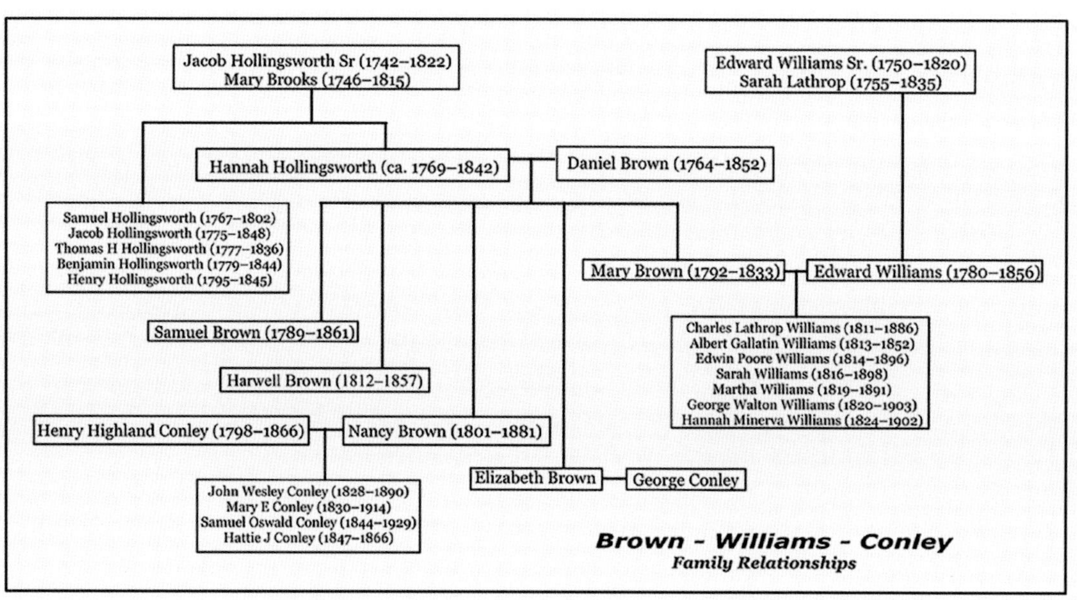

A family tree of some of the relationships among early white residents of the valleys. Image by author, 2023.

First Families

Lewis W. Richardson (1904–88), who claimed no family relationship to Rev. Jesse Richardson, attempted to better organize the scattered versions of the traditional story of the settlement of Nacoochee in a paper he compiled for the White County Historical Society. His narrative adds historical details that may have been gleaned from interviews with older residents of the valleys, including perhaps Victor Bristol.

Richardson turned the story into a manuscript that perpetuated, mostly without reservation, some of the myths of the valleys' resettlement in the early 1820s. Surnames are alphabetized and organized into two groups, one from Burke County and one from Rutherford County. He has both groups arriving in early March 1823, while other versions of the story use a date of March 1822. He named thirty-nine families in the group from Burke County but noted that three other names had been forgotten. Seventeen families in the Rutherford County group were identified, while another four names could not be remembered. In spite of its flaws, this typescript manuscript is the most coherent version of the story, and some of the families most relevant to Sautee Nacoochee are extracted here.[24]

According to tradition, the largest group of forty or so families was from Burke County and led by Daniel Brown. The same tradition also has Brown accompanied by his sons-in-law Edward Williams and Henry Conley, although the former may not have arrived until 1825, and the latter did not arrive until 1828. All three of these men and their families would play a major role in the resettlement of Nacoochee. The second party, which had only twenty-one families, left Rutherford County under the leadership of Rev. Jesse Richardson, his son-in-law Abraham Littlejohn, and Adam Pitner. They and their descendants would also leave indelible marks on the valleys.

BROWN

The list of sixty-one families begins with the Brown-Williams family, who led the first party from Burke County. No other family had more of a share in shaping the Sautee Nacoochee we know today. It was no accident that this prosperous family acquired the greatest part of the Nacoochee Valley, since Edward Williams, as did many others, searched out the best land that would be distributed in the lottery.

Daniel Brown (1764–1852) and his wife, Hannah Hollingsworth (ca. 1773–1842), were married in Randolph County, North Carolina, in August 1788. Her parents, Jacob and Mary Brooks Hollingsworth, were Methodists but had roots among the Quaker community in Chester County, Pennsylvania. The Hollingsworths married in Pennsylvania in 1768 but moved to what became Guilford County, North Carolina, before the Revolution. In 1792 Jacob Hollingsworth, his brother Samuel, and their families were part of the ill-fated Wofford Settlement on the Georgia frontier. The house Jacob Hollingsworth built about 1793, near what is now Alto, Georgia, still stands and is listed in the National Register of Historic Places.[25]

Daniel Brown's ancestors were among the Quakers who settled along the Delaware River in the late seventeenth century. His parents married around 1750, probably at his mother's parents' home in Lower Alloways Creek Township, New Jersey. They had several children before moving to Guilford County in the North Carolina Piedmont, where Daniel was born around 1764. Daniel's father was a Quaker, while his mother was Presbyterian; after the Revolution they were all stalwarts in the Methodist Church. The lives of Daniel's siblings remain obscure, but his sister, Mary (1761–1829), married Edward Carter (1761–1838) around 1781. In the early 1820s, around the time Daniel and his family moved to Nacoochee, Mary and Edward moved to Rabun County, along with at least one son and his family. Mary Brown and Edward Carter are both buried at the Head of Tennessee Baptist church in Dillard.

Not long after their marriage, Daniel and Hannah Brown moved to Burke County, North Carolina, and in 1793 bought property in an area known as the North Cove, a few miles south of Linville Falls and near Hannah's parents and other Hollingsworth relatives. The Browns built up a large farm, and Daniel is also believed to have operated a store. They had two sons—Samuel (1789–1861) and James Harwell (1812–57)—and six daughters: Mary (1792–1833), Sarah, or Sallie (1798–1872), Nancy (1801–81), Patsy (1806–60), Edith (1808–1900), and Elizabeth (1810–84).

In February 1810 Mary Brown married Edward Williams (1780–1856), and in 1822 her sister Nancy Anne married Henry Highland

Conley (1798–1866), both in Burke County, North Carolina. The Conleys would play a major role in the early resettlement of the valley around Helen, but, contrary to the traditional tale, they appear to have remained in North Carolina until 1828. A few years later, the youngest of the Brown daughters, Elizabeth, married Henry Conley's brother George, but they remained in Burke County, where they occupied the old Brown home place in North Cove. The youngest of the Browns' children, whom they called Harwell, came to Nacoochee with his parents but never married.[26]

If the federal census is correct, Daniel Brown had acquired eleven slaves by 1810, and, by the time he moved to Nacoochee, he owned fourteen or fifteen. He continued to own slaves into the 1840s, when he apparently gave them to his son Harwell, who was enumerated with sixteen slaves in 1850. The Browns are believed to have acquired parts or all of Land Lots 57 and 72 of the Third District of Habersham County (now in White County) in the 1820s, but dating those transactions is problematic since Daniel Brown may not always have recorded his deeds in a timely manner. Their house was in Land Lot 72 near the present Nichols-Hunnicutt-Hardman House but, if Moffat's map is correct, on the south side of Unicoi Road. It may have been located west of the present house in an area now on the north side of SR 17. He is said to have built a saw- and gristmill on the Chattahoochee near the present site of Nora Mill in Land Lot 57 as early as 1824.[27]

Detail from Moffat's map of Nacoochee in 1837, showing most of Daniel Brown's farm in Land Lots 57 and 73. Courtesy of Sautee Nacoochee Community Association.

WILLIAMS

Daniel and Hannah Brown's daughter, Mary, married Edward Williams in Burke County, North Carolina. Williams was a descendant of Richard Williams (1606–92) and Frances Deighton Williams (1611–1706), who immigrated from Gloucestershire to Taunton, Massachusetts, in 1636 or 1637. Edward was born in Easton, a town about ten miles north of Taunton in southeastern Massachusetts, one of twelve children born to Edward Williams (1750–1820) and his wife, Sarah Lathrop Williams (1755–1835). The younger Edward Williams grew up on the family farm but got "tired of the rocks" and in 1799 boarded a ship for Charleston, South Carolina. He was like many young men of his generation who escaped the confines of New England to seek their fortune on the southern frontier in the early years of the republic. He spent two years in Charleston, perhaps working for a mercantile company, but in 1802 decamped to Burke County in western North Carolina. Georgia historian E. Merton Coulter, who wrote the definitive biography of Edward Williams's son George Walton Williams, suggests that the elder Williams may have been working for a mercantile company that had branches in Charleston and Morgantown in western North Carolina. Edward Williams was apparently always addressed as "Major Williams," but none of the sources suggest how he came by that title. Since he was too young to have fought in the Revolution, the title may, as Coulter suggests, date to undocumented service in the North Carolina or Georgia militia during the War of 1812 or the contemporaneous wars with the Creek and the Seminole.[28]

After their marriage Mary and Edward set up housekeeping near her parents, and seven children were born in Burke County: Charles Lathrop (1811–86), Albert Gallatin (1813–52), Edwin Poore (1814–96), Sarah L. (1816–98), Martha (1819–91), George Walton (1820–1903), and Hannah Minerva (1823–1902). There were also twins who did not survive infancy. All but two of the children who survived childhood lived and died in White or Habersham County. Daughter Martha married Robert Junius Gage (1810–82) in 1845, and they raised a family of several children in Union, South Carolina, where both are buried.

The youngest of the Williams sons, George Walton Williams, left home before he was eighteen, moving to Augusta and then in 1852 to Charleston, where he built a wholesale grocery business into a career as a merchant and banker. He was married twice and father of thirteen children, but of the seven by his first wife, Louisa A. Wightman, none survived beyond the age of nine, and of the six by his second wife, Martha Fort Porter, two died as infants. He was an important figure in Charleston before, during, and after the Civil War and his is said to have been the first "mercantile house" to reopen after the war. He continued to make his main residence in Charleston, but in 1876 he built a summer house, which he called Mountain Home, at Nacoochee. His memoirs, if sometimes overly sentimental, remain one of the richest sources of information on the early history of Sautee Nacoochee.[29]

The other Williams sons—Charles, Albert, and Edwin—and daughter Hannah remained at Nacoochee, and each received a house and land from their father. Charles married Jacob Hollingsworth's granddaughter Hannah Hollingsworth (1810–87) in 1836, and they had seven children who survived infancy. Both are buried at Nacoochee Methodist church cemetery. Albert and his wife, Elizabeth Stephenson (or Stevenson), married in 1843 and had four children before his untimely death in 1852. Edwin Williams married Jane Elizabeth Perkins (1818–1910) in 1838, and they raised eight children at Nacoochee. They too are buried at Nacoochee Methodist church cemetery.

The youngest of the surviving Williams children, Hannah, married Dr. Fletcher Starr (1820–98) in 1845 and with him had seven children, only to see four of them die of typhoid fever, contracted while attending a family gathering in the late summer of 1893. The oldest daughter, Sarah, was the last to leave home, when she married John S. Dobbins (1800–86), a widower with four young children, in 1851. They had no children of their own. He is buried at Clarkesville, she at Nacoochee.

Edward Williams referred to "my mountain lands" in his will, which probably included the site of his dairy on Tray Mountain. In November 1825 Williams took title to Land Lots 56 and 73, adjacent to his father-in-law's property along the river. It is believed that he built a temporary house, perhaps a log house, before constructing their final residence in Land Lot 74 in 1830. Inherited by their daughter Hannah Williams Starr, who christened it Starlight, the house burned to the ground in 1959.[30]

Like his father-in-law, Edward Williams owned slaves. Two slaves are counted in the 1820 census, three in the 1830 census, seven in 1840, and five in 1850. Most were probably house servants. His son Charles had eleven slaves in 1850 and twenty-one in 1860, making him one of the largest slave owners in White County. When Edward Williams died in 1856, his obituary in the *Southern Christian Advocate* in Charleston noted that "for a long time he had been regarded as the model farmer of upper Georgia" and had won prizes in corn, wheat, and cheese at the state's annual agricultural fairs. Williams also established "the first cheese-dairy in the South" and continued to manage it personally until "his advanced years forbade him longer attending to it."[31]

Starlight, the residence of Edward and Mary Brown Williams, photographed here shortly before it burned in 1959. The rooms above the front porch were a later addition. T. Lumsden, *Nacoochee Valley*.

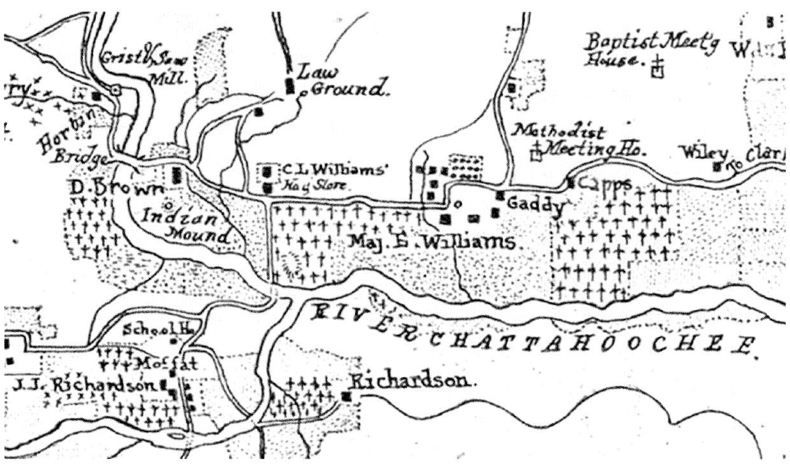

Detail from Moffat's map of Nacoochee in 1837, showing the location of Charles Williams's house and store (*left of center*). Edward Williams's house was at the sharp curve on Unicoi Road (*near center*). The structures on the north side of the road opposite his house may have been slave quarters. Courtesy of Sautee Nacoochee Community Association.

BELL

The tenth family on the numbered list was that of William Bell, who "settled on what is now known as Bell Branch of Buckhorn Mountain. It was named for him." The list also notes that he was the grandfather of Thomas Montgomery Bell (1882–1944), who served as the Ninth District representative in Washington from 1905 to 1931 and for whom Tom Bell Road, three or four miles southwest of Nacoochee, is named. All of that suggests that William Bell, born about 1815, who was enumerated along with his wife in the 1850 Habersham County census, was a nephew or son of the Montgomery Bell whom Gedney documents in the Helen Valley. A family by that name, along with three other Bell families, is shown in the 1820 Rutherford County census, and it is likely that the family was that of Montgomery Bell and his wife, Comfort Brittain. They had eight children, including three sons, Montgomery Jr. (1832–97), James Cicero (1834–1908), and William Brown (1839–99), who is said to have lived in the White County courthouse while building a new house in 1860. The Bells, who were Scots-Irish, were in what is now known as Helen Valley by 1827, and some of the family are thought to have been buried in the England Cemetery in Helen, but the truth of that remains uncertain.[32]

CONLEY

As noted earlier, Daniel and Mary Brown's daughter Nancy Anne married Henry Highland Conley (1798–1856), a fourth-generation American descended from one John O'Connelly, who immigrated from Ireland in 1743 and eventually settled in Burke County, North Carolina. By the time Henry was born, the family was known as Connelly, but he himself adopted a more phonetic spelling, Conley, after coming to Georgia. Henry and Nancy were married in Burke County in 1822, but they apparently did not join her parents in Nacoochee until 1828, when they acquired property in Helen Valley from Elijah England.[33]

ENGLAND

The eleventh party was "William England, wife, and family," but this may be another instance of the list conflating and confusing generations. The Englands also were a Scots-Irish family, descended from a William England, who was born in Ireland around 1720 and came to Pennsylvania with his parents in 1733. Around 1760 the family, which included five sons, moved from Maryland to North Carolina. One of those sons, Daniel (1752–1818), married Margaret Gwinn (ca. 1752–1849) in the late 1770s. After the Revolution, William England's grandchildren moved to western North Carolina, where the 1820 census found thirteen separate England households in Burke County and another three in Rutherford County. Some of them show up in the 1830 Habersham County census, including Richard, Elisha, Joseph, Jonathan, and Martin England, who were likely all siblings or cousins. Gedney has well documented their time in Helen Valley.[34]

PITNER

The third "leader" of the party of Nacoochee immigrants from Rutherford County was Adam Pitner (1789–1873), who has also been well documented by Gedney. His grandfather immigrated from Germany in the early eighteenth century and settled in Bucks County, Pennsylvania, where there were several Pitner families in the 1790 census. After 1800 some of the family moved to Maryland, Delaware, and Virginia, including the immigrant's son John Pitner (1762–1842) and probably Adam as well. In 1814, in North Carolina, Adam married Rachel Campbell (1782–1852), a Quaker who was promptly dismissed for marrying outside the church. They had at least six children. By the 1820s John Pitner and some of his relatives were in Tennessee, where the 1830 census shows two Adam Pitners, one in Sevier County and one in McMinn. The Georgia gold rush was then in full swing, which may have been a factor in Adam's decision to buy Land Lot 39 in the Helen Valley. John Pitner, Adam and Rachel Pitner, and many of their descendants are buried at Nacoochee Methodist church cemetery.[35]

HARSHAW

Moses Harshaw and his family are among the traditional number of families coming from North Carolina, with the patriarch remembered mainly for being "the meanest man in the valley." Moses Harshaw (1795–1859) married Nancy England (ca. 1797–after 1860), the daughter of Daniel and Margaret England. Harshaw was among the several slave-owning families in Burke County in 1820 but was notorious for his poor treatment of them. By all accounts, including public records, he was an arrogant and lawless man. He and Nancy had four children, but it was not a happy marriage. They divorced in 1857. In May 1825 Harshaw paid William B. Wofford $1,500 for Land Lot 44, which encompasses 250 acres northwest of the intersection of Bean and Sautee Creeks. Harshaw is credited with having built the house, which is listed in the National Register of Historic Places, in the 1830s, but the relatively high price paid in 1825 suggests that there was already a house or other improvements on the property at that time.

Harshaw was a rich man by the 1850s, when he claimed $20,000 in real estate, the bulk of it agricultural land in Sautee Valley. It is not clear how much he and his family actually resided in the valley; in 1850 they were enumerated as residents of Clarkesville, but he died in 1858. His widow continued to live in Clarkesville until her death sometime before 1870. In addition to property in Clarkesville, he acquired other acreage, including Land Lots 21 and 22 along Bean Creek, where he is reported to have had eleven gold mines, which he worked using slave labor. When he died in 1859, those land lots were subdivided into forty-acre gold lots by his heirs. It is possible that housing for Harshaw's slaves was the genesis of the Bean Creek community.[36]

RICHARDSON

Rev. Jesse Richardson (1763–1837) was a stalwart of the white resettlement of the valleys in the 1820s and 1830s. He was born in Virginia, but nothing is known of his heritage. As a young man caught up in the Second Great Awakening, he was "converted and soon began to call sinners to repentance." In the spring of 1788, he began a long career as a circuit rider, deacon, and elder in what was then known as the Methodist Episcopal Church. In 1790 he and Daniel Asbury (1762–1825) formed the first Methodist circuit to cover Lincoln, Rutherford, and Burke Counties in western North Carolina, but Richardson rode that circuit only a year. In January 1794 he married Rutha Jones in Surry County, North Carolina, but continued his career as a Methodist minister. Their first child, John L. Richardson, was born in October 1794. A daughter Sarah was born in 1796, followed by Thomas in 1798, Joseph in 1805, and twins Betsy and Nancy in 1810.

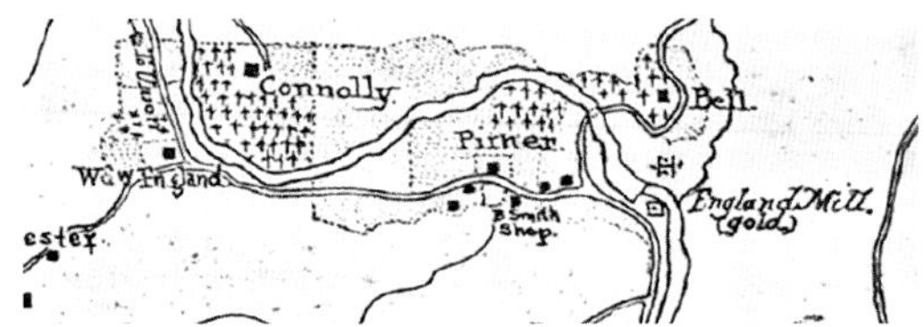

Detail from Moffat's map of Nacoochee in 1837, showing the families living in Helen Valley, all names included in the traditional tally of families that migrated from North Carolina. Courtesy of Sautee Nacoochee Community Association.

John L. Richardson (1794–1850) married Susan Simmons (1797–1886) in 1815 and was himself an ordained Methodist minister. As Methodist ministers, the Richardsons were naturals to lead the cohort of migrants that set out from Rutherford County. They acquired several hundred acres that encompassed most of the rich bottomland along Dukes Creek in Land Lots 88 and 89, where Jesse and John L. Richardson and Abraham Littlejohn built houses. John L. Richardson was one of the largest slaveholders in the valleys in 1850. Thomas, Joseph, and John L. Richardson, as well as their sister Sarah Richardson Littlejohn, all came to Nacoochee with their parents in the 1820s. Jesse Richardson and many of his descendants are buried at Nacoochee Methodist church cemetery.

LITTLEJOHN

Rev. Jesse Richardson's daughter Sarah (ca. 1796–1867) married Abraham Littlejohn (ca. 1796–after 1850) in Rutherford County, where at least five households named "Littlejohn" can be found in the 1820 census. The family appears to have originated in Virginia in the early eighteenth century, and Littlejohns are thought to have migrated to Orange County, North Carolina, around 1760, before moving into western North Carolina after the Revolution. Very little documentation for this family has been located beyond what can be gleaned from the federal census of Habersham County from 1840 to 1860.

Detail from Moffat's map of Nacoochee in 1837, showing the Richardson and Littlejohn farms along Dukes Creek, south of the Chattahoochee River. Courtesy of Sautee Nacoochee Community Association.

They owned no slaves but had five or six children. Their house is depicted on Moffat's map of Nacoochee in 1837. Sarah Littlejohn died in 1867, and Abraham was gone before 1870. Both are probably buried at Nacoochee Methodist church cemetery.

LEONARD

According to the traditional account, "Capt. Wilbur Lenard" and his family "first settled across the river near what is known as the Robertston Mine place, but they later moved and settled on what is known as the Dover Mill place, where he built a mill today known as the Stovall mill." The surname is a misspelling of "Leonard," and the entire name is a corruption of the name of Rev. Wilkes Timothy Leonard (1804–62). He was born in Washington, North Carolina, and was one of five siblings, who all became Methodist ministers. In December 1828 he married Jane Lovelady (born about 1805) in Habersham County. She was likely the daughter of Jesse Lovelady, who, the traditional narrative states, "settled on Mauldin Creek," just downriver from Sautee Nacoochee.

Wilkes Leonard spent only ten years as a Methodist minister, probably in the 1840s; the 1850 census gives his occupation as "farmer." When White County was created in 1857, he was the first county ordinary. When war broke out, he resigned his position and joined the Confederate army as captain of Company G, Twenty-Fourth Regiment, Georgia Volunteer Infantry, "White County Marksmen," when the unit was organized on 24 August 1861. Captain Leonard was killed at the Battle of New Bern, North Carolina, in March 1862, but his body was brought back to Nacoochee for burial at the Methodist church cemetery.[37]

Captain Leonard was probably the son of James Leonard, who purchased Land Lot 85 in 1824. James was born in the 1760s and appears next door to his son Wilkes Leonard in the 1830 Habersham County census. By 1840 James and Wilkes Leonard had moved their families to Cherokee County, where James died in 1855. Wilkes Leonard returned with his family to Nacoochee before 1850 and, with his wife and their seven children, lived on 127 acres around the junction of Nester Branch and Chickamauga Creek in Land Lot 13.

The fourth family listed was that of John Trammell, who, along with his wife, six children, and several slaves, "settled where Robertstown now stands." There are two "Jno. Trammells" in the Burke County census in 1790, both of whom must have been born before the Revolution, since they are listed as heads of households, as well as "Big Denis Trammell" and "Little Denis Trammell." By 1800 the Trammells dispersed to upstate South Carolina.[38]

Habersham and White Counties historical documentation names both "John Trammell" and "Jehu Trammell," but the research of some genealogists suggests that both names refer to the same person. Although his name has been transcribed as "John" in some indices of the 1850 and 1860 censuses, it appears that the correct identification is Jehu Jasper Trammell (1793–1869), born in Union County, South Carolina, the son of William Trammell (1752–1842) and Mary Zelpha (Zilphia) Lynch Trammell (1759–1836). He is believed to have married Elizabeth Fain (1791–1870) in Habersham County in 1818, but no marriage record has been located. Several families with the Fain surname can be identified in the early censuses of Habersham County.[39] Herbert Kimzey's transcription of early Habersham County land records documents Trammell's purchase of Land Lot 21, Third District, in March 1821, encompassing the southern half of Grimes Nose. In the fall of 1823, he also bought Land Lot 7, located on Smith Creek just south of what is now Unicoi State Park. That was the beginning of the thousand or so acres that the Trammells owned in and around what is now Robertstown.[40]

"Jahue" Trammell appears in the Habersham County census in 1830 and was likely the same Jehu Trammell who witnessed Edward Williams's will in 1855. He was one of the five justices of the inferior court elected when White County was organized in 1857. Jehu and Elizabeth Trammell apparently moved to Bartow County, Georgia, after the Civil War; both are buried at Oak Hill Cemetery in Cartersville.[41] The federal census shows the family owning nine slaves in 1860, some of whom must have stayed in Nacoochee after the Civil War. There are now three dozen African Americans with the Trammell surname in the Bean Creek Cemetery, and Trammell Drive is one of the local roads in the Bean Creek community.

Among the list of sixty-one families, the twenty-fourth was said to be brothers Job and Abraham Sosebee and their mother, "Aunt Patsy Sosebee, who had seven boys and four girls, who came with the party. They were the ancestors of the Sosebee families in this county." Another version of the story adds that they settled near Sautee Creek "on the Chimney Mountain Road," which would have been in the upper watershed of the creeks that are the sources of Chickamauga Creek, north of SR 356. Precisely where that might have been has not been researched. The Sosebees had a penchant for biblical names, including Isaac, Solomon, and Job, but no Abraham or Job Sosebee was documented in Habersham County in the first half of the nineteenth century. Nevertheless, the Sosebees were certainly among the early settlers in northeastern White County.

The first Sosebee in the area appears to have been Sampson Sosebee (1786–1863), who was enumerated in 1820 with his wife and young son in the federal census of Habersham County. He was the eldest of ten children of Job Sosebee (1759–1821) and Elizabeth Tankersly (1763–1849) and was born in Spartanburg, South Carolina, where his parents had moved before 1790. Around 1808 he married Elizabeth Abbot (1786–1880) in Spartanburg, and they had four children. The youngest, Noah (1820–83), is reported to have been just a few weeks old when they moved to Habersham County, although the census records his birthplace as Georgia. Sampson's brother Thomas (1793–1874) was also in Habersham County by 1830. They were joined by their younger brother Job Jr. (1799–1876) by 1840, and in 1850 their brother Isaac (1797–1870) was also enumerated in the Habersham County census.

It is believed by some that Sampson's parents were intending to move to Georgia as well, but his father died suddenly and was buried in South Carolina. Job may have sold his farm in South Carolina in 1821 and moved to Habersham County, but in his widow's will, filed for probate in 1849, she refers to herself as "widow and relict of Jobe Sosebee of Franklin County, deceased."[42] Sampson and Elizabeth Sosebee and many of their descendants are buried at Cool Spring Methodist church cemetery, a few miles east of Sautee Nacoochee.

A newspaper clipping reporting on the Williams family's centennial reunion at the old homestead, Starlight, in 1922 outlined Edward

Williams's life and noted that he "came from North Carolina in 1822, purchased land in Nacoochee Valley, and comfortably settled about 40 of his tenants, who came with him, upon his newly acquired property." A hundred years later, some of those tenants remained, prompting George Walton Williams Jr. to present his estate manager, John Martin Sosebee (1862–1937), grandson of Isaac Sosebee, with a Bible inscribed, "Presented to John M. Sosebee in commemoration of the 100th anniversary of the association as landlord and tenant of the Williams and Sosebee families, 1822–1922, Nacoochee, Ga., September 6, 1922, by George W. Williams."[43] The Sosebee who was Edward Williams's farm manager in the antebellum period has not been identified, but the most likely candidate is Isaac Sosebee, who may have been followed by his son Thomas Jefferson Sosebee (1833–94).

The identities of the forty tenant families that Williams is said to have brought with him from North Carolina have not been found, but a few of the sixty-one families might have been tenants, including family 28, which the list noted only by his first name. Jim "settled on Edward Williams's lands," and John Carroll (26) and Theron Wheeler (39) were both noted in the traditional list of the sixty-one families as having settled on "Daniel Brown's land." These people were among the obscure, barely named families in the list who never had the advantages of the Brown and Williams families but who formed at least a plurality of their neighbors.

STOVALL

In his numbering, the author of the list of sixty-one families inexplicably skipped the twenty-eighth family but then included two family names in the next entry. "The twenty-ninth family was John Stovall and wife and Isaac Baker, who settled in the Blue Creek District," which is southeast of Mount Yonah. John Stovall (ca. 1804–after 1880) was born in North Carolina, the eldest son of Frederick J. Stovall (1784–1838). In the 1820 federal census, there are two Stovalls, Herod and Thomas, listed in Rutherford County, North Carolina, and at least two John Stovalls in Franklin County, Georgia. There were none in Habersham County. It is unclear how these Stovalls might have been related, but William Isaac Stovall (1863–1945), who was a major figure in the history of Sautee Nacoochee beginning in the late nineteenth century, was one of John Stovall's grandsons.[44]

John Stovall married Elizabeth McCollum (ca. 1812–after 1880) about 1830. She was the daughter of William McCollum (1789– 1876) and his wife, Susan (1788–1824), who married in Pendleton, South Carolina, in 1810 and had several children before they moved to Habersham County in the early 1820s. Hers was one of the first burials in Blue Creek Cemetery. William McCollum remarried and moved away, but his father, Daniel McCollum (1760–1850), spent his last years with the Stovalls at Blue Creek. He was a New Jersey–born veteran of the Revolution and is buried at Blue Creek along with his wife.[45] The Stovalls, like many of their neighbors, worked a small farm themselves, without slaves. Their descendants were still farming along Blue Creek as late as the 1930s, and several generations of Stovalls and McCollums are buried in the old Blue Creek Baptist church cemetery in southern White County.

CAPPS

Moffat's map of Nacoochee in 1837 depicts the residence of a family named Capps, which is almost certainly "the ninth family." Elisha Capps "never did drink any water. Later it was told of him that he left and went to California in 1852, where he married and lived until his death. His place is still known as the Capps place on the river." There are Cappses in western North Carolina in the 1820 census but none in Burke or Rutherford Counties. No Cappses appear in Habersham County until the 1850 census, and even then none are named Elisha. Once again the traditional story seems muddled. Moffat's map locates a Capps family next door to Edward Williams, although the federal census does not appear to corroborate that. The 1850 census does record one Franklin Capps, but in another district. He died prematurely in the 1850s, leaving his widow, Elizabeth (ca. 1822–ca. 1875), to finish raising their five children. The Cappses may have been among the tenants who, in addition to enslaved African Americans, kept Edward Williams's enterprises operating. That Elisha Capps did not drink water might not seem so idiosyncratic when one remembers that typhoid, cholera, and a host of other waterborne diseases took a terrible toll. It is only within the past hundred years that Americans began taking clean water more or less for granted.

CRUMLEY-MONROE

According to the legendary account, "the eighteenth party was the Rev. William Crumley and family, grandfather of Rev. Howard Crumley. He settled up near the foot of Chimney mountain and the land is still owned by his descendants." Rev. William Smith Crumley (1794–1882), who was a Methodist minister, and his wife, Elizabeth Jones Monroe (1790–1836), are buried in the Monroe Cemetery near Sky Lake Road and SR 255. He is reported to have organized a Methodist community in that area in 1832.[46]

"Jim Monroe and family" were recorded as the fifty-seventh family on one version of the list of early settlers. The census suggests that James Monroe Sr. (1760s–before 1840); his wife, Elizabeth (ca.1765–1850); and their son Jesse (1794–1875) came from Rutherford County in the 1820s, and it is likely that Reverend Crumley's wife was their daughter Elizabeth. Jesse Monroe was one of the first trustees of the Nacoochee Methodist church when the land for it was deeded in 1834. The elder Monroe apparently died in the 1830s, and Jesse took his family, including his mother, to Cherokee County in the 1840s. In 1843 he deeded "two acres more or less" in Land Lot 28 of District 6, including the graves of his parents and sister, to be used for a church building and cemetery.[47]

WIDOW BENNETT

Identification of this widow and her children "who settled on Sautee Creek on what is now known as the Harry Williams place" is uncertain. No Bennetts in Habersham County have been identified in the federal census of 1830–50 or in the local cemeteries. However, Moffat's map of Nacoochee in 1837 depicts the Widow "Burnett" farm on the west side of Sautee Creek just south of Ben [*sic*] Creek, and it has become clear that the name was mistakenly recorded as "Bennett" when the list of sixty-one families was put together. Available evidence suggests she was the former Frances Bell, daughter of Montgomery Bell, but then the young widow of L. B. Burnett, one of the original trustees of the Nacoochee Methodist church, who died in 1835. James R. Wyly administered Burnett's estate, which included Lots 55 and 56 of District 3, which were later acquired by Moses Harshaw.[48]

BAKER

Isaac Baker, who was listed with John Stovall, might have been a relative, but no proof of that has been located. Nine Baker families are listed in the 1820 census for Burke County, North Carolina, but none of them are named Isaac, a name that does appear, however, twice in the Duplin County census and once in the Iredell County census that same year. There were also Bakers in Habersham County, Georgia, in the 1820 and 1830 censuses, but none named Isaac. Some of the family moved to Cass (now Bartow) County in northwest Georgia in the 1830s, including an Isaac Baker, but again no firm connection to Habersham County has been documented.

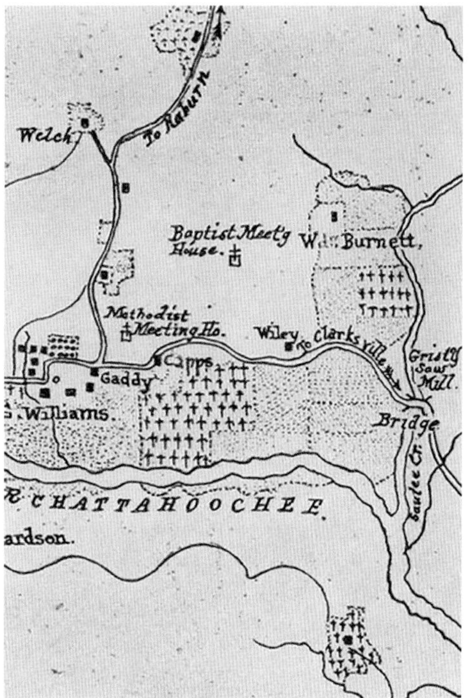

Detail of Moffat's map of Nacoochee in 1837, showing the Widow Burnett's farm along Sautee Creek. Courtesy of Sautee Nacoochee Community Association.

6

Community Development

The community of families at Sautee Nacoochee, often connected by both blood and marriage, rebuilt old Cherokee farms and homesteads and constructed new ones, as they continued the centuries-old agricultural use of the valleys and developed their own cultural institutions. The discovery of gold on Dukes Creek and elsewhere in 1828 precipitated the nation's first real gold rush and sent waves of prospectors into northern Georgia. Since it was quickly realized that the gold belt was not easily mined in the Nacoochee and Sautee Valleys, they escaped much of the environmental devastation seen on the upper reaches of Dukes and Bean Creeks and in the hills and valleys to the west. Fortunes were made by a few in the mines, and others found their fortune as merchants or simply by continuing to grow wheat, corn, and oats, for which there was always high demand.

ROADS

When Sautee Nacoochee was being resettled in the 1820s and 1830s, travel was slow and difficult and mostly on foot, over roads that were sometimes little more than footpaths through the wilderness. Most of the earliest roads in Sautee Nacoochee were simply improvements on the system of prehistoric trails discussed in chapter 2. While many early trails were eventually widened to accommodate wagons, roads remained in generally poor condition across rural Georgia until well into the twentieth century. In 1829 James Silk Buckingham, traveling in the eastern Georgia Piedmont, described a likely typical condition of these early roads: "The road itself was the worst, being formed apparently by the

A view of the Unicoi Road around 1920, looking southeast with the Nichols-Hardman House visible at left. Brooks and Greear, *Images of America*, 26.

mere removal of the requisite number of trees to open a path through the forest, and then left without any kind of labour being employed, either to make the road solid in the first instance, or to keep it in repair."[1]

Whatever their condition, roads were extremely important in the antebellum South, where waterborne transportation was often impractical. Roads were not just a means of transportation either: to live on a well-traveled public road, or later a railroad line, was considered a great thing, not only for convenience but also for the chance of social interaction. As cultural historian John Stilgoe notes, "Southerners treated their roads as extensions of church, courthouse, and store, seeing in them the potential for excitement that northern city dwellers found in streets. Strangers, especially Europeans and 'Yankees,' failed to understand the extraordinary importance of the road in southern culture because they searched for the towns or hamlets so uncommon south of Pennsylvania and ignored the roads and waterways that substituted for towns."[2] Charles Lanman, author of one of the earliest descriptions of "the vale of Nacoochee," wrote of a night he spent at Tesnatee Gap in the 1840s. His host told him, "I have always desired that I could live on some public road, so that my girls might occasionally see a civilized man, since it is fated that they will never meet with them in society." Such attitudes meant that antebellum houses were almost always built within sight of the road, if not within a few feet of it, and so it was with many of the early residences around Sautee Nacoochee.[3]

In most counties created prior to the Civil War, the formal designation of roads was one of the first orders of business for the county inferior court, since that ensured some level of public maintenance of the thoroughfare. In Habersham County several roads were already in place, including the Unicoi Turnpike, which opened in 1819. The earliest minutes of the Habersham County Inferior Court do not record designation of any public roads; it can only be assumed that the network of trails and roads left by the Cherokees remained in use.

Construction and maintenance of public roads and bridges were civic responsibilities under the supervision of the inferior court, but the duty was often shirked. In 1868 the state legislature authorized counties to elect county commissioners of roads and revenues, and in 1872 it passed a law that required the labor of all able-bodied males between the ages of sixteen and fifty for maintenance work on public roads under the supervision of the county road commissioners. Ministers were exempt, and no one could be required to work more than nine days in any one year.[4]

In 1849 George White published his ambitious *Statistics of the State of Georgia*, which included commentary on each county. In writing of Habersham County, he noted that "for a mountain county, the roads are fair," and he greatly admired "an elegant covered bridge resting on three arches" over the Tugaloo River and another covered bridge, 183 feet long, over the Chattahoochee River at Nacoochee.[5]

Thirty years later a newspaper reporter from Gainesville wrote much the same thing: "The road we traveled is one of the smoothest and least hilly in all this Piedmont region. If it be true that the civilization of a people may be judged by their roads, those living between this point and Nacoochee are above criticism in that regard, and need no higher evidence of the exalted type they have attained."[6]

There is virtually no mention of roads or bridges in early Habersham County records, but historical documentation corroborates much of what is shown in the Moffat and Hollis maps. The scale and orientation of these maps are awkward and sometimes difficult to interpret, and a number of features depicted have since disappeared. Yet most of the roads can be correlated with roads shown on Georgia State Highway Board maps of White County produced before World War II and on

The covered bridge over the Chattahoochee at Nacoochee. Brooks and Greear, *Images of America*, 79.

the 1937 U.S. Forest Service map of the area, which also shows land-lot lines. Much of Georgia SR 17 and 255, SR 75 north of Nacoochee, Bean Creek Road, Rabun Road, Lynch Mountain Road (probably the best preserved of the early roads), Garland Bristol Road, and the recently closed Hardman Road retain their historic nineteenth-century alignments, albeit with some twentieth-century alterations.

As the rural population declined after World War I, many of the county's smaller roads began to disappear. One of the most significant, since it almost certainly predated white resettlement of the valleys, left Unicoi Road in front of Charles Williams's store, a half mile east of the modern intersection with SR 75, and ran roughly south to what Moffat showed as a ford in the Chattahoochee River. Last depicted on the 1937 U.S. Forest Service map, it had disappeared by 1940, along with the road that paralleled the Unicoi Road on the south side of the river.

Detail from the 1937 U.S. Forest Service map, annotated by others, showing Sautee Nacoochee. All the roads are depicted as unpaved, with the dashed lines representing roads that the State Highway Board would label "primitive." Many of these roads correspond with those shown on the Moffat map. Georgia Secretary of State, Office of Surveyor General, Historical Map File, 1937, Georgia Archives, Morrow.

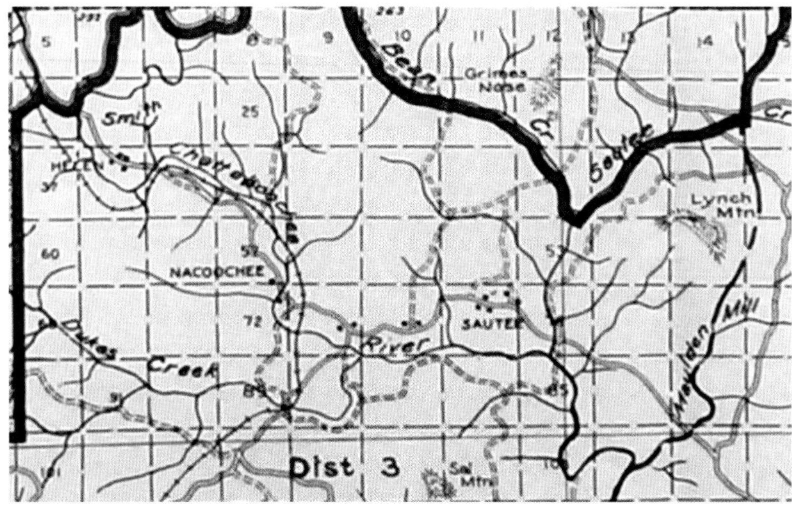

More recently, Hardman Road has been abandoned as a public road. The early militia district courthouse, or "Law Ground," for Nacoochee was located on this road, and it likely predated white resettlement of the valleys. Its closure is a loss to the community's historic fabric. Lynch Mountain Road, which flanks the Sautee Valley on the east, remains a well-preserved road, one of a diminishing number of unpaved roads that characterized travel all over White County prior to World War II.

Natural fords in rivers and streams were generally stable, so the route of roads leading to them were stable as well. The location of bridges, however, did not always remain the same. As long as bridge building remained a private endeavor, the loss of a bridge from flooding or general deterioration might lead to a new bridge in a new location. Although expensive to build, covered bridges were common across Georgia in the nineteenth century as a way to protect the structure's framing and joinery and prolong its useful life. With good maintenance a covered bridge could last almost indefinitely, while a wooden structure and deck exposed to the elements might last only ten or fifteen years.

Moffat's map depicts all the river and stream crossings as natural fords, except for a bridge where Unicoi Road crossed Sautee Creek and another where it crossed the Chattahoochee River at the western end of Nacoochee Valley. The first bridge at Sautee Creek may have been built by James R. Wyly. Moffat's map stops short of the upper Sautee Valley, but at an early date there was likely also a bridge where the much-traveled Rabun Road crossed Chickamauga Creek. The present Stovall Mill Bridge at that location was built in 1895 to replace an earlier bridge that washed away in a flood.

RAILROADS

The first railroads in the southeastern United States connected ports like Savannah and Charleston with the cotton-producing plantation belt farther inland. After the Civil War, the railroads began to venture into the Upper Piedmont and, in the early twentieth century, into the mountains, the latter often built to access the forest timber. Along the way there were the Luddites resistant to the new iron rail, worried about the noise, the encroachment on private property, and fields and houses set afire by sparks from the engine. Some towns, such as Decatur east of Atlanta, even prohibited railroad construction within their corporate limits.

One of the first regularly scheduled railroads in the South began operations from Charleston to Hamburg in 1833. That same year the Georgia Railroad Company was chartered to build a railroad from Augusta to Athens with a branch to Madison, which was completed in 1841. By 1845 the Madison branch line was a main line that extended to meet the state's Western & Atlantic Railroad to Tennessee and the Central of Georgia Railroad from Savannah at a junction six miles from the Chattahoochee River, around which the city of Atlanta would soon grow. Other railroads were chartered and a few completed before the Civil War, but none in northeastern Georgia, where the closest railroad remained Athens, sixty miles away, until the 1870s.

In 1873 the Atlanta and Richmond Air-Line Railway was completed, bringing a railroad through Cornelia and within seventeen miles of Nacoochee. At the same time, construction had begun on the Northeastern Railroad of Georgia, chartered in 1870 to build a line from Athens to Clayton. It was completed as far as Clarkesville, ten miles from Nacoochee, in 1874 and on to Tallulah Falls in 1882. The entire line was open to Franklin, North Carolina, in 1907. Finally, the Marietta and North Georgia Railroad, which had been chartered in the 1850s, was completed as far as Canton in 1879, Ellijay in 1884, and Murphy, North Carolina, in 1887.

People in rural areas with rich agricultural and natural resources did all they could to encourage railroad development. Not only would they have a ready outlet for their products, but a vast array of manufactured goods and cheap foodstuffs would become available as well. In areas such as White County, where rail service was deemed unprofitable, private investors often tried to fill the gap. In 1888 Jesse R. Lumsden (1848–1933), grandson of John L. Richardson, and Robert A. Williams (1860–1947), grandson of Major Williams, incorporated the Nacoochee Valley Railroad Company with the help of three Charleston investors: George B. Edwards, J. Lamb Johnston, and S. V. Stewart. The company was authorized to build a rail line "from some point on the Richmond and Danville Railroad between Mount Airy [Cornelia] in Habersham County and Lula in Hall County, the nearest and most practicable route . . . to some point in Nacoochee Valley in White County." As with several other proposed rail lines into northern Georgia, the Nacoochee line did not attract enough investors to be built. It would be nearly twenty-five years before a rail line passed through Nacoochee.[7]

STATE HIGHWAYS

The first real efforts toward a national highway system emerged during the Progressive Era of the early twentieth century: the dawn of the automobile age and of a growing awareness of the need to improve the wretched condition of most rural roads. Private auto clubs began cobbling together "auto trails" before World War I, and major stretches of the east-west Lincoln Highway (dedicated in 1913), the north-south Dixie Highway (begun in 1915), and other long-distance roads were in use by 1920.

In July 1916 the Federal Aid Road Act was enacted, creating the nation's first highway-funding mechanism. Especially important was a provision for federal funding for rural post roads, provided there were no tolls. Since dealing with federal grants at the county level was not feasible, the law stipulated that all states have a highway agency staffed by professional engineers who would administer the federal funds and ensure that all roads were properly constructed. A month later, in August 1916, Georgia created a State Highway Board charged with planning, constructing, and maintaining a publicly accessible, toll-free state highway system. Over the next few years, improved roads began to take shape, funded then, as now, by federal grants and a state gasoline tax. The latter was raised during the 1920s; by 1931 half of the state's revenue was being spent on roads.

Early in 1922 construction began on a new road from Cleveland to Blairsville and on into North Carolina. Completed in 1926 and now known as U.S. 19, it paralleled the old Logan Turnpike, the county's last toll road, which shut down when the highway opened. Crossing Neel's Gap, it was one of the first state highways in White County. A state highway was designated between Cleveland and Nacoochee in 1930 and two years later extended over Unicoi Gap. Convict labor was used for highway work, including improvements in the road from Cleveland to Helen, later designated SR 75, work that included construction of a concrete bridge in Helen.[8] The State Highway Board (which would be reorganized as the Department of Transportation in 1972) produced at least six "general highway" maps of White County, dated 1934, 1940, 1951, 1960, 1964, and 1985.

Although the map from the 1930s is not as reliable as later ones, they are all helpful to understanding the evolution of White County's modern road system. A postal map also exists from 1931 but is not accurately drawn. The highway board updated their statewide highway maps monthly throughout the 1930s. By the time the 1940 State Highway Board map was compiled, many of the "primitive" roads designated on the earlier maps, including those of Moffat and Hollis, had disappeared or at least were no longer being delineated.[9] A major statewide campaign to pave and improve rural roads began in the 1930s and resulted in an overhaul of the state highways through Sautee Nacoochee—17, 75, and 255—the latter having been designated a state road in the 1940s. Completed in 1950, the work included paving as well as significant realignment of parts of those roads to make them more suitable for automobile traffic.

Highway 75 north of Nacoochee had undergone some realignment in the 1930s, especially around Helen, where a concrete bridge was built over the river. After World War II, the portion of the highway south of Nacoochee was completely rebuilt, using the roadbed of the defunct Gainesville and Northwestern Railroad, which had ceased operation in 1931 and offered an excellent foundation for a highway. Along with that came a major alteration to the intersection of Highways 17 and 75.

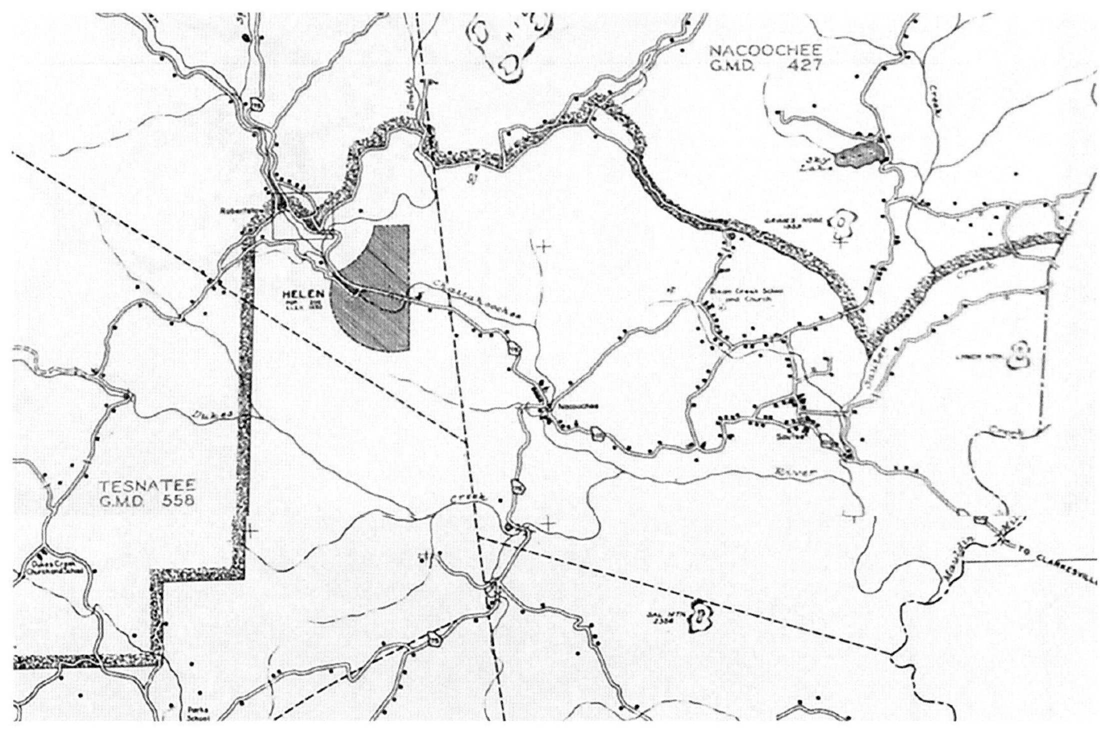

Detail from the 1940 State Highway Board map of White County. Most of the small black squares represent residences. Georgia Department of Transportation, "Archive Maps."

The original route of Unicoi Road included a turn to the northwest just west of where the Nichols-Hardman House stands today, continuing a quarter mile to an intersection with Highway 75 south of the river crossing. As part of the 1950 project, the present T-shaped intersection was constructed, leaving an alley of trees on Hardman Farm to mark the route of the old Unicoi Road.

Parts of Highway 17 were also realigned in 1950 to eliminate the worst of the curves in the old road, which had become a major hazard in the days of fast automobile travel. A mile east of Highway 75, Wright Road survives, like an oxbow lake along a river, where the historic route of the Unicoi Road was straightened in 1950. A sharp curve in Highway 17 must still be negotiated west of Sautee, but the 1950 project eliminated an almost ninety-degree turn at that location. Highway 255 underwent minor realignment during its improvement as a state highway. Finally, the State Highway Board's project in 1950 replaced the covered bridges at Sautee Creek and the Chattahoochee River with modern reinforced-concrete bridges. The small covered bridge over Chickamauga Creek survives because the road was rerouted when it was paved and improved in the 1970s.

Post Offices

In 1775 the Second Continental Congress appointed Benjamin Franklin as the first postmaster general. In 1789 the Constitution authorized Congress to establish post offices and post roads, and President George Washington created the U.S. Post Office in 1792. The first stamps were issued in 1847, and in 1872 Congress established the Post Office Department as an official executive department. In 1971 the U.S. Post Office began operating as the U.S. Postal Service.[10]

The first post office in northeastern Georgia was established in 1796 at Carnesville in Franklin County. Located forty miles southwest of Sautee Nacoochee, it remained the only post office in that county until after the War of 1812. A post office designated Habersham was opened

Detail from Merrill, Finegan, and Riemann's *Map of Northern Georgia* in 1864. Post offices were established at all these communities in the 1820s and 1830s. Item 2006458675, Geography and Map Division, Library of Congress, Washington, D.C.

in 1820, but its location is uncertain. It was replaced by the Clarkesville Post Office in 1823, where the mail was delivered by a post rider from Augusta once a week.

NACOOCHEE POST OFFICE

On 26 September 1826, John Cleveland was appointed postmaster at "Naucooche."[11] His life is poorly documented, but he appears to have been the eldest son of Habersham County pioneer Benjamin Cleveland. If so, he died in September 1828. The poor quality of early post office records may obscure other documentation, but it is likely that the date of Cleveland's appointment marks the founding of the Nacoochee Post Office. No documentation for its original location has been found.

On 4 March 1830, Thomas B. Cooper was appointed postmaster of the "Naucoochy Valley" Post Office. His tenure was apparently only a little over two weeks. The post office record of his replacement names Edward Williams and Henry Albright, leaving it unclear whether they were postmasters together or in succession, since only one date (20 March 1830) is recorded. Henry Albright was enumerated in the 1830 Habersham County census but otherwise remains undocumented. Post office records support the family's belief that the post office had been located in Charles Williams's store "continuously" since at least 1830.[12]

On 24 January 1837, Charles Lathrop Williams was appointed postmaster of the Nacoochee Post Office. He would continue in that capacity until the Civil War. After the war Williams regained his citizenship in October 1865, no doubt intending to continue as postmaster; but in August 1866 he was replaced by Josiah R. Dean Sr. After the latter's death five years later, Williams was again appointed postmaster of the Nacoochee Post Office.

Joseph A. Richardson served as interim postmaster for a few months after Williams's death, until April 1887, when Charles Lee Williams (1849–1930) was named postmaster at Nacoochee. He was followed by his brother Edward G. Williams (1836–1901) for a brief period before November 1890, when Ella F. Cunningham became Nacoochee's first female postmaster. Her maiden name is unknown, but, as early as 1875, she married William H. Cunningham. Running a farm encompassing over 150 acres of land (Lots 54 and 75 along Unicoi Road, just east of SR 255), they had at least two sons. The post office was relocated there when she became postmaster.

In October 1892 Alfred P. Williams (1839–1913), eldest son of Edwin Poore Williams, replaced Cunningham as postmaster. A few weeks later, he requested that the post office be relocated from the Cunningham house, which was next door to his own, to where it had operated "continuously for over sixty years," at his uncle Charles's old house and store. On 19 January 1893, Thomas C. Williams (1838–96) was named postmaster, and the post office was returned to the store.

In August 1896 Thomas Williams died, and Dr. John H. Alley (1856–1914) was appointed postmaster. When he retired, his wife, Katie (1877–1951), was named postmaster on 5 April 1912. The post office remained in the old Williams store, as indicated by the post office location report she filed in July 1913. The following spring Katie Alley applied to move the Nacoochee Post Office "one mile air-line distance west" of its existing location to a point "35 yds" east of the new Gainesville and Northwestern Railroad, most likely the brick commercial building Gov. Lamartine Hardman had built near the Nacoochee depot. Dr. Alley

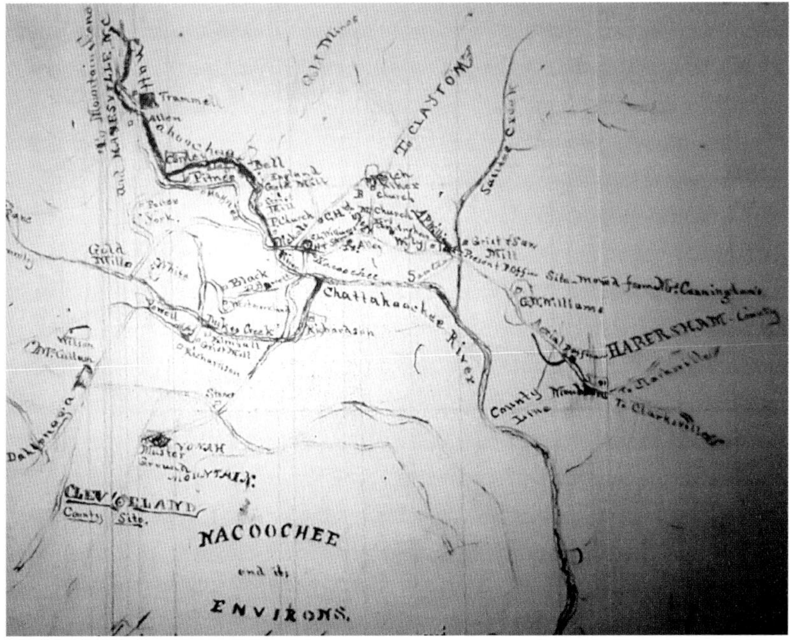

Map for the relocation of the Nacoochee Post Office in 1893. U.S. Post Office, "White County," M1126, roll 124, Records of Site Locations, National Archives and Records Administration, Washington, D.C.

died in 1914, but Katie married Jess Cannon by November 1917, when she received another appointment as postmaster for the Nacoochee Post Office. She continued to run it until she retired in September 1942. Her successor was Fannie Ruth Thurmond, who served until the Helen Post Office was established in 1952 and mail addressed to Nacoochee was routed there. Seven years later the Nacoochee Post Office was discontinued entirely with the establishment of the present Sautee Nacoochee Post Office.

SAUTEE POST OFFICE

On 27 February 1893, a few weeks after his cousin Thomas was appointed Nacoochee postmaster, Alfred P. Williams (1839–1913) completed an application to establish a new post office. Named Sautee, it was located on Unicoi Road a quarter mile or so west of Sautee Creek and near the junction of the Nacoochee and Sautee Valleys, which probably made it more convenient for a larger number of people (the application states that the post office would serve "about 200").[13] Presumably since its inception, the post office was located in what is now the Old Sautee Store, but contrary to some reports, the store, which was built about 1872, was not built as a post office. Most rural nineteenth-century post offices were not purposefully built but rather occupied a part of an existing commercial structure or sometimes a private residence.

Alfred P. Williams was appointed Sautee's first postmaster on 7 April 1893 and served until his death in February 1913. In October of that year, his son Robert A. Williams (1874–1952) was named postmaster, which he remained until he retired in March 1948. Lula L. Logan (1874–1949) was his successor, an unmarried daughter of Lula and Simpson Logan, who had deep roots in Nacoochee. She was still the Sautee postmaster when the name of the post office was changed to Sautee Nacoochee, effective 30 April 1959, a change reflecting the elimination of the Nacoochee Post Office. Logan died less than three months later and was succeeded by her younger sister, Florence C. Logan, who was appointed postmaster in July 1961 and served until her retirement in the 1970s.

LYNCH POST OFFICE

In May 1893 James E. Hood applied to establish the Lynch Post Office only two miles north of the new Sautee Post Office; in January 1894 he was named postmaster. Hood was probably operating the store in front of the Harshaw-Stovall House, which is where the post office was located. His son James C. Hood was installed as postmaster in 1896 and remained until May 1905, when William S. Allen applied to have the post office moved about a mile north, near what is now the intersection of SR 255 and Sky Lake Road. It remained there until it was discontinued in 1909.

RURAL FREE DELIVERY

That the Lynch and Sautee Post Offices were established in such proximity to each other may have been part of the Post Office Department's early efforts to implement the free delivery of mail in rural areas. Before rural free delivery, patrons often had to walk for hours to the nearest post office and back or, if one could afford it, hire a private carrier. With the growing popularity of mail-order catalogs, there was a demand for better service. In 1893 Georgia's Rep. Thomas C. Watson introduced legislation mandating rural free delivery, and, although the legislation was signed into law, Congress was slow to make necessary appropriations.

Standard mailboxes were prescribed by the post office in 1901 and soon were a fixture of the landscape. Boxes were to be metal, weatherproof, and six by eight by eighteen inches and mounted so mail carriers could make deliveries without leaving their vehicle.[14] Establishing routes and post offices was an enormous undertaking, but by 1902 nearly everyone living outside cities and towns received free postal deliveries. Walter B. Lumsden Sr. (1876–1973) began delivering mail at Sautee before 1910 and became the longest-serving rural letter carrier for the community by the time he retired after World War II.

Churches

The second edition of Adiel Sherwood's *Gazetteer of the State of Georgia*, published in 1829, includes a section titled "Religious Denominations in the State." Overwhelmingly Protestant, the state had only three Roman Catholic churches, in Savannah and Augusta, and a single Jewish synagogue, also in Savannah. There were a few Episcopal congregations, but most were Baptist, Methodist Episcopal, and Presbyterian. Quakers from Orange County, North Carolina, established a colony at Wrightsborough in 1768. Some early settlers in northeastern Georgia in the 1820s had roots in that colony, but by 1810 the Quakers had disbanded and left the South because of their bitter opposition to slavery.

At Sautee Nacoochee, as with much of the southern Piedmont, every nineteenth-century church was Baptist, Methodist, or Presbyterian. Across the state there were significantly more Baptists than Methodists, but in Habersham County Methodists outnumbered Baptists by a small margin. In 1850 the federal census of Habersham County reported nineteen Methodist churches, collectively valued at $19,000 and with seating for 7,700 people. There were also in the county eighteen Baptist churches valued at $18,000 and seating 7,200 and a single Presbyterian church valued at $1,000 and seating 400 in Clarkesville. These figures may be exaggerated, as the total population of the entire county in 1850 was less than 9,000.[15]

Sherwood neglected to mention the state's Lutheran churches, several of which were begun in the eighteenth century. There were not enough Lutherans in Habersham County to form a congregation in the nineteenth century, but there were Lutherans and other denominations who joined Methodist and Baptist churches, including those at Nacoochee. Many churches of any denomination did not have weekly services, and so attendance at monthly services was typically large. Charles Lanman, who visited northeastern Georgia in the 1840s, noted that "the people attend the Sunday meetings from a distance of ten and fifteen miles" on horseback; they often made it a daylong affair.[16]

BAPTIST CHURCHES

For most of the past two hundred years, there have been more Baptist congregations in the state than any other denomination, and a large majority of the churchgoing public counted themselves as Baptist. In 1845 the denomination split into Northern and Southern Baptist conventions over the issue of slavery, and the sides have never reconciled. In 2016 there were over sixteen million Southern Baptists, making it by far the country's largest Protestant denomination. Today there are at least twenty-seven separate Baptist denominations, each sharing a belief in the ability of the individual to decide his own faith within the framework of the Bible. Originating in Holland and England in the early seventeenth century, the Baptists were known for their literal interpretation of the rite of baptism (i.e., total immersion in water) and their opposition to infant baptism.

The first Baptist church in Georgia was established at Kiokee in Columbia County in 1772. Believing as they did that every member was a minister of the Gospel and that all that was needed was a "call" from God, the Baptists were able, unlike the Presbyterians with their requirement for formally educated clergy, to respond quickly to the demand for new churches on the rapidly expanding southern frontier of the late eighteenth and early nineteenth centuries. Although the churches were independent, they typically joined in loose associations, and by 1817 there were already enough Baptist churches in northeastern Georgia to organize the Tugaloo River Association.

A bitter divide in the Baptist church emerged during the Second Great Awakening, a period of intense religiosity that began in the 1790s and lasted through the first quarter of the nineteenth century. The rift was primarily over whether Baptists should support the missionary work and temperance leagues of the period. In 1829 Sherwood found "a sluggish indifference in many to the promotion of 'missionaries' and a bitter hostility in others; but since the revival of 1827–28, its prospects are brighter."[17]

By 1836 churches were typically either Missionary Baptist or Primitive Baptist. As missionary work became part of mainstream Baptist associations, most, but not all, of the missionary churches dropped the word from their formal name. The Primitive (first or original) Baptists were sometimes called "Hardshell Baptists" or "Old School Baptists," as they continued to carry out foot-washing ceremonies, sing a cappella, and enjoy an occasional dram of whiskey. There is also an African American Primitive Baptist church, distinct from and more conservative than the white congregations. Today there are, nationwide, perhaps one hundred thousand members of Primitive Baptist denominations.

In 1829 Adiel Sherwood reported in his *Gazetteer* that Georgia had 356 Baptist churches with 28,268 members, an increase of 66 churches in two years. He noted, however, that "few, not more than seven or eight ministers [out of two hundred], have shared the advantages of a collegiate education; nor do the Baptists think such education indispensable to preaching."[18] Twenty years later he found 879 Baptist churches, along with 809 Methodist churches, 97 Presbyterian churches, 20 Episcopal churches, 8 Catholic churches, and a scattering of other denominations. The earliest Baptist church in Habersham and White Counties appears to have been a Bethlehem Baptist church, which was organized soon after the county's founding, at Walter Adair's old schoolhouse five miles north of Clarkesville.[19]

Old Nacoochee Missionary Baptist Church

Moffat's map of Nacoochee in 1837 depicts the site of the "Baptist Meet'g House," which Dr. Tom Lumsden located a half mile north of the Lumsden-Couch House. The congregation claims to have organized in 1822 and built their first church in 1826, the first church built at Nacoochee. In 1851 the congregation relocated to the intersection of Lynch Mountain Road and Hub Tatum Road, where they remain today. The congregation is now known as the Old Nacoochee Missionary Baptist Church. Why the congregation relocated is not clear, but it is quite possible that the Missionary Baptist Church was actually a splinter group from an original Primitive Baptist congregation that subsequently disappeared.[20]

There are 275 known graves at the Old Nacoochee Missionary Baptist church cemetery, with numerous Holcombes, Sosebees, Wilsons, and other names prominent in the early history of the valleys. When the congregation moved, they left behind an older cemetery, a not unusual occurrence among the Baptists, who typically abandon old church

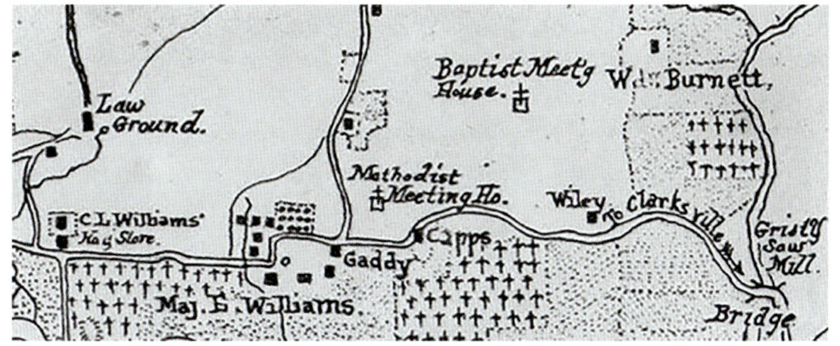

Detail from Moffat's map of Nacoochee in 1837, depicting the locations of the Baptist and Methodist "meeting houses." Courtesy of Sautee Nacoochee Community Association.

The old Nacoochee Missionary Baptist church. Photograph by author, 2021.

sites more readily than the Methodists and Presbyterians. Lumsden counted fifty headstones in the late twentieth century, all but two of which were typical, unmarked field stones.

Two graves are marked by marble monuments, one of them commemorating Rachel Foster (1773–1851), mother of the congregation's first minister, Robert Foster (ca. 1795– before 1880). Both he and his wife, Margaret (ca. 1807–after 1880), were born in South Carolina, and both are documented in the federal census of Nacoochee 1830–70, along with several of their children. He was a farmer of modest means, and, as with many Baptist preachers, the church was not his only occupation. He died before 1880, and his wife sometime after that, but exactly when and where has not been recorded. Both are probably buried in unmarked graves.[21]

The other marble monument memorializes Mary Farris (1742–1827). She may have been the wife of William Farris (1750–1835), a veteran of the Revolutionary War who died in Rabun County. William "Faris" and a female who was probably his wife, both over forty-five years old, along with two adult males and three adult females, appear in the 1820 census of Habersham County but cannot be located in the 1830 census. However, one of the first officers in Rabun County, when it was organized in 1819, was Samuel Farris (1792–1862). He married Rebecca Pinson (1791–1848) in Habersham County in 1822, and it seems probable that he was one of William Farris's sons.[22]

Bean Creek Missionary Baptist Church

No certain date for the establishment of Bean Creek Missionary Baptist church has yet been documented. Some sources believe the date to have been as early as 1862, but most African American churches were not organized until after the Civil War. During the war some of the enslaved African Americans may very well have begun holding prayer meetings in an old log cabin across the road from the present church, but formal organization as a Baptist church had to wait until after the war.[23]

Tradition holds that the cabin, which no longer exists, was built by Cherokees, which would not be surprising. If it were in fact that old, the cabin might well have served as the Nacoochee Militia District courthouse too, since that also seems to have been approximately where the Moffat map locates the "Law Ground." In an interview published in "The Mayor of Bean Creek," in *Foxfire* in 2014, Lena Dorsey states that the African American church at Bean Creek "was a Methodist church to begin with." Whether she means that the congregation actually switched denominations is not clear, but that may have been the case. Shortly after the Civil War, probably in 1866, several African American churches in northeastern Georgia are said to have gathered at Bean Creek church and formed the Union Middle River Missionary Baptist Association. Sometimes known as the Middle River Union Baptist Association or Middle River First Union Baptist Association, the organization championed education in the African American community in the late nineteenth and early twentieth centuries. The association, which historically included churches in southwestern North Carolina as well, remains an active organization today.[24]

Baptism through bodily immersion was a cornerstone of the Baptist faith and was originally conducted in an artificial pool in the branch across the road from the church. Until the church built its own baptismal pool in 2003, the congregation sometimes used the baptismal pools at other churches and even the swimming pool at Alpine Crest Resort in Helen.[25] It is not clear when the congregation was able to construct a new wood-framed building, but by 1883, when they were finally able to purchase an acre of ground around the building for ten dollars from Capt. James H. Nichols, the church was apparently already standing. Trustees listed in the deed are James Alston, Prince Wyley, Green Alston, Reuben Jarrett, and Wiley Jones.[26]

James Alston was born about 1817 in North Carolina and probably married his wife, Rosetta, or Rose, before the Civil War. Documented in the 1870 and 1880 census, they had at least six children, including Reuben Jr., who was born about 1883. Reuben Sr. died before 1900, and his wife before 1910; if they are buried at Bean Creek, the graves are unmarked. Henry Jarrett, born in Alabama and ten years older than Reuben, was most likely his brother; they were recorded living next door to each other in 1870 and 1880. Also recorded in the 1870 census and possibly buried at Bean Creek was Julius Jarrett, born about 1773 in Virginia, who was likely Henry and Reuben's father.

Wiley Jones was born in 1844 in South Carolina. Shortly after the Civil War, he married Roseann, who was born in Georgia in 1845. They lived first at Cool Springs, five or six miles west of Clarkesville, before moving to Nacoochee in the 1870s. They had nine children, but by the time the family appeared in the 1900 census, all but three of the children were dead. Wiley and Roseann probably died at Nacoochee and may be buried in unmarked graves at Bean Creek. The Bean Creek school, which is discussed in more detail in the following chapter, was located on the site of the present church until it closed in 1944. In 1955 the congregation tore down the school and the old church building and built the present building.[27]

The cemetery at Bean Creek has irregular rows, running in a generally north-south direction across the hillside north of the church. Typical of many rural nineteenth-century graveyards, especially those used by African Americans, many of the burials at Bean Creek are unmarked. Some are marked with field stones or cast concrete, with the

The Bean Creek Missionary Baptist church, built 1955. Photograph by author, 2021.

name incised in the wet concrete; a few have professionally engraved stone monuments. Some of the temporary markers installed by the funeral home at burials have been lost or relocated, as have many of the field stones that marked the oldest graves. The precise number of burials is uncertain, but recent professional surveys suggest that there are as many as 230, and more will likely be documented in a survey using ground-penetrating radar.[28] Marked or not, the cemetery contains the graves of many individuals significant to the history of Sautee Nacoochee in general and to the African American community at Bean Creek in particular. Prominent among the surnames represented there are Austin, Brown, Dorsey, Jarrett, Jones, Lowery, Nicely, Richardson, Trammell, and Wyley.

Crescent Hill Baptist Church

In April 1921 a group of Baptists approached Dr. Lamartine Griffin Hardman (1856–1937) for his assistance in forming a Baptist church using the old Presbyterian church on property that had reverted to the heirs of Captain Nichols the previous year. Being a devout Baptist himself, Dr. Hardman bought the property from Nichols's daughter and in October 1921 conveyed the old church, complete with "electric light fixtures, pew Bibles, song books, organ, and heater," to the newly organized Crescent Hill Baptist church. Although Hardman kept his letter of membership at the Baptist church in his hometown

The Crescent Hill Baptist church, built by Captain Nichols in the 1870s for a Presbyterian congregation. Photograph by author, 2021.

of Commerce, his brother Thomas Colquitt Hardman, who was himself a Baptist minister, was one of the first deacons in the Crescent Hill Baptist Church.[29]

Other Baptist Churches

In 1879 a group of Baptists organized Union Baptist Church and built a church on property donated by Captain Nichols. The church is located on SR 356, just west of Unicoi Lake. Bethel Baptist Church, which was founded in 1901 by members of Union Baptist Church, is located less than five miles northeast of the Union Baptist church on SR 356. The Baptists were prone to schisms over one thing and another, but the organization of Bethel may have simply been a matter of a growing population and a desire to make attendance more convenient.[30]

The newest Baptist church in the valleys was the Community Baptist Church, organized in 1927. The Nacoochee Institute was a Presbyterian school, but, as part of an outreach program, they built a small building "five miles back in the hills," near the site of the old Monroe Methodist church. Known as the Monroe Community House, it was used for educational and entertainment purposes as well as religious services. In 1955 a pastor was called and the Community Baptist Church was organized. The congregation has since relocated to Hall County.[31]

METHODIST CHURCHES

The Methodist Church began as a movement within the Church of England and was so called for the "methodical" way in which John Wesley (1703–91) and his followers approached the practice of religion. In Savannah, in April 1736, he and his brother Charles (1707–88) set up a society that was, essentially, the first Methodist congregation. Although that congregation did not survive long, the Wesleys returned to England and established the first self-sustaining Methodist society in London.

Concerned with the plight of the poor in England, the Methodists were influenced by the fiery rhetoric of George Whitefield (1714–70). Whitefield roiled the established church, as he almost single-handedly created a new style of fervent evangelism designed to reach out to the underclass that he, the Methodists, and others felt had long been ignored by the Church of England. His revivals throughout the colonies, beginning in Savannah in 1738, continued the Great Awakening, begun by Jonathan Edwards in New England a few years earlier. In spite of the Anglican attempt to control it, Methodism spread rapidly in America. With its natural appeal to the underclass, the Methodists made many converts among the colonists, who were beginning to feel ignored by both the Anglican and the Presbyterian Churches. This was especially true in the South, where by 1773 two-thirds of the Anglican "Methodists" were located.

After the American Revolution, it was inevitable that there would be a separate church, and at the Christmas Conference in Baltimore in 1784, the Methodist Episcopal Church was formally organized. In 1844 the Methodist Episcopal Church split over the issue of slavery, bringing into being the Methodist Episcopal Church South. Although a few southern congregations refused to leave the original church, that split lasted until reunification as the Methodist Church in 1939. In 1968 other splinter groups from the denomination rejoined as well, forming the modern United Methodist Church.

During the Second Great Awakening, the Methodist Episcopal Church began to flourish, and Methodists became associated in the South with "camp meetings." Usually held in August, traditionally a slow time in the agricultural seasons, these gatherings for preaching, singing, and communion continued the tradition of the "holy fairs,"

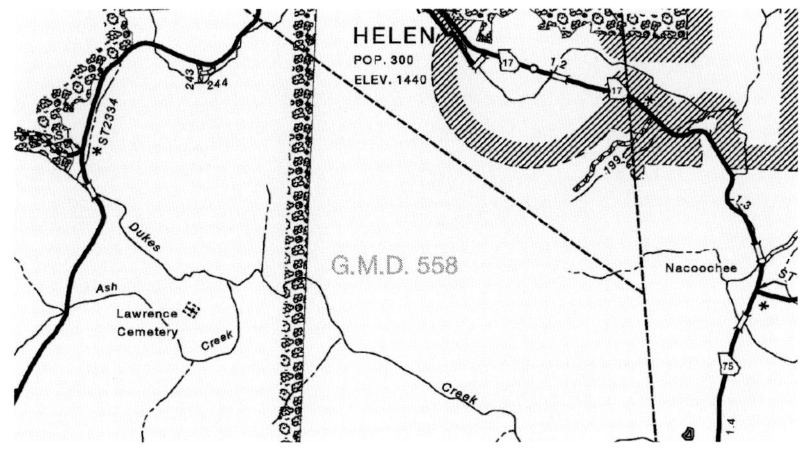

Detail from 1985 GDOT highway map, showing Highway 75 and Nacoochee (*right*) and the Lawrence Cemetery and Highway 75A just south of the intersection of the Richard B. Russell Scenic Highway (*lower left*). Smithgall Woods now encompasses much of the area between the two highways. Georgia Department of Transportation, "Archive Maps."

a medieval tradition adopted by the Presbyterians in Scotland in the seventeenth century. By the 1830s permanent campgrounds with a central "arbor" or pavilion were being established, including the ones that remain in White County at Mossy Creek (1833) and at Loudsville (ca. 1835). Both of those are associated with Methodist churches, but the meetings were essentially nondenominational and would have been familiar to the people at Sautee Nacoochee in the nineteenth century.[32]

Rock Springs Campground was founded in 1887 by former slaves who had been attending the Mossy Creek meetings and is one of the very few African American campgrounds in existence. Union Grove Campground was established in 1925 by the Congregational Holiness Church, one of a number of Pentecostalist denominations that emerged in the late nineteenth century. Today no other Georgia county has as many extant religious campgrounds as White County.[33]

In 1821 the Methodists established their first church in what is now White County, at Mossy Creek, six miles south of Cleveland. By the time the Methodist conference sent Nathaniel Rhodes, the first of its circuit-riding ministers in Habersham County in 1826, Methodist churches had also been established at Mount Zion and Nacoochee. There was a Methodist church with an African American congregation on the Dahlonega road west of Cleveland in the late nineteenth and early twentieth centuries. Methodist records show that Nelson Martin (ca. 1830–after 1880) and Wyatt Oliver (born about 1820) were ordained ministers in the Methodist Church, and one or both of them may have ministered at this church. The church's name has not been discovered.[34]

The Baptists have been credited with building the first church at Nacoochee in 1826, but there was a Methodist church by that time as well, which is not surprising, given the residence in the valley of several Methodist ministers. Most prominent was Rev. Jesse Richardson (1763–1837). Richardson was one of the pioneers of Methodism in Burke and Rutherford Counties, North Carolina, after the Revolution, beginning a long career as a circuit rider, deacon, and elder in the Methodist Episcopal Church in the spring of 1788.

His son John L. Richardson (1794–1850) and his son-in-law Abraham Littlejohn (ca. 1790–after 1860) were also ordained ministers in the Methodist Church, and all three came to Nacoochee, probably around the same time in the early 1820s. In addition, Rev. Josiah Askew (1765–1845) was from Burke County, North Carolina, and came to Habersham County in the early 1820s. He was a grandfather of renowned Methodist minister and bishop Atticus Green Haygood (1839–96) and his sister Laura Askew Haygood (1845–1900), who was one of the great Methodist missionaries in the late nineteenth century.

Lawrence Cemetery Methodist Church

According to Dr. Lumsden, the first Methodist church at Nacoochee was located "at the site now known as the Lawrence Cemetery," which is about three air miles west of Nacoochee and a few hundred yards east of Georgia SR 75 Alt. The church is poorly documented, but, as noted earlier, it appears to have been in existence by 1826.[35] No church building remains, but it is believed that some of the early settlers built a "little log schoolhouse" there that was occasionally used for church services. In June 1828 James Gilliland and his wife, Elizabeth, donated an acre of land in Land Lot 67 of the Third District for the purpose of building a Methodist church there, but it is not clear if one was ever built. The trustees named in the deed were Ebenezer Fain, Benjamin Crumley, David Fain, Francis Bird, Benjamin Allison, Melchizedek Charles, John Richardson, John Harwell, and Thomas Hughes.[36]

The origin of the cemetery's present name is uncertain. The oldest marked grave in the cemetery is that of one S. Harwell and dates to 1826. Numerous members of the Allison family are buried there as well, along with Holcombes, Ledfords, Abernathys, and Allens, all surnames found among the lists of the legendary sixty-one families. The only Lawrence in the antebellum census of Habersham County is Rhody Lawrence, who was apparently a widow in her twenties, with three small children in 1830. No other Lawrences can be located until after the Civil War. Perhaps her husband was the first burial in the cemetery, or perhaps it took its name from the Lawrence family who lived at Nacoochee after the Civil War.

Monroe Methodist Church

As noted in the previous chapter, Rev. William Smith Crumley (1794–1882) and his wife, Elizabeth Jones Crumley née Monroe (1790–1836), were among the legendary pioneers in Nacoochee. Dr. Tom Lumsden credits him with organizing, as early as 1832, a Methodist congregation near the junction of current SR 255 North and Sky Lake Road. Elizabeth Monroe Crumley died in 1836 and was buried on property east of Sky Lake Road that belonged to her family. In 1843 her brother, Jesse Monroe, deeded to Methodist trustees land around her grave and, across the road, a building referred to as Pleasant Hill Meeting House "that they may . . . build thereon . . . a place of worship for the use of the Methodist Episcopal Church." In recognition of this gift, the church on the site was afterward known as the Monroe Methodist church, and the cemetery as Monroe Cemetery. Trustees named in the deed were Ephraim McClain, Jesse Regans, John Heath, Wilkes T. Leonard, Benjamin Crumley, and William S. Crumley. William Smith Crumley died in 1882 and is buried next to his wife. The congregation apparently disbanded after the 1918 departure of its last pastor, Rev. Roy Pierce Etheridge (1889–1979). Today the cemetery survives, but there is no sign of a church.[37]

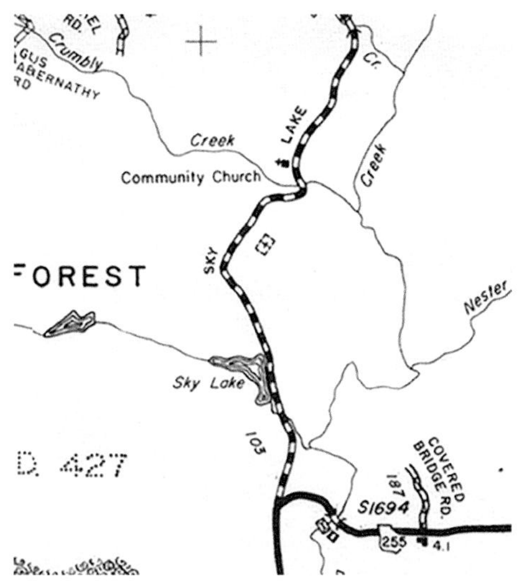

Detail from a 1985 highway map, showing Sky Lake Road, with the symbol for Monroe Cemetery placed just south of Crumbly Creek. Georgia Department of Transportation, "Archive Maps."

Nacoochee Methodist Church

In May 1833 thirty-year-old Mary Brown Williams, Edward Williams's wife, died and was buried on the hillside across Unicoi Road from their house. In April 1834 Williams deeded six acres to the trustees of the Nacoochee Methodist church—Abraham Littlejohn, John L.

Richardson, Daniel Brown, Soloman Leonard, Henry Conley, L. B. Burnet, Jesse Monroe, Reuben Philips, and James Quillian. The deed states that the land was given "together with all and singular the houses, woods, waters, ways, privileges, and appurtenances," suggesting that there were already improvements on the property.[38] The deed for the property, which stipulated there were always to be nine trustees, was witnessed by Rev. Jesse Richardson and Edward Williams's son Charles L. Williams. Trustees Littlejohn, Richardson, Brown, Leonard, Conley, and Monroe were among the sixty-one families discussed in the previous chapter; the other three trustees were not.[39]

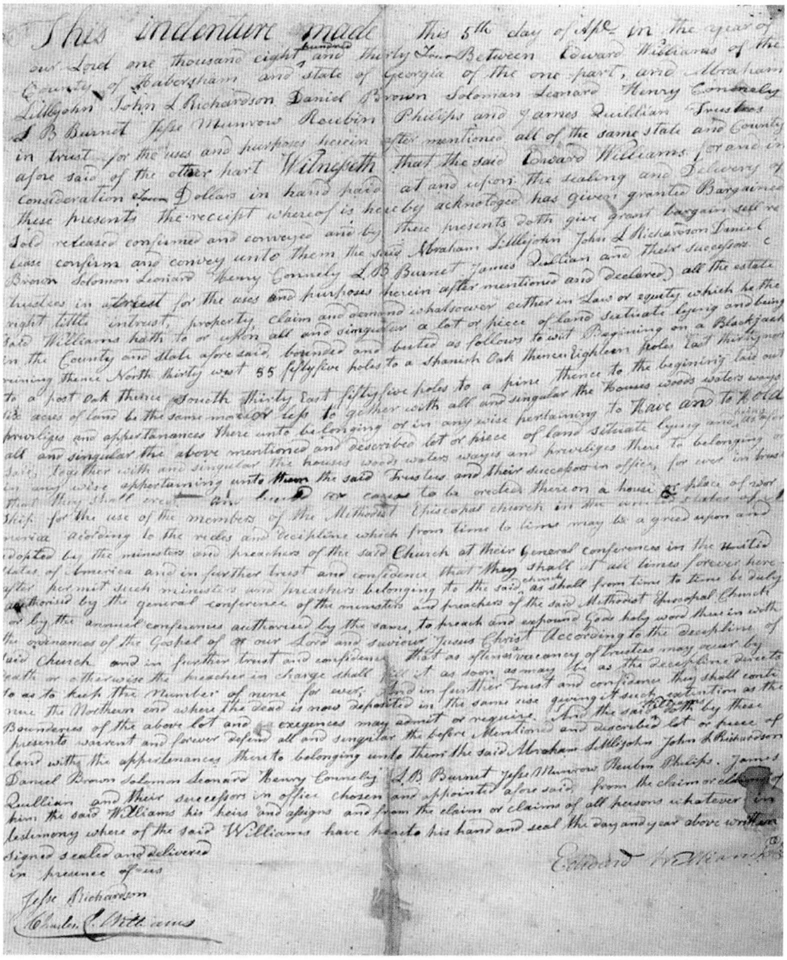

This indenture made this 5th day of Apl in the year of our Lord one thousand eight hundred and thirty four Between Edward Williams of the County of Habersham and state of Georgia of the one part, and Abraham Littlejohn John L Richardson Daniel Brown Soloman Leonard Henry Connely L B Burnet Jesse Munrow Reubin Philips and James Quillian Trustees in trust for the uses and purposes herein after mentioned all of the same state and County afore said of the other part Witnesseth that the said Edward Williams for and in consideration of [illegible] Dollars in hand paid at and upon the sealing and delivery of these presents the receipt whereof is hereby acknowledged has given granted Bargained Sold released confirmed and conveyed and by these presents doth give grant bargain sell release confirm and convey unto them the said Abraham Littlejohn John L Richardson Daniel Brown Solomon Leonard Henry Connely L B Burnet James Quillian and their successors trustees in trust for the uses and purposes herein after mentioned and declared all the estate right title interest property claim and demand whatsoever either in Law or equity which he the said Williams hath to or upon all and singular a lot or piece of land situate lying and being in the County and state afore said, bounded and butted as follows to wit Beginning on a Blackjack running thence North thirty west 55 fifty five poles to a Spanish Oak thence Eighteen poles East [illegible] to a post Oak thence South thirty East fifty five poles to a pine thence to the beginning [illegible] [illegible] …

Edward Williams

Signed sealed and delivered in presence of us
Jesse Richardson
Charles L. Williams

The 1834 deed conveying land for the Nacoochee Methodist church from Major Williams to the church's trustees. Courtesy of Nacoochee United Methodist Church.

Trustee L. B. "Berry" Burnet was Littleberry (or Littleberry B.) Burnett, born in 1804 to Swan Pritchett Burnett of Burke County, North Carolina, and his wife, Frances Bell. Swan's parents were Thomas Burnett and Elizabeth Littleberry. Frances's parents were early Nacoochee settlers Thomas Bell and Jane Montgomery Bell, which may account for Littleberry's presence in the valleys before the 1830 census, where he appears as household head Berry Burnet. In 1833 he married Maria Louisa Hamilton of Clarkesville. Littleberry died in 1835, leaving his young widow with two daughters. James R. Wyly was appointed administrator of his estate, selling five hundred acres, parts of Land Lots 52 and 53, to George Edmondson for $522 on 7 March 1837. Five days later Edmondson sold it for $700 to Moses Harshaw. The deed was recorded on 5 December of that year, by which time Burnet's widow and two daughters had departed with her parents for Nacogdoches, Texas.

Trustee Reuben Philips was probably Rev. Reuben Philips (1795–1887). He was enumerated in the 1830 Habersham County census, with his wife, four children, and a woman born in the 1750s who may have been his mother or mother-in-law. Several households named Phillips, or Philips, are listed in the 1820 census of Burke and Rutherford Counties, North Carolina. However, neither Reuben Philips nor any similar name can be found in that census, and he has not been certainly located in any of the later censuses.

Finally, the trustee James Quillian was Rev. James Milton Quillian Sr. (1793–1869), buried in an unmarked grave at the Mossy Creek Methodist church. He married Sarah Pricket in Franklin County in 1815, and, although they have not been located in the 1830 Habersham County census, his brother Clemond and his father, James Quillian Sr. (1757–1858), were both listed as head of a household in that census. His son William would go on to be president of Wesleyan College in Macon.

The new Nacoochee Methodist church was finished in 1836. Besides being more conveniently located, the new site had the added benefit of removing the church from the upper reaches of Dukes Creek, which were being ravaged in the ongoing search for gold. The new church was a plain, wood-framed, front-gabled structure, like most churches of the period, but it was the first building in the valley to be painted white, gaining it the moniker "the white church." In 1878 the church deeded that building and two acres to John Jasper Methvin for use as a

Monument commemorating enslaved African Americans at Nacoochee buried in unmarked graves in the Nacoochee Methodist church cemetery. Photograph by author, 2021.

The Nacoochee United Methodist church. Photograph by author, 2021.

schoolhouse and built a new building, which stood until it burned and was replaced by the present structure in 1943. There have been several significant additions since then.[40]

The Williams and Brown families were stalwarts of the Methodist church. In 1849 one observer noted, "Mr. Williams is a zealous, though not a bigoted member of the Methodist Episcopal Church." Nevertheless, other sources say that he was not baptized and thus unable to take communion until shortly before his death.[41] Many of the earliest and most prominent people in the history of Sautee Nacoochee are buried at this Methodist cemetery, including Edward Williams and his wife, Mary Brown; her parents, Daniel and Hannah Brown; Rev. Jesse Richardson and his son John L. Richardson; and Adam and Rachel Pitner and Adam's father, John.

The graves of Abraham Littlejohn, one of the church's early pastors, and his wife, Sarah, who was Jesse and Rutha Richardson's daughter, have not been located. Many families continued to use family cemeteries, and the Littlejohns may have as well. It is also possible that they are buried in unmarked graves in the Nacoochee Methodist church cemetery. As of 2018, 620 graves had been inventoried in the cemetery, but there are certainly dozens more in unmarked graves. In particular, local historians have documented at least 120 slave burials at the northern end of the cemetery.[42]

Chattahoochee Methodist Church

In May 1860 Jehu Trammell deeded three acres in Land Lot 28 to the trustees of the newly organized Chattahoochee Methodist Church. The first church building was a log structure, but in 1889 the congregation built the present wood-framed building, which was made famous when it was used in the filming of *I'd Climb the Highest Mountain* in 1951. Among the hundreds of graves in the church cemetery are those of several generations of families with roots in the earliest days of white settlement. Among the most numerous surnames are Abernathy, Adams, Allison, Cantrell, Fain, Robinson, Sims, Vandiver, and Westmoreland.[43]

Bean Creek Methodist Church

Although none of the standard published histories of White County mention this congregation, and it does not appear on any of the state highway maps, several local residents confirm its existence. In an interview published in *Foxfire* in 2014, Lena Dorsey stated that the African American church at Bean Creek "was a Methodist church to begin with." When this might have been is not clear. Recent research suggests that the church was founded by Byas (or Byus) Richardson Sr. (1861–1933), the son of Shadrack and Emily Richardson, who were thought to have been slaves of Rev. John L. Richardson. Born at Tesnatee, Byas Richardson is buried in the Mount Yonah Cemetery. Located near the intersection of Rabun and Bean Creek Roads, the Bean Creek Methodist church was always a small congregation, since most of the African American community were Baptist.[44]

PRESBYTERIAN CHURCHES

Although vastly outnumbered by Baptists and Methodists today, the Presbyterians played an extremely important role in the early settlement of the Carolina and Georgia frontier. Although a latecomer to Sautee Nacoochee, the Presbyterians would have a major impact on the community in the first quarter of the twentieth century. A Calvinist cousin to the Dutch Reformed Church and the French Huguenots, the Presbyterian Church was established as the Church, or Kirk, of Scotland by John Knox in the late 1550s. Before authorizing formation of a new congregation, the Presbyterians demanded an educated clergy and insisted on a list of subscribers committed to the support of the minister and the church. The Presbyterians' strong support of education had a tremendous influence on the Scots-Irish before and after emigration to America. Of the 207 enduring colleges founded in the United States before the Civil War, for example, 48 were organized by Presbyterians, 34 by Methodists, 25 by Baptists, and 21 by Congregationalists. The Presbyterian impact on the development of education in the United States, including Sautee Nacoochee, was significant.

The Presbyterians held the first of their camp meetings in Kentucky in the 1790s, but, by the early 1800s, camp meetings had become a fixture of the Methodist Church. The Presbyterians were appalled as fervent, emotional preaching and singing at these meetings overwhelmed their sense of propriety. The Presbyterians' lack of enthusiasm for the revival experience helped ensure that the religion-hungry pioneers of the upcountry would turn to the Methodists and the Baptists and away from the Presbyterianism that had nurtured the Scots-Irish for two and a half centuries.

As a result, the Presbyterian Church in America began a long period of slow decline. While a few congregations managed to flourish, many simply ceased to exist. When Levi Willard, for instance, joined the Decatur Presbyterian Church in 1826, it was noted that his letter of membership from his old church in Eatonton could not be gotten "as it [the church] had become nearly extinct."[45] Even the congregation of the great Mount Zion Presbyterian church in Hancock County, which had been the center of a flourishing Presbyterian community in the early 1800s, sold their building to the Methodists and merged with the Sparta Presbyterian church in the 1840s.

Nacoochee Presbyterian Church

When Capt. James H. Nichols (1835–97), a devout Presbyterian, came to Nacoochee after the Civil War, the closest Presbyterian church was the one founded at Clarkesville in 1832. He soon found that some of his neighbors were Presbyterian but had been worshipping with the Methodists, most notably among them Massachusetts-born Josiah Robinson Dean (1796–1871), who had come to Nacoochee to mine for gold in 1857. After some initial difficulties, the presbytery in Augusta authorized a new Presbyterian church at Nacoochee and assigned a minister, Rev. Joseph Washburn. Because his circuit also included churches at Blairsville, Brasstown, and Dahlonega, services were held at Nacoochee only one Sunday each month. The first elders were Captain Nichols, Josiah Dean, and John Glen (1798–after 1880).[46]

Captain Nichols, whom George Walton Williams called "that liberal-hearted Christian gentleman," donated three acres and built the Nacoochee Presbyterian church in 1872.[47] It flourished until the death of some of its founding members in the 1880s. After the death of Captain Nichols, there were so few members that the church was dissolved, although it was reorganized in 1901. Two years later the Athens Presbytery organized the Nacoochee Institute, and, since the church was an integral part of the school, the Nacoochee Presbyterian Church began worshipping in the institute's auditorium. Little used, the old church was finally allowed to revert to the Nichols heirs. After the school's auditorium was destroyed by fire in 1926, the Nacoochee Presbyterian Church built a new building and continues to worship there today.

EPISCOPAL CHURCHES

The Church of England originated with Henry VIII's challenge to the supremacy of the pope in the 1530s and was introduced into the Americas with the first communion at Jamestown in 1607. During the Revolution northern Anglican churches tended to support the Crown, while those in the South tended toward support of the patriot cause. All Anglican churches in the colonies suffered at the end of the Revolution, when much of their state funding, once provided by the Crown, disappeared with the newly established principle of separation of church and state.

With some difficulty the American congregations reorganized themselves to ensure apostolic succession of their bishops; they met in Philadelphia in 1789 to adopt a constitution, canon law, and an American version of the *Book of Common Prayer*. However, the church never had a large membership in Georgia. Sherwood counted only four Episcopal congregations in the state in 1826; a fifth had been formed in Macon in 1824, but the congregation was unable to maintain a minister.

One of the main reasons that Grace Episcopal Church was established in Clarkesville in 1838 was to give the state a sixth parish, thereby making the Episcopal Church in Georgia eligible for its own diocese and bishop. By 1840 the Clarkesville congregation had completed a fine church building, which was apparently used by both Episcopalians and Presbyterians, until the latter completed their own building in 1847. Many of the earliest members were summer refugees from fever-ridden plantations on the coast. The Clarkesville church remained the only Episcopal church in Habersham or White County until 1993, when the Church of the Resurrection was founded two or three miles south of Sautee Nacoochee.[48]

Education

Beginning in New England in the seventeenth century, strong support for literacy and education was a characteristic of colonial America, in large part because most Protestant denominations, since the time of Martin Luther, put a premium on an individual's ability to read and interpret the Bible. From a secular point of view, Thomas Jefferson spoke for many when he wrote in 1816, "Where the press is free, and every man able to read, all is safe." As a result, by 1800 the United States had a higher literacy rate than any European country except Scotland, where the Kirk of Scotland helped make literacy nearly universal. Primarily because of racial animosity, the overall literacy rate in the South lagged behind the rest of the nation until late in the twentieth century.[49]

Through most of the antebellum period, especially in the South, African Americans were prohibited by law from being taught to read and write, and Georgia adopted such a law in 1829. In spite of that, some slave owners defied the law, and many slaves simply taught themselves. In 1870 well over three-quarters of African Americans in the state were illiterate, while less than 10 percent of the white population was so. Thirty years later about half the Black population remained illiterate, while the proportion of illiterate whites had increased to nearly 20 percent.

Outside the middle and upper classes, who could afford private academies, "old field schools" formed the beginnings of public education in the United States. Often located on worn-out fields, thus the name, they were generally privately supported within the community, members of which would donate the land, construct the building, hire the teacher, and bear the expense of running the school. Students were generally those whose parents could afford to support the school, although nearly every county had "poor school funds" required by state law to support a few indigent scholars. The quality of the schooling, then as now, depended largely on how much the community was willing to spend.[50]

These schools provided only an elementary education—reading, writing, and arithmetic—and were often known as "blab schools," for the teaching that was conducted through the students' oral recitation of lessons. Still, it was a great thing in the days before the birth of public school systems in the 1870s. George Walton Williams (1820–1903) remembered as much when he wrote later in life, "Persons who have been blessed with a classical education cannot appreciate too highly their advantages. Those who have not been so favored, keenly feel the difficulties under which they labor."[51] More colorfully, Charles H. "Bill Arp" Smith (1826–1903), the famous Georgia writer and politician, wrote around the same time, "But for my town raising [in Lawrenceville, Georgia] and old field school education, I too would have made a very respectable cracker."[52]

These early schools were almost always one room, frequently in a log building with a dirt floor, and are poorly documented. Early Habersham County records mention Walter Adair's schoolhouse, which was probably on the Unicoi Road a mile or two north of Clarkesville, and the county courts may have held their first meetings there before the county seat was formally designated. There is no indication of how long that school operated, but by 1829 Clarkesville had a "neat two-story academy," suggesting the kind of education possible for the more affluent citizens of the county. Twenty years later George White wrote that "considerable attention is paid to the subject of education, and common schools are numerous."[53]

In 1850 the federal census reported that 1,022 children, all of them white, had attended school in Habersham County over the previous year. Most of those would have attended private schools; there was a single public school in Habersham County, the least of any county in the state except for Floyd County. The public school had one teacher and twenty students. The county contributed $350 annually for its support, augmented by $200 from "other sources." It is not clear where this school was located.[54] Moffat's map of Nacoochee in 1837 locates a schoolhouse south of the Chattahoochee River, on the west side of the land lot between the river and Dukes Creek. The school almost certainly stood on property owned by the Richardsons, and some sources credit Rev. Jesse Richardson with its establishment. Edward Williams provided substantial support for the school, as did Jesse's son John L. Richardson.

As noted in the previous chapter, Adeline Moffat (1815–80) came to Nacoochee with her parents in the 1830s and taught school prior to her marriage in 1841. The map of Nacoochee as it existed in 1837 that her brother drew in 1891 depicts the schoolhouse as well as two Moffat residences near the Richardsons on Dukes Creek. How long the family

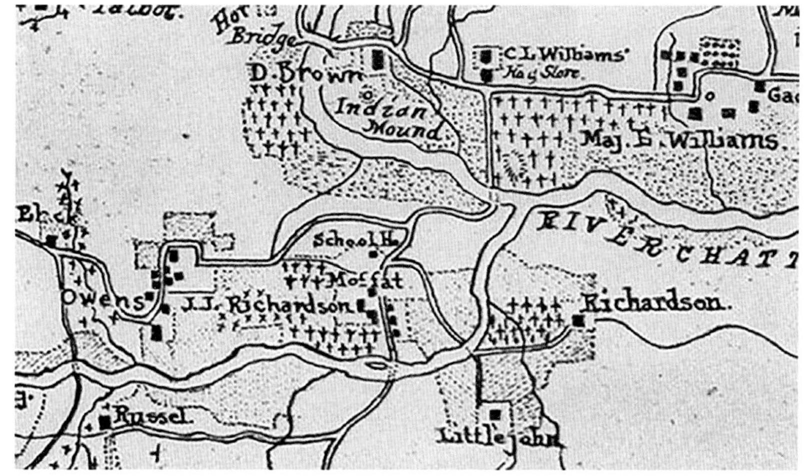

Detail from the 1837 Moffat map, with the "School H." near the center of this image. Courtesy of Sautee Nacoochee Community Association.

remained at Nacoochee is not clear, since no Moffats have been identified in the federal census of Habersham County in 1830 or 1840. The schoolhouse probably continued in use through the remainder of the antebellum period, but nothing is known of any other teachers who may have taught there.[55]

There had been abortive efforts to establish a state-supported public school system in the late 1830s and again in the late 1850s, but both failed for lack of funding by the legislature or local government. In October 1870 Gov. Rufus Bulloch signed a bill that reestablished a public school system. Elections for county school boards were held in January 1871, but efforts were hampered by public resistance to anything coming out of the state's Republican government, since the state was still under military occupation. Not until the election of a Democratic governor in 1872 did the board of education approve the first textbooks for use in local schools, a significant step in establishing a public school system in the state.[56]

African Americans freed from bondage were nearly half the state's population, and they began setting up schools for African Americans immediately after the Civil War. With assistance from northern philanthropists and educators, and little, if any, assistance from state or local government, Atlanta University (1865) and Clark College (1869) were organized in Atlanta. At the same time, work began to address the massive lack of educational opportunities for African Americans throughout the state. At the local level, literate African Americans began teaching, while others donated land for a building and provided financial support for teachers. After the Freedmen's Bureau closed in 1870, the northern aid societies began limiting their efforts to the state's larger cities, where they focused on educating generations of African American teachers.[57]

When Georgia established its first public school system in 1870, "separate but equal" facilities were mandated for white and African American students, a policy that was upheld in the landmark U.S. Supreme Court case *Plessy v. Ferguson* in 1896. Inevitably, the facilities were separate but almost never equal. Some of the inequality of conditions is suggested in the fact that White County had African American grammar schools at Rock Springs near Mossy Creek, Oak Springs at Cleveland, and Bean Creek at Nacoochee but had no high school. Until after World War II, African American students who wanted a high school education had to go to Habersham County to find it.[58]

WHITE COUNTY BOARD OF EDUCATION

In October 1870 the Georgia legislature enacted a law calling for elections of county boards of education and local school trustees. Accordingly, in January 1871 each militia district in White County elected one representative to the county board of education and three "trustees" to oversee the district schools. Nacoochee chose Elijah Fletcher Starr for the board of education seat. The first trustees were Thomas Watson Fain (1822–99), Capt. James H. Nichols (1835–97), and James H. Williams (1845–1909).[59]

Hampered by the political confusion at the state level, the county still did not have a functioning school system in 1872, although its first superintendent, Rev. John Jasper Methvin (1846–1941), was appointed that year. Born and raised in Wilkinson County, a few miles east of Macon, he was the son of a prosperous farmer. A Confederate veteran, he studied law and was admitted to the Georgia bar before being ordained a Methodist minister in 1870.

The Panic of 1873 and the subsequent nationwide depression may have slowed progress too, and the early history of public education in White County remains murky. The county's two high schools, at Cleveland and Nacoochee, appear to have been privately funded when they first opened in 1877, and there is no evidence of a public high school at Nacoochee prior to 1928. Reverend Methvin was "in charge" of both schools, "alternately." The six-month school term that he advertised in 1877 suggests that he taught six months at Nacoochee and six at Cleveland.[60] In addition to the grammar schools at Shoal Creek and White Creek, the schools at Cleveland and Nacoochee taught grammar and high school grades. As the population grew, additional grammar schools were established, and by World War II there were twenty-three public grammar schools in White County.

HIGH SCHOOLS AT NACOOCHEE

At the Nacoochee Male and Female High School he advertised in 1877, Reverend Methvin was the principal, as it were, but he had an "assistant" in Amelia Starr (1858–93), daughter of Dr. Elijah Fletcher Starr and his wife, Hannah, and so Edward Williams's granddaughter. Hers was a typical occupation for a young woman before marriage, and she probably quit teaching when she married Frank Logan Asbury in 1881. She would be the first of four siblings who died of typhoid fever in August and September 1893. All are buried at Nacoochee Methodist church cemetery. The music teacher at the high school was M. B. Butt. She cannot be located in the federal census, and no other details of her life have been documented.[61]

The circumstances of Methvin's departure from Nacoochee are unknown, but in 1881 he was appointed president of Gainesville College, a short-lived school in Hall County, where he served for two years. From there he went to Butler Female College (1883–85) at Butler, Georgia, about halfway between Macon and Columbus. He ended his career in Oklahoma, where he worked mostly with the Kiowa, Comanche, and Apache. He was a prolific writer, especially in publications of the Methodist Church, and was inducted into the Oklahoma Hall of Fame in 1932. He died in Anadarko, southwest of Oklahoma City, in 1941.[62]

In 1882 some in the local community banded together to organize a high school at Nacoochee and bought four acres of land on the west side of Sautee Valley. The original trustees were Dr. Elisha F. Starr, James G. Glen, Samuel O. Conley, John L. Richardson, Alfred P. Williams, and Herbert H. Dean.[63] It was around this time that Captain Nichols's daughter, Anna Ruby, is thought to have used what is now the

Crescent Hill Baptist church as a schoolhouse, probably for primary grade students. Many churches, including or maybe even especially African American churches, doubled as schoolhouses. As late as 1940, the Georgia State Highway Board's map of White County designated the Bean Creek church "Bean Creek School and Church."[64]

NACOOCHEE INSTITUTE

The Nacoochee school that Dr. Starr and the others organized in 1882 continued, but, by the early 1900s, it was offering no more than a ninth-grade education. Jesse Richardson Lumsden (1848–1933), great-grandson of John L. Richardson, who had founded the first school at Nacoochee in the 1820s, is credited with being the catalyst for the creation of the Nacoochee Institute, a landmark at Sautee Nacoochee. In the fall of 1902, Lumsden is reported to have offered the old Nacoochee school building and the four acres on which it sat to the Athens Presbytery of the Presbyterian Church if they would establish

The main building at Nacoochee Institute, originally constructed around 1870. Courtesy of Sautee Nacoochee Community Association.

a full-fledged, coeducational high school at Nacoochee. It was a bold suggestion, made possible by the fact that the Nacoochee school was privately owned and governed by a board of trustees. After a series of negotiations, the presbytery agreed and the Nacoochee Institute was established in 1903.

Only the first floor of the two-story school building, which measured thirty-five by seventy-five feet, was finished, and it had "one ordinary box stove" to heat a single giant room. The building was apparently renovated over the spring and summer of 1903, when "a temporary partition was put in, dividing this room into two and a large stove and two drums were purchased and installed, which added much to our comfort." The coeducational Nacoochee Institute opened its doors for its first session on 2 November 1903.[65]

Until 1909 the school was taught by Joel Taylor Wade (1862–1957), a Presbyterian minister. He was assisted by Minnie Amelia Asbury (1882–1975), the daughter of Amelia Starr Asbury and great-granddaughter of Edward Williams. Wade has not been located in the federal census during the period, and his life remains undocumented. In the summer of 1904, the institute finished the second floor and divided it into five rooms: "a music room, practice room, and three recitation rooms and arranged for their heating." The following year they built a three-story addition, and in 1906 a twenty-nine-room dormitory was built by neighbor William I. Stovall for $6,000.[66]

About 1908 Wade resigned, and Prof. John R. Long of Columbus took over for a year before the board hired Rev. John Knox Coit (1872–1945), who led the Nacoochee Institute for the next twenty years. During that time it grew to encompass 301 acres of land, three school buildings, four dormitories, eight cottage homes, and sixteen farm and service buildings. There were six grammar school grades, and two junior-high grades, offering home economics, music, and commercial departments. There were fourteen teachers and 262 students.

A contemporary of the Nacoochee Institute was the Rabun Gap Industrial School, founded in 1905 thirty miles northeast of Sautee Nacoochee. Its purpose was "to help educate and give hope to the isolated people of the region." Students attended classes and also "worked the school farm, growing and preparing their food and maintaining the school's buildings, including dormitories."[67] In February 1926 the Rabun Gap School was destroyed by fire; six weeks later the Nacoochee

Dormitory at the Nacoochee Institute.
Courtesy of Sautee Nacoochee
Community Association.

The burning of the Nacoochee Institute, 1926. Courtesy of Sautee Nacoochee Community Association.

Institute burned as well. As a result, in 1928 the two schools merged and built new buildings at Rabun Gap, where the Rabun Gap–Nacoochee School continues to operate today.

GEORGE W. WILLIAMS HOME FOR CHILDREN

Before his death George Walton Williams Jr. (1860–1923) gave $6,000 to the Presbyterian synod at Athens for the purpose of establishing "a home for orphan and dependent little girls." Williams and his father had been perennial supporters of the Charleston Orphan House, which was the nation's first municipal orphanage when it was established in 1790. After his death in 1923, George Jr.'s widow, Margaret Milligan Adger (1863–1941), and their children provided funds that enabled the synod to purchase Mountain Home, the summer residence that George Sr. had completed in 1876. In 1923 it was owned by George Jr.'s sister, Martha Williams Carrington (1867–1930), who had inherited it when her father died twenty years previously. It was a "beautiful Country Estate" on sixty acres of land that included "a large 13-room house built of the finest materials, still in good state of repair, furnished throughout and supplied with water from mountain springs above."[68]

As noted in the previous chapter, Victor Hollis was superintendent of the orphanage and with wife, Mamie, lived onsite. They were in charge of day-to-day operations, over the years giving "real Christian home training" to as many as thirty children at a time, many of whom also received the advantage of an education at the nearby Nacoochee Institute. The orphanage continued to operate after the Nacoochee Institute moved to Rabun Gap. In 1930 the Hollises, who had three daughters and a son of their own, were caring for eleven girls and seven boys there. It was finally closed in 1938, when Hollis accepted the position of superintendent at the Georgia Industrial Children's Home in Macon.[69]

THE NEW NACOOCHEE SCHOOL

At the old site of the institute at Sautee, the rubble was cleared away, and in 1928 the county built a new school on the site to serve as both grammar and high school. Ten years later a gymnasium was constructed. The school closed in 1978. In 1986 the Sautee Nacoochee Community Association took possession of the old school building, which they refurbished and converted into a thriving arts and community center. It remains the home of the Sautee Nacoochee Community Association.

BEAN CREEK SCHOOL

The origins of Bean Creek school are not well documented. Although the school may have been more or less contemporaneous with the Bean Creek church, the records of the Freedmen's Bureau, which typically identified church schools in its field reports, suggest that no African American school was in existence at Nacoochee in 1868. The trustees of the Bean Creek Missionary Baptist church did not acquire the property until 1883, at which time the church was already standing, and the deed states that the property could be used only for "religious worship or educational purposes," suggesting that the building may already have been used as a school. A separate one-room schoolhouse was built in the late nineteenth century, perhaps as early as the 1880s. There was one teacher for the school, which went through seventh grade. Only one of the teachers has been identified, Eugenia Jarrett. African American, she was born Eugenia Dennis in September 1896 in Texas but moved to Georgia sometime before 1920. She married Robert L. Jarrett in 1923 and continued to teach at the Bean Creek school throughout the 1920s. During the Depression she and her husband moved to Asheville, where she died in June 1996, at the age of one hundred.[70] The school continued to operate until 1944, when it and the two other African American schools in the county were consolidated, and students were bussed to a new, segregated school in Cleveland.[71]

The Vale of Nacoochee

By 1850 the gold rush was moving on to California, although the creeks and the rivers in northern Georgia would continue to suffocate from runoff from hydraulic mining for decades to come, and gold mining would remain an important aspect of the local economy into the twentieth century. The destruction of the mountain forests from timbering was still decades away, and the large, productive farms that the Brown, Williams, Wyly, and Harshaw families established in the valleys created a bucolic setting that was appreciated by artists, poets, and sentimentalists of all stripes.

The so-called Legend of Nacoochee had also taken shape as early as 1840, at a time when many Romantics were searching out "authentic" Indian legends. There may have been a single local source to the Nacoochee tale, but, as one observer noted in 1843, the legend of star-crossed lovers was "like the snow-ball, capable of adding greatly to its bulk in rolling from lip to lip, and had I not as mortal a horror of Indian stories, as of revolutionary tales, Fourth-of-July orations, and temperance addresses, what a nice little piece of tragedy I might weave up from the groundwork which the popular legend affords."[72] All of that aside, the natural beauty of the valleys was a real attraction in and of itself.

GEORGE COOKE (1840)

George Cooke (1793–1849) was born in Maryland and began his career as a portraitist in Virginia before he left to study and work in Europe in 1826. After his return in 1832, he traveled the South, making his living primarily as a portraitist. He was also a painter of romantic landscapes but is remembered primarily for his gigantic painting *Interior of St. Peter's, Rome*, which hangs in the University of Georgia Chapel. He produced five essays for the *Southern Literary Messenger*, the last titled

"Sketches of Georgia," in which he described a trip through northeastern Georgia in the late summer of 1840. Climbing Mount Yonah, he described the magnificent view, concluding that "the most pleasing object from this mountain is the valley of Naucooche, through which the Chatahooche winds its way, skirted with verdant plains for many miles, and closed in between high hills, with here and there a farm-house."[73] He also noted the mound that stood "in the midst of the vale" and recounted a version of the legend of Nacoochee, in which an Indian maiden is sacrificed as a burnt offering to appease the spirits. Cooke painted several scenes in northeastern Georgia, possibly including Nacoochee, but, if so, it has not been located.[74]

THOMAS ADDISON RICHARDS (1843, 1850)

Another of the early admirers of Nacoochee was the noted artist Thomas Addison Richards (1820–1900), one of the first to paint southern landscapes. He was born in England, but his family emigrated to the United States in 1831. In the mid-1830s they moved from Savannah to Penfield, in Greene County, Georgia, where Richards's father was one of the original trustees for what became Mercer University. In 1838 Richards moved to Augusta, where he taught classes in painting and wrote short travelogues of his journeys around the state. In 1841 he set off to search out and draw picturesque scenery all across the South. The following year he and his brother William Cary Richards published *Georgia Illustrated in a Series of Views: Embracing Natural Scenery and Public Edifices*. The volume featured eleven steel engravings of Addison Richards's sketches from the field.

From 1842 to 1844 the brothers collaborated in the publication of *Orion: A Monthly Magazine of Literature and Art*, with William as editor and Addison as illustrator. In October 1843 the *Orion* featured "The Valley of Nacoochee, with the Legend of the 'Evening Star,'" authored by Addison Richards. The issue included his illustration *The Vale of Nacoochee*, along with the earliest of what would be many romantic paeans to the beauty of the valleys.

> Now, reader, lend me your eye—fix it in the direction of the setting sun;—there! you see that extent of slightly undulating plain, some seven miles in length and about a mile and a half in breadth? You see the ranges of hills which environ it, and jealously seclude it from all the outer world;

T. Addison Richards, *The Vale of Nacoochee*, lithograph 1842, depicting a westerly view from where Unicoi Road enters the valley from the east. *Orion* 3 (1843).

hills and mountains of luxuriant verdure, in all the varied tints which close proximity and gradually growing distance impart! You see the gorgeous carpet which covers the plain; the emerald grass, the tassellated maize, the ripening grain of every nature! You see the cottages, with "the smoke's blue wreaths ascending with the breeze," which sprinkle its surface! You see the fickle shadows cast by the changing clouds?—ah! and you see too, I can tell it by the bright sparkle of your eye—you see the serpentine windings of an infant stream, through the bosom of the vale-the waters of the young Chattahoochee: this, reader, is the valley of the "Evening Star."[75]

Richards had visited Nacoochee in 1842 and was shown the sights by Edward Williams, "the hospitable proprietor of a great portion of the valley." It is not clear who told him the legend of Nacoochee, which is supposed to mean "Evening Star," but his account was only one of many published versions of the tale.[76] In 1854 Richards published *American Scenery, Illustrated*, which included descriptions and engravings of Toccoa and Tallulah Falls and noted that Clarkesville was "a convenient place from which to reach all the surrounding points of interest," including "the winsome vale of Nacoochee and the noble Yonah." There was also a poem by his brother that took the legend of Nacoochee a romantic step further:

> Enshrined in my heart is the vale of Nacoochee,
> And memory often makes pilgrimage sweet
> To the beautiful haunts of the bright Chattahoochee
> Where its silvery fountains in melody meet.
> The poets may boast if they will of Wyoming,
> Of peerless Avoca, and lovely Cashmere;
> My fancy, contented without any roaming,
> Shall find in Nacoochee a valley more dear.[77]

CHARLES LANMAN (1849)

Charles Lanman (1819–95)—"author, government official, artist, librarian, and explorer"—was born in Monroe, Michigan, at the western end of Lake Erie. His narrative accounts of his travels in the eastern United States published in the 1840s established his reputation. One of those travelogues was his *Letters from the Alleghany Mountains*, published in 1849, which included a long narrative and illustrations of his travels in the Southern Appalachians from northern Georgia to western Virginia. In addition to "letters" from Clarkesville, Track Rock Gap, Toccoa Falls, and Tallulah Gorge, Lanman's third letter was titled "Valley of Nacoochee, Georgia, April, 1848."

> The valley is perhaps a mile wide, and, as the surrounding hills are not lofty, it is distinguished more for its beauty than any other quality; and this characteristic is greatly enhanced by the fact, that while the surrounding country remains in its original wilderness the valley itself is highly cultivated, and the eye is occasionally gratified by cottage scenes which suggest the ideas of contentment and peace. Before the window where I am now writing lies a broad meadow, where horses and cattle are quietly grazing, and from the neighboring hills comes to my ear the frequent tinkling of a bell, which tells me that the sheep or goats are returning from their morning rambles in the cool woods.[78]

Lanman also described two Indian mounds at Nacoochee, both similar in size and about a half mile apart. One mound was cultivated, while "the other is covered with grass and bushes, and surmounted, directly on the top, by a large pine tree." He went on to relate two legends that he was told by "the 'oldest inhabitant' of this region." Neither of them was the now traditional version. Instead, in the first, the two mounds were erected over the remains of Nacoochee, the daughter of a Yamasee chief, and Kostoyeak, a Cherokee chief, to commemorate their long and exemplary life together. In the second legend, Nacoochee, a beautiful Cherokee maiden, died in her fifteenth summer, and the "newly born" Evening Star appeared in the sky on the night after her burial. He noted even then that one of the mounds had been plundered and that he had been told that "pipes, tomahawks, and human bones were found in great numbers."[79]

HENRY ROOTE JACKSON (1849)

Lanman concluded his account of Nacoochee with the poem "Mt. Yonah—Vale of Nacoochee," which had been written by Henry Roote Jackson (1820–98), a prominent lawyer in Savannah before the Civil War, who unsuccessfully attempted to prosecute the owners of the notorious slave ship *Wanderer* in 1859. Better known as a major general in the Confederate army, Jackson published a volume of sentimental poetry, titled *Tallulah and Other Poems*, in 1850. The volume includes a version of the poem used in Lanman's book, titled simply "The Vale of Nacoochee."[80]

SAMUEL JONES CASSELS (1851)

Born in Liberty County, Georgia, Samuel Jones Cassels graduated from the University of Georgia in 1828 and was ordained as a Presbyterian minister the following year. In the early 1830s, he served as minister to the Presbyterian church in Washington, Wilkes County, where he married Mary Eliza Winn and where their son was born. In 1836 they moved to Macon to take charge of the Presbyterian church there, only to lose their twenty-two-month old son in 1837.

In the spring of 1838, he published *Providence and Other Poems*. It is notable chiefly as the first work of a Georgia poet published inside the state. Six months later his wife died. After her death Cassels continued his ministerial work and in 1841 moved to Norfolk, Virginia, to take charge of the Presbyterian church in that city. In 1850 he published a second volume of poetry, which was a single epic poem, *America Discovered*. In 1851 Cassels published the third and final volume of his poetry, *Liberty Poems*, which included "Nacoochee: An Indian Legend." He prefaced that poem with the note that, while visiting the "beautiful valley" of Nacoochee in 1837, he heard the traditional tale from "one of the early settlers" and based his poem on that. Cassels died in 1853.[81]

7

African Americans at Nacoochee

One of the great histories written about Atlanta is Gary Pomerantz's *Where Peachtree Meets Sweet Auburn: A Saga of Race and Family*. In an epic story, Pomerantz illustrates the parallel strands of history, separated by race, in the stories of the Allens and Atlanta's last white mayor and of the Dobbses and the city's first African American mayor. Much of that Black/white dichotomy is evident across most of the South, including White County. Yes, people white and Black had many, many common experiences, but African Americans had many that would have been unthinkable to a white person.

African Americans have been living in and around Sautee Nacoochee for more than two hundred years. A few African Americans were enslaved by rich Cherokee planters and so became a part of their unhappy story. Many more were part of the slow-motion, white invasion of the Cherokee Nation in northeastern Georgia in the first quarter of the nineteenth century. Some of these African Americans worked in the fields, but slave labor was also critical in operating many of the small industries and mines established in and around Sautee Nacoochee before the Civil War. There were even a few "free people of color" in northeastern Georgia.

The Civil War brought freedom to enslaved African Americans, but the early promise of a so-called soft reconstruction was soon replaced by military rule, as it became clear that much of the white population would resist accepting African Americans as equals in any way. As they negotiated their freedom, a very few African Americans were able to acquire some land, but it was seldom the best. Most of them were forced into sharecropping or tenancy and, like many poor whites of the period, barely scratched out a living.

As Jim Crow made life miserable for most African Americans in the 1890s and early 1900s, the African American population in rural areas, including White County, began a precipitous decline, as those who were able sought better opportunities in the state's larger cities or in the big northern cities. By 1910 the African American population in White County had fallen to less than 8 percent of the total and to less than 6 percent in 1940. By the time schools were finally desegregated in 1964, there were fewer African Americans in White County than there had been in a hundred years. Today there are more people of Hispanic origin than there are African Americans in White County.

Free People of Color

As the Cherokee knew when they rejected the status for themselves, "free people of color" were anything but free. Defined as anyone with one or more great-grandparents who had African ancestry, free people of color were required by state law to register with the inferior court in their county of residence or risk being sold into slavery again. Many of the laws passed restricting the movement and rights of slaves applied to free people of color as well, including a prohibition against being taught reading or writing and engaging in independent real estate transactions. Additional taxes were levied as well, including a five-dollar annual poll tax, which was twenty times what a white person paid.[1] In spite of these restrictions, a few free people of color prospered, none more so than the celebrated James Boisclair (ca. 1804–50). He moved to Lumpkin County in 1829, opened a store, and went on to work what is still known as the Free Jim Mine, just east of Dahlonega.[2]

Free people of color were never numerous in Georgia; the federal census counted only 3,500 in the entire state in 1860. In northeastern Georgia 9 were counted in Habersham County in 1840 but only 2 in 1850. One of the latter was a fourteen-year-old girl named Mariah, whose last name was not given, living in the household of her guardian, John E. Craig, a carpenter in Clarkesville. The other free person of color in Habersham County in 1850 was Samuel Roman, age nineteen, in the household of farmer David Washburn. Neither one of these free people of color, both of whom were designated "mulatto" (i.e., mixed race), have been located in later censuses.[3]

By 1860 Habersham County registered forty-three free people of color, and White County, which was created out of the western part of Habersham County in 1857, counted eleven free people of color. In White County they were in two households, both headed by women, living next door to each other at Mossy Creek.[4] Mary Ann Oglesby was born in South Carolina around 1810 and was enumerated in the 1860 census with a twenty-seven-year-old woman, also named Mary and possibly a daughter, who was also born in South Carolina. By the late 1840s, they were in Georgia, and by 1860 the household included four children—Eliza, John, Martha, and an unnamed infant only six months old. The three youngest children are designated "Mu" for mulatto; the remainder of the family was designated "B" for Black. Both Mary Ann and Mary listed their occupations as day laborers, probably as farmworkers. They may have also worked as laundresses or cooks, two other occupations open to African American women in this period. They could even have worked in some capacity at the Mossy Creek ironworks.[5]

The household next door was headed by Sophiah Thomas, who was thirty-six years old, a farmer, and a Georgia native. Also in the household were two teenagers, John Thomas, age twenty, and Katharine Thomas, age sixteen. Both worked as day laborers, but, whereas Sophiah Thomas's race was indicated as mulatto, the two younger Thomases were both shown as Black. How that distinction was made is not clear. The federal census in 1870 enumerated John Thomas working as a blacksmith and living with his wife and two children in Cleveland, Georgia, but none of the other family members shown in 1860 have been documented after the Civil War.[6]

Slavery

The first African slaves in the British colonies arrived at Jamestown in 1619. Their importation continued for nearly two centuries, primarily at first to provide labor for the tobacco plantations that proliferated in Tidewater Virginia. By the time of the first federal census, in 1790, there were some 760,000 enslaved African Americans in the United States, representing nearly 20 percent of the population. Slavery existed in every county south of the Mason-Dixon Line and, regardless of the myth of a "pure" Anglo-Saxon population in the southern highlands, African Americans, enslaved and free, played an important role in the history of Appalachia.

The morality of slavery in America was always challenged. The Puritans in Massachusetts passed a law prohibiting slavery in 1621, but it applied only to Puritans themselves and placed no restrictions on slaveholding by "strangers" (i.e., non-Puritans) in the colony. Throughout the colonial period, German Protestants and Quakers, who had a significant presence on the southern frontier, were particularly outspoken in their opposition to slavery, and the Presbyterian schools established across the South generally espoused an antislavery ethic as well.

In spite of the fact that Pennsylvania and other northern states had already abolished slavery, the country failed to outlaw the practice when the Constitution was ratified in 1789. Instead, in the notorious "Three-Fifths Compromise," the constitutional requirement for a decennial census to determine congressional representation was framed such that three-fifths of the slave population would be included in determining the state's congressional representation, even though none of the slaves could vote. This gave the southern, slaveholding states a large structural advantage in the House of Representatives until the Civil War.

For a brief period in the 1820s, East Tennessee was a center of the abolitionist crusade in the United States, but by the 1830s the expression of abolitionist sentiments almost anywhere in the South was violently opposed. Nat Turner's Rebellion in Southampton County, Virginia, in August 1831, killed sixty white people and frightened the white population throughout the slaveholding districts. The residents of Burke County, North Carolina, some of whom were friends and relatives of the Nacoochee pioneers, were particularly concerned, since many of their own slaves were from Southampton County.[7] Some historians have argued that there was "greater social harmony" between the races in the southern highlands, with smaller numbers of African Americans perhaps making them less of a threat to the status quo. Carter Woodson (1875–1950), a prominent African American historian, thought that in Appalachia there "was more prejudice against the slaveholder than against the Negro."[8]

Frederick Law Olmsted (1822–1903) traveled extensively across the South between 1852 and 1857 and talked to all sorts of southerners, from rich planters to poor whites, and even a few slaves. Beginning with *A Journey in the Seaboard Slave States* in 1856, he provided a powerful indictment of slavery. He argued not only the obvious harm done to the enslaved human being but also the effect the "peculiar institution" had on slaveholders and common white laborers, whose exposure to slavery made them "even more indifferent than negroes to the interests of their employers." One of the up-country farmers told Olmsted that he saw how slavery "worked on the white people. It made the rich people who owned the niggers passionate, proud, and ugly, and it made the poor people mean."[9]

SLAVERY IN GEORGIA

The colony of Georgia was established in 1732, in part as a bulwark against the Spanish in Florida. In 1735, fearing that slaves would undermine from within and be too prone to go over to the Spanish at the earliest opportunity, Gen. James Oglethorpe and the other trustees of the colony succeeded in getting a royal decree banning slavery in Georgia. Slaves had a strong incentive to escape to Florida: although they might remain slaves, their legal status would be very different. Spanish law, derived from Roman law, recognized that slaves had certain rights, including the right to sue their owner, buy their freedom, and own property. English common law, in contrast, treated slaves (and English wives, for that matter) no differently than cattle or any other chattel property. In the English colonies, slaves had no legal or civil rights and were always subject to arbitrary punishment or even death.

In 1751 political pressure, no small part of it coming from Carolina planters, forced the king to rescind the ban on slavery in Georgia, and by 1790 slaves composed 35 percent of the state's population, working primarily on the rice and cotton plantations of the coastal counties. Development of the cotton gin and of short-staple cotton, which could be grown in a shorter growing season, precipitated a boom in cotton production across the Piedmont and a corresponding boom in the demand for slaves. By 1820 there were nearly 145,000 enslaved African Americans in Georgia, representing 42 percent of the state's population.

In 1798 Georgia prohibited importation, but not ownership, of slaves, and Congress passed an act to end importation nationwide in

Emblem created by Josiah Wedgewood and adopted in 1797 by the Society for Effecting the Abolition of the Slave Trade. Metropolitan Museum of Art, Wikipedia Commons.

1807, effective 1 January 1808, the earliest possible date. In spite of that, illegal foreign trade in slaves continued until the eve of the Civil War, with one of the last slave ships, the *Wanderer*, landing on Jekyll Island with more than 400 slaves in November 1858. As many as 48,000 Africans were forcibly brought to Georgia in the early 1800s, and many more were moved or sold from worn-out tobacco plantations in Virginia and Maryland. With natural increase, by 1860 there were 462,198 enslaved African Americans in Georgia, representing 44 percent of the total population, making Georgia second only to Virginia in the number of slaveholders and slaves. Even so, in 1860 less than one-third of Georgia's adult white male population of 132,317 owned slaves. Of that group, only 6,363 of them owned 20 or more; nearly half owned fewer than 6.

In rural Georgia the largest numbers of slaves were found on plantations along the coast, in southwest Georgia, and in a "Black Belt" across middle Georgia. In those regions, where the soil, terrain, and climate lent themselves to large-scale cotton or rice production, the African American population often outnumbered the white population three to one. Even in the Piedmont, more than two-thirds of the population of Troup, Greene, Wilkes, Lincoln, and Columbia Counties were listed as slaves in 1860.

The topography of northeastern Georgia, and much of the rest of the Southern Appalachians, limited agriculture to relatively small parcels of land that produced a variety of crops, mostly for local consumption. In that environment the economics of slave labor typically could not justify the expense of purchasing another human being just to work in the fields. As a result African American slaves made up progressively smaller proportions of the population as one journeyed up the Piedmont. From half of the population of Elbert and Clarke Counties, the proportion dropped to a third in Jackson and Madison Counties and a fifth in Franklin and Gwinnett. In Habersham and Hall Counties, an eighth of the population was enslaved in 1860, while in Lumpkin and White Counties less than a tenth were slaves. In the rugged terrain north and west of the Blue Ridge, less than 5 percent of the population of Rabun, Towns, and Union Counties were slaves.

Edward W. Phifer Jr.'s study of five counties in western North Carolina, one of which was Burke County, found that, unlike the slaveholding planters in the Piedmont and southern Georgia, who might depend entirely on agriculture for their livelihood, the largest slaveholders in the mountains were notable for the diversity of their income.[10] Frederick Law Olmsted noticed this when he traveled the Southern Appalachians in 1853: "Of the people who get their living entirely by agriculture, few own Negroes; the slaveholders being chiefly professional men, shopkeepers, and men in office who are also land owners, and give a divided attention to farming."[11]

As a result, slaves in northeastern Georgia were not necessarily or even usually, perhaps, engaged in agriculture. Slaveholders might lend or rent a slave's labor and often did when a neighbor was in need. Some contracted out their slaves to work in the gold mines in the 1830s or the railroads in the 1850s. As one advertisement stated, slaves could be had as "cooks (meat and pastry), washers and ironers, house servants, and seamstresses, blacksmiths, carpenters, field hands, shoemakers, plow boys and girls, body servants, waiters, drivers, and families."[12]

The Library of Congress's collection *Born in Slavery: Slave Narratives from the Federal Writers' Project, 1936 to 1938* includes 2,300 first-person audio and transcribed accounts of slavery and more than five hundred black-and-white photographs of former slaves, giving names, faces, and experiences to individuals, which the census and tax records do not. For most of the four million African American slaves in the South in 1860, there is no such personal documentation. One of the best sources is the federal census records, although even those provide only the barest of details. In the 1850 census, all adults and children, including slaves, were enumerated individually and named for the first time. Unlike the entries for white people, the entries for African American slaves almost never included anything but their sex and age.

Very occasionally, slaves were named in deeds of sale, such as when Moses Harshaw sold three people in 1833 and described them as "3 Negroes, Dice, Linda, and Katherine." The four slaves that went to his wife when they divorced were also named: "Melissa, a woman about 25, Ester, a girl about 12, Hannah, a girl about 9, and Bill, a boy about 7."[13] With few exceptions, however, many, if not most, of the slaves in northeastern Georgia remain anonymous figures. Few individual slaves can be traced to named individuals in the 1870 census.

Slavery in Habersham and White Counties

In 1820 and 1830, the federal censuses of Habersham County, which then included White County and northern Stephens County, counted around 8 percent of the population as African American slaves, which amounted to 274 people in 1820 and 909 in 1830. Ten years later the census counted 942 slaves in Habersham County, or nearly 12 percent of the population, one indicator of the economic boom that came with the gold rush. By 1850 there were 1,218 slaves, which was nearly 14 percent of the total population, although there were many more in the eastern part of the county, where there was much more arable land.[14] In 1860 the census of White County, which was created out of western Habersham County in 1857, counted 263 slaves, which was just under 8 percent of the total population. Most of them lived around Cleveland, Mossy and Shoal Creeks, and the valleys of northeastern White County, where the land was most fertile and the largest slaveholders were located. There were forty-seven white slaveholders, a number that included four women. Nearly a third of the slaveholders owned only a single slave, who may have worked as the family cook or simply as an extra hand for the farm or business. Two of the three doctors in Cleveland, both married with children, each had a teenaged African American boy who no doubt worked closely with the rest of the household.

Thomas Barton and his wife, Dicey, were an elderly white couple living alone near Mossy Creek. They had a single enslaved African American, a thirteen-year-old girl, whose presence no doubt allowed them to continue to live in their home. Whether the child had relatives nearby is unknown, although several other slaveholding households were in the area. William B. West was a clerk of the superior court in White County and lived with his wife, Mariah, and their six children in Cleveland. They had a single slave, an eighteen-year-old African American man, who most likely did much of the work on their small farm. He was not isolated, however, since West's elderly parents, James and Eunice, lived next door and claimed ownership of eleven slaves, most of them children, who lived in two slave houses on the elder Wests' farm.

Sometimes, the stark data of the census is difficult to interpret. Martin Trammell, a forty-two-year-old farmer, lived with his wife, Mary Louise Vandiver, on 250 acres, where Unicoi State Park is now located. They apparently had no children but claimed as a slave a three-year-old African American girl. She has not been named or located in later censuses. Similarly, their neighbor George W. Carodine and his wife, Darcus, both about forty years old and also apparently childless, did not own taxable property in White County in 1860, but he made their living as a "grocery keeper." The 1860 slave schedule includes a four-year-old African American boy in their household. He also has not been named or located in later censuses.

It is clear that many of the slaves were part of family groups. Half of the county's slaveholders counted more than one but fewer than ten slaves, many with small children, and typically housed in a single slave dwelling. Unfortunately, the census takers did not always or even usually make family groups obvious in their enumeration of slaves. Jesse Holcombe, who had been among the first pioneers in the resettlement of what became White County, and his wife were elderly and had a single son, who was a schoolteacher, so most of the work on the farm would have fallen to their slaves. He claimed ownership of three adults and eight children, five of them under the age of ten, and all occupying a single house on their farm.

Jehu and Elizabeth Trammell, who were also among the "first" families of Nacoochee, owned nearly eight hundred acres, bisected by Smith Creek and encompassing most of what is now Robertstown. Both were elderly in 1860, and much of the farmwork would have depended on their four adult African American slaves, who lived in a single house along with four children under the age of ten. In 1860 Henry Conley, who lived in what became known as Helen Valley, had twelve slaves, eight of whom were children. There were two dwellings for his slaves. His neighbor Martha England had only one slave dwelling for her seven slaves, who were apparently a couple and their five children.

Frederick Dover was a thirty-two-year-old farmer, millwright, and house carpenter who lived with his wife, Sarah, on Land Lot 103, encompassing the present intersection of SR 75 and Duncan Bridge Road. She was ten or twelve years older than he, and they apparently had no children. They owned nine slaves in 1860, all adults except one twelve-year-old boy, living in three slave dwellings. It is possible that at least one of the enslaved men worked with Dover and learned carpentry or other skills.

Only nine slaveholders in White County in 1860 owned ten or more slaves. In addition to James West, mentioned earlier, Christopher Columbus Meaders (1808–87) owned ten slaves who lived in two houses on his property near Mossy Creek. He had moved there from Banks County in 1848 and become a prosperous merchant. One of his sons, John Milton Meaders (1850–1942), would found the renowned Meaders Pottery in the early 1890s. Only three of his ten slaves were adults; half were ten or younger. John T. Carter owned several hundred acres along Shoal Creek. He claimed eleven slaves, only four of whom were male, all living in two slave dwellings on his farm.

Slavery at Nacoochee

Some seventy-three enslaved African Americans, well over a quarter of the slaves in the entire county, were enumerated at Nacoochee in 1860. Over half of those were owned by three children of Major Williams: Charles, Edwin, and Hannah. Charles L. Williams was the largest slaveholder in White County in 1860, claiming ownership of twenty-one African Americans. They included a fifty-year-old woman, who might have been his cook, and five men and four women between the ages of twenty-one and forty. Four teenagers and seven younger children completed the list of Williams's slaves, who lived in five houses. Unfortunately, the census taker made no attempt to record what clearly must have been several family groups. Charles's brother Edwin Poore Williams claimed eighteen slaves. The oldest was a sixty-year-old man, and there were six other African American men between the ages of twenty and forty, along with three teenaged boys, who were probably worked as men. There were only three women and four children, all living in three slave dwellings in 1860.

Dr. Elijah Starr and his wife, Hannah, the major's daughter, were living in the old Williams family home, christened Starlight, on Unicoi Road with their four children. Two slave dwellings were declared, one of them occupied by what appears to have been two enslaved African Americans in their early twenties and their three children. The other was the residence of a sixty-year-old African American woman, who was most likely the Starrs' cook. According to the 1860 census population schedules, living in the same house with the Starrs that year was

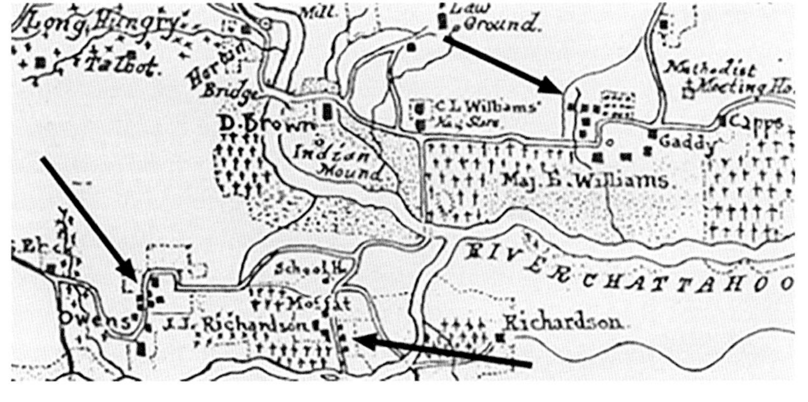

Detail from Moffat's map, annotated with arrows to indicate the clusters of buildings that could have been slave quarters for the Richardson and Williams slaves. A similar cluster was probably present on Moses Harshaw's property along Bean Creek, which was not part of Moffat's map. Courtesy of Sautee Nacoochee Community Association.

a seventy-five-year-old white man, Ephraim McClain, whose occupation was listed as "old." He claimed ownership of two thousand acres and a family of slaves that included two adults and six children. He also listed a single slave dwelling, which was presumably in addition to the two that Starr claimed. Elijah Starr was McClain's guardian, but if there was a familial relationship, that has not been documented.

Across the river was the Williams's neighbor John L. Richardson, who came from Burke County with his father, Jesse. They probably brought slaves with them from North Carolina. The younger Richardson owned twenty slaves, housed in five dwellings: five African American men and five women over the age of twenty, along with two teenaged boys and eight children under ten. Here again family groups were not recorded.

Moses Harshaw's house on Rabun Road remains a landmark as the Harshaw-Stovall House listed in the National Register. Harshaw was also a large slaveholder prior to his death in 1859. Many of his family members, including his brothers Jacob and Abraham in Burke County, North Carolina, and his nephew Sidney, over the Blue Ridge in Union County, were slaveholders. A tally in the 1860 federal census shows that as many as 225 African Americans in Burke and Cherokee Counties in western North Carolina and in Union, White, and Habersham Counties in northeastern Georgia were enslaved by members of the Harshaw family.[15]

Moses Harshaw owned several hundred acres on the lower reaches of Bean and Chickamauga Creeks, along with land in the Sixth District in extreme northern White County. He claimed twenty slaves in the 1850 census slave schedules, at least some of them at his residence in Clarkesville. In the census that year, alone among White County slaveholders, he reported two slaves as being "fugitive from the state," which is perhaps another sign of his reputation as an egregiously mean person, especially to his slaves. Like Richardson and a number of others, he is thought to have used slave labor for the gold mines on his land lots.

At Harshaw's death in 1859, he left only eleven slaves. It appears that his ex-wife, Nancy Harshaw, who kept the surname after their divorce in 1857, and daughter Adeline had seven of them, including a young African American man, two African American women, and four small children, all at their home in Clarkesville. In addition, Alonzo Harshaw, who was thirty years old and unmarried, was living in the family home in Sautee Valley and had four slaves of his own (two African American men and a seventeen-year-old woman with a child six months old, both mixed race). They were living in two slave houses on the property. As suggested by some of the situations and statistics cited here, slavery was not a monolithic institution but rather varied from place to place and in as many ways as there are human beings, white and African American. The experience of the Williams family's house "servants," the common euphemism in the South for slaves working in the household, for example, was likely far different from that of the African Americans unfortunate enough to be slaves of Moses Harshaw.

While there is no mitigating the wrong of slavery, there is some evidence that where their numbers were fewer, and so less of a perceived threat, slaves were often managed less stringently. At Nacoochee they invariably lived in proximity to white families, and many African American slaves attended the white Methodist and Baptist churches and were even buried in the church cemeteries, albeit in separate sections. Overall the closer personal contact and intertwining of lives may have created a marginally less toxic environment than existed on one of the huge cotton plantations in South Georgia, even if the fundamental racism of nearly every white Southerner toward African Americans, slave or free, never varied.[16]

SLAVE HOUSING

In 1860 the federal census slave schedules recorded dozens of slave dwellings in White County. In Nacoochee Valley the Williams brothers and their sister had a total of eleven houses for their slaves, while the Richardsons had at least six along Dukes Creek. The Harshaws had four or five at the northern end of Sautee Valley, and there were others as well. The precise location of these dwellings remains uncertain, but Moffat's map depicts a number of small unnamed dwellings in proximity to one another, similar to the clustering typical of most slave housing.

The double-pen log house moved and restored by the Sautee Nacoochee Community Association in 2005 may be one of the three slave dwellings reported by Edwin Poore Williams on the 1860 census. Thought to have been built around 1850, although it could have been built ten years earlier, the house originally stood not far from the main house and was probably used as a residence for the family's household servants. How typical the cabin was of the Williamses' other slave dwellings is uncertain, but there is no reason to think it was necessarily

The log house moved from its original location on Unicoi Road, where it is thought to have been built by Edwin P. Williams around 1850. Photograph by author, 2021.

atypical. Many wealthy planters wanted household servants close at hand, but they also took care in the appearance of their surroundings. In addition, as one historian has pointed out, by the 1850s slaveholders were providing better-built slave quarters, either through renovation of older quarters or new construction, which reflected "a new attitude among masters that combined Christian duty, paternalism, scientific agricultural reform, and a sharp business approach to slave management."[17]

A typical improvement might be the replacement of simple wooden window shutters with glazed sash, which not only improved the health of the inhabitants but also made it possible to do more work indoors in inclement weather. The two front doors and the dimensions of the building, sixteen by twenty-eight feet, suggest that it might have been built for two enslaved families, but, with a single fireplace, it is possible that it was built as a kitchen and living quarters for the cook.

The earliest slave dwellings were probably log, with packed earthen or puncheon floors, not unlike houses occupied by white yeoman farmers in the early days of white resettlement of the valleys. By the 1840s, as sawmills became more common and lumber more affordable, braced-frame buildings with wide-board flooring became more common. The better of these houses were finished with lapped siding, while others were finished with cheaper, board-and-batten siding. Often the interior walls were left unfinished, and the only ceiling might have been a partially floored loft accessed by a ladder.

Civil War

The rancorous debate over slavery, which was often acted out politically in arguments over states' rights in a federal relationship, culminated in threats of secession and a four-way race for the presidency in 1860. The election of Abraham Lincoln on 6 November sent many white southerners into a rage, and on 20 December South Carolina proclaimed its secession from the Union. Also that month the Georgia legislature ordered a special election in which each county would choose two delegates to a convention to decide the issue of secession. There was a strong pro-Union sentiment across North Georgia, and when the election was held on 2 January 1861, the majority of delegates elected from the counties in northern Georgia were pro-Union, or "co-operationist." The vote in White County was 320 to 48 against secession. By the time the delegates convened in Milledgeville two weeks later, three more states had left the Union.[18]

New Jersey–born Horatio Hennion, who bought the Mossy Creek Iron Works in 1854, was an outspoken Unionist who later remembered, "When the results of the election declared Lincoln elected, the secesh were wild with joy, but there were very few non-union men in Hall and White Co. A great many union men were elected by their constituents but were bribed or talked over, and voted for and signed the secesh ordinance."[19] The final vote, on 19 January 1861, was 208 to 89 in favor of immediate secession, and $1 million was appropriated for the governor to establish and arm a ten-thousand-man state militia. On 4 February 1861, representatives of the states that had seceded—South Carolina, Georgia, Florida, Alabama, Mississippi, Louisiana, and Texas—met in Montgomery, Alabama, and organized the Confederate States of America. North Carolina, Tennessee, Arkansas, and Virginia would follow.

The Confederate firing on Fort Sumter on 12 April 1861 began hostilities. Throughout the spring and summer, men were volunteering and joining new militia units, with both sides thinking the war would be over in a matter of weeks. On the Fourth of July 1861, President Lincoln addressed Congress, requesting and getting a call for five hundred thousand men. He called the war a "People's contest" and one with a lofty purpose: "a struggle for maintaining in the world, that form, and substance of government, whose leading object is, to elevate the condition of men."[20]

Edward Williams's youngest son, George Walton Williams, knew the awful loss that war would surely bring. As he wrote from Charleston in September 1861, "I think there is a fair prospect for a long and bloody war. I have no fear but that the South will ultimately succeed, but we shall suffer dreadfully. I have made up my mind never to live under Abe Lincoln." When he visited Nacoochee around the same time, he found that someone had managed to climb the great pine tree growing on what was probably the natural mound across the valley from Starlight and that "the Secession flag was thrown to the breeze before the State of Georgia seceded. Long may it wave o'er Nacoochee and the home of the brave."[21]

In late June and July 1861, the Twenty-Fourth Georgia Volunteer Infantry Regiment was organized, with ten companies from northeastern Georgia. One of those was Company C, the White County Marksmen, and in its roster was Dr. Elisha F. Starr, who enlisted on 20 July and was appointed surgeon on 1 November 1861. In the same company were Moses Harshaw's oldest son, Hemphill, and four of Major Williams's grandsons—Charles L. Williams's son James, Edwin Poore Williams's sons Alfred and Church, and James Edwin Williams, son of Albert Gallatin Williams, who had been buried at sea in 1852. All of them enlisted on the same day. The regiment was officially mustered into Confederate service on 24 August 1861, a month after the celebrated Confederate victory at Manassas, Virginia, in the First Battle of Bull Run. The regiment was eventually transferred to the Army of Northern Virginia, where they fought the rough campaign from the Seven Days Battles to Gettysburg before joining Longstreet in 1864.[22]

Another regiment with several Nacoochee volunteers was the Fifty-Second Georgia Infantry, organized at Atlanta in April 1862. Many of its members were from the counties of northeastern Georgia, including White County. Its regimental roster included two sons of John L. Richardson, one son of Robert Westmoreland, and one son of John Stovall from Blue Creek.[23] In White County and elsewhere, a low-grade guerrilla war was fought between Union and Confederate sympathizers. Union men were ambushed, beaten, and killed and sometimes saw their houses burned down by Confederate sympathizers. A few moved away rather than face such peril; others simply learned to hide their true sympathies. The mountains of North Georgia also became a haven for lawless gangs of "Tories" and "bush-whackers," who fought their own guerrilla war against the Confederacy. As one historian wrote, White County was a "dangerous place" during the war years.[24] In April 1862 nearly 110,000 soldiers from both armies fought for two days at Shiloh and Corinth, suffering casualties of nearly 24,000, which was more than the casualties from all previous wars combined. In August almost as many fell in a single day of fighting at Antietam.

EARLY CASUALTIES

At least 620,000 people would ultimately die in the war, more than the combined total of all of the United States' wars before Vietnam. As the military made great advances in its ability to kill, they sometimes neglected the sanitary conditions in their own camps. As a result, for every two deaths on the battlefield, three more died of diseases, including typhoid fever, smallpox, measles, diarrhea, pneumonia, malaria, and tuberculosis.

The Confederate soldiers at Nacoochee suffered through a litany of death early in the war, beginning with the death of Edwin Williams's son Church from disease in a Confederate camp at Goldsboro, North Carolina, in May 1862. That was followed by the death of his cousin James Edwin Williams, killed by typhoid fever in a camp at Richmond, Virginia, in late June. Williams's parents, Charles and Hannah, may not have even gotten that news when their daughter Georgia died on 30 July 1862. Robert Westmoreland was discharged after a severe case of measles, only to die at home in White County in November 1862. In late December John and Elizabeth Stovall found that their son James had died in an Atlanta hospital. Alonzo Harshaw did not die, but he was discharged for medical reasons in 1863.[25]

THE BLOCKADE

The Union blockade of Southern ports, which began in April 1861, effectively strangled international trade with the Confederacy. The Union eventually commissioned five hundred ships, which destroyed or captured three times that many blockade runners over the course of the war. Some runners got through but could carry only a small fraction of what the South needed. By 1862 inflation was rampant in the South, as supplies of formerly imported staples, such as salt and flour, virtually disappeared. People began to go hungry, especially the poorest, who often lived hand to mouth anyway.

One of the most widely felt shortages was that of salt, which everyone needed as a preservative, if nothing else. In early November 1862 Georgia's governor Joseph E. Brown, "knowing how indispensable [salt] is to health and comfort," announced that the state had contracted for the procurement of salt from the Virginia Salt Works in what is now West Virginia. County inferior courts managed distribution, beginning with a half bushel free to every war widow, of whom there were already twenty-two in White County by the end of 1862. The rest was to be sold at cost, and virtually every family, including the Richardsons, Williams, Starrs, and Harshaws, appear on the salt lists in 1862 and 1863. In spite of that, there was never enough salt to meet demand; some people were reduced to trying to retrieve salt by boiling the dirt from their smokehouse floors.[26] George Walton Williams was a leader in dealing with the food crisis among the poor in Charleston. By the end of the war, he and the Subsistence Committee that he organized were responsible for the distribution of $2 million in supplies to the city's poor.[27]

Port Royal fell to the Union gunboats in November 1861, and Jekyll and Saint Simons Islands off the Georgia coast were occupied in March 1862. With the fall of Fort Pulaski the following month, the Union foothold on the coast was unassailable. Panicked planters began moving what they could inland, including their slaves. If they could not be moved, many were sold in what remained a strong market for much of the war. Probably in 1862 George Williams bought around a hundred enslaved African Americans, including a dozen or more families, "partly from humanity, and partly from gain," he later wrote. There is no record of what Williams paid for the slaves, but it could not have been a small amount. His goal was to move them to Nacoochee so that they could "make bread and meat for themselves," while at the same time providing supplies for soldiers' families in Charleston.[28]

How he housed these people is not known, but construction of additional slave dwellings was likely necessary. During the war George Williams and his brother Edwin were taxed on hundreds of acres of land in dozens of separate parcels in the Nacoochee District alone, so the slaves could have lived almost anywhere. Among those parcels of land were three encompassing over 120 acres in Land Lots 23, 43, and 44, which are at the heart of the Bean Creek community. It is certainly possible that Williams's slave housing was part of the genesis of the African American community at Bean Creek.[29] At war's end and emancipation, according to his biographer, Williams was disappointed that all of them returned to the Low Country, although he claimed not to have regretted what he had done.[30]

WAR INDUSTRY

During the Civil War, there were still farms to operate, especially since the Confederacy needed desperately to maintain its agricultural production. On Sautee Creek at Unicoi Road, Edwin Williams had a tanyard and operated a gristmill, a sawmill, and a blacksmith shop, all probably at least partially dependent on slave labor. Williams was also one of several contractors producing pikes and knives for the state. Sometimes called "Joe Brown" pikes, after Georgia's wartime governor, Joseph E. Brown, they were part of similar efforts elsewhere in the South to address a severe shortage of matériel of all kinds. Several hundred of the pikes were shipped to Charleston, but none were ever used for their intended purpose before they were confiscated by the Union army. George Williams remembered that they "sent the pikes North as a barbarous relic of the Lost Cause."[31]

Both of John L. Richardson's sons were captured at the Battle of Champion Hill in May 1863, the pivotal battle leading up to the siege of Vicksburg. Jesse, named for his grandfather Rev. Jesse Richardson, died in a Union prisoner-of-war camp in May 1864. The slaughter at Gettysburg on 1–2 July 1863, followed by the fall of Vicksburg two days later, left the South in shock. Worse was to come as the two armies fought each other to a standstill at Chickamauga in September. Northeastern Georgia was not directly affected by the fighting, but Gen. William Sherman's cartographers did go to the trouble of mapping the area as far as Clarkesville and Blairsville. Then, in May 1864, Sherman led his army into Georgia and, on 15 November, left Atlanta in flames. After his ruinous March to the Sea, he was able to present Savannah as a Christmas present for President Lincoln on 21 December 1864.

The war was every bit the disaster that George Williams feared it would be, as he wrote in 1869: "The lofty pine which withstood the pelting storms of centuries, the tree that marks the grave of Nacoochee and bore aloft the Confederate flag during the late bloody war, died with our Lost Cause. It is now a blighted tree—fit emblem of the tempest-tossed Confederacy."[32] In early April 1865 Richmond fell, and six days after that Robert E. Lee finally surrendered to Ulysses S. Grant at Appomattox courthouse. Another week and President Lincoln was dead from an assassin's bullet. Sporadic fighting dragged on until the last Confederate forces surrendered in Texas in June. The CSS *Shenandoah* continued its raiding until it was surrendered in Liverpool in November. Not until 20 August 1866 did President Andrew Johnson issue a proclamation that formally ended the war.

Emancipation

Lincoln had issued his Emancipation Proclamation on New Year's Day 1863, broadening the goals of the war to include abolition of slavery. Although the proclamation was immediately applicable only to states then in rebellion, it provided a legal framework for the military to free the thousands of slaves who were captured or escaped through Union lines. Declared as contraband in the early war years, these African Americans were freed and, not incidentally, allowed to join the Union army. As Union control in the South expanded in 1863, more and more slaves found that the shortest path to freedom was the closest Union army, and slavery began, in the words of one historian, "to unravel from within." With the Union occupation of East Tennessee in late 1863, freedom for the slaves at Nacoochee was sixty miles away, although it is unclear if any took advantage of that opportunity.[33]

The growing number of freed slaves who followed Sherman's army through Georgia became such a logistical burden that he was forced to destroy the bridges at the Ogeechee River behind him to keep the freed slaves from following. Many other Union commanders had a similar problem. For obvious reasons African American slaves had a single goal in mind, which was freedom, and what would come after that was seldom thought through. For too many the euphoria of freedom was soon replaced by a simple, desperate search for food and shelter. A myriad of private efforts tried to meet the need, but not until 3 March 1865 did Congress pass an act establishing the Bureau of Refugees, Freedmen, and Abandoned Lands, more commonly known as the Freedmen's Bureau.

In the first year or two after the war, African Americans were experiencing the exhilaration of freedom, while some white Southerners were coming to grips with the fact that $2 billion in slave assets had evaporated. Many of both races were just trying to cope with the almost total collapse of the South's agricultural economy. There is no record of when the enslaved African Americans of White County learned of their emancipation. Henry McNeal Turner, one of the few African American Freedmen's Bureau agents, was shocked to find African Americans who felt that they could not claim their freedom because "Mas ain't tol' me so yit."[34]

The Williams family has a long tradition that, when the war ended, Edwin Williams read the Emancipation Proclamation to his freed slaves from the "Emancipation wall," a stone wall near his house. That may be so, but emancipation may have been old news to them, since house servants and drivers quickly relayed information via an astonishing grapevine of communication among the slaves. In a place like Nacoochee, with well-connected slaveholding families, African American slaves had a rich source of information, never mind the enslaved people from the Low Country who appear to have lived at Nacoochee for two or three years during the war. In any case formal acknowledgment of the fact of freedom left much else to be negotiated.

Reconstruction

With the collapse of the Confederacy in April 1865, state government collapsed too. On 17 June 1867, President Johnson appointed James Johnson of Columbus as provisional governor of Georgia, and military occupation of the state began. The total number of troops in the entire state was usually around one thousand, but it was enough to preserve a semblance of order as freed African Americans worked out the terms of their freedom with their none-too-happy white neighbors.

Lincoln's lesser-known Proclamation of Amnesty and Reconstruction, issued in December 1863, provided for near-complete amnesty for Southerners, except for military leaders, Confederate politicians, and a few others. For most all that was required was an oath of allegiance to the United States. Even that would prove difficult for some, including Edmund Ruffin, who had fired the first shot at Fort Sumter. Declaring his "unmitigated hatred to Yankee rule—to all political, social and business connections with Yankees, & to the perfidious, malignant, & vile Yankee race," he committed suicide on his front porch two weeks after the last Confederate troops surrendered. Closer to home Robert Toombs refused to sign a loyalty oath and never regained his citizenship. Nevertheless, he was able to dominate Georgia's 1877 state constitutional convention, which began to undo what little Reconstruction had been able to accomplish.

On 29 May 1865, President Andrew Johnson issued proclamations that pardoned all participants in the late rebellion except Confederate officers, people with more than $20,000 in property, and a few others. However, those excluded could apply individually for a pardon, and, by the spring of 1866, the president had issued seven thousand individual pardons. One of those was Charles L. Williams, who had probably been excluded because of his wealth but who was pardoned on 25 October 1865.

The presidential proclamation also authorized the formation of new state governments, with new constitutions. Delegates for the constitutional convention were elected on 24 October 1865, but only fifty thousand Georgians were eligible to vote, all of them white, amounting to less than half of what the electorate had been before the war. In the end the convention created a constitution that portended worse to come. As historian E. Merton Coulter commented on the new constitution, "The ordinance of secession was repealed, not declared null and void as required, slavery was abolished in a reluctant article of little grace, and the wartime debt was repudiated by a narrow margin only after agonizing debate. All in all, the resulting constitution was not greatly different from the old one."[35]

A rising tide of violence against African Americans was being reported across the South in 1866, symbolized by the rise of the Ku Klux Klan and similar groups. The general intransigence of the defeated white population and the Democratic party is perhaps best exemplified by the Georgia legislature's appointment of former Confederate vice president Alexander Stephens to the U.S. Senate in 1866, although he was never seated. Rejection of the Fourteenth Amendment by the Georgia legislature and the rest of the South that year confirmed the problem. In spite of President Johnson's campaign to preserve his program for a lenient approach to reconstruction, the fall elections delivered a crushing blow to his supporters, culminating in the passage, over the President's veto, of the Reconstruction Act of 2 March 1867, quickly followed by a second act. "Radical Reconstruction" had begun. The act divided the old Confederacy into five military districts, with Georgia falling into the Third District, along with Florida and

Alabama. All the constitutions passed by Southern states since the war were nullified, and new constitutional conventions based on universal manhood suffrage were called.

In 1867 and 1868, Gen. John Pope, commander of the Third District, headquartered in Atlanta, supervised voter registration for the election of representatives to a constitutional convention. He divided Georgia into forty-four districts, each with three registrars, two white and one African American. The Radical Republicans in Congress pushed for withholding the voting franchise from all white Southerners for five years, but in the end only a few thousand in Georgia were actually excluded from the voter rolls, and some of those perjured themselves and registered anyway.

Across the state 102,411 whites registered to vote, along with 98,507 African Americans. In White County 522 white males and 80 African American males registered to vote, which was approximately the same proportion as found in the general population in 1870. An additional 24 white males registered by March 1868. The only member of the Williams family to register was Alfred P. Williams, eldest son of Edwin Poore Williams. Of the Richardsons only John L. Richardson's brother Thomas and Thomas's son Joseph registered to vote. Among the first African Americans registered to vote in White County were Baltimore Alston, Calvin Nicely, Alphonzo Williams, Shadrack Richardson, the brothers George and John Trammell, and the Welches: Charles, Anthony, Henry, and David.[36]

Many whites boycotted the entire process, and in the end only 39,000 white Georgians voted, compared to 110,000 before the war. By some estimates as many as 20,000 African Americans were intimidated into not voting. Delegates were chosen that included 169 white delegates and 33 African American delegates, a lopsided proportion that did not reflect the state's population.[37] The convention went forward in spite of Governor Jenkins's efforts to halt it, which included refusing to use state funds to hold the convention and absconding with $400,000 from the treasury. He also took the state seal, which was not returned until the state was "redeemed," as conservative Democrats styled the transition from Republican rule in 1872.

The new constitution provided for universal manhood suffrage, prohibited slavery and whippings for punishment, abolished imprisonment for debt, established a poll tax, and even granted women the right to own property. It was approved in April 1868, at the same time that a coalition of Black and white Republicans elected the most democratic legislature that the state had ever had. Among those elected were three African American state senators and thirty-nine African American state representatives.

The new legislature convened on 4 July 1868, and among its first acts was to pass the Fourteenth Amendment to the U.S. Constitution, which guaranteed citizenship to anyone "born or naturalized" in the United States as well as equal protection under the law. Its passage allowed Georgia's reentry into the Union, but that progress was undone by the state's continued recalcitrance in coming to terms with free African Americans. In September the legislature expelled all of its African American members, except four whose African heritage was called into question because of their light skin, on the grounds that the Fourteenth Amendment did not necessarily allow African Americans to be elected to public office. As a result, Georgia's U.S. representatives in Washington were seated in the late fall of 1868, but in March 1869 were expelled. Georgia's U.S. senators were never seated, and, by the end of the year, the state was again under military rule. A purge of ex-Confederates in the state legislature allowed passage of the Fifteenth Amendment in February 1870, and in July Georgia was readmitted to the Union for a second time.

The Transition to Freedom

The war freed four million slaves but left them with little more than the clothes on their backs and a wall of resentment from most white Southerners. For their part the white South was being forced to adjust to the facts of a free-labor market, and many were convinced that the agricultural economy would collapse without slavery. In the summer of 1865, rumors were rampant in the African American community that the government would be giving each household "forty acres and a mule" at Christmas that year.[38] White southerners knew this was patent nonsense, but they also believed that there would be a terrible slave uprising when the nation's largesse was not forthcoming.

Such rumors and the desire to establish their autonomy prompted many of the freedmen to delay coming to terms with white farmers, who needed African American labor just as desperately as the African Americans needed to make a living. By January 1866 most freed slaves had signed some sort of agreement with either their former owner or with someone else to provide labor on the farms that supported the southern economy. Housing was an immediate point of leverage, since none of the former slaves owned land, and all needed a place to live. Housing, therefore, was a critical part of sharecropping and tenancy arrangements.

For millions of freed slaves and landless white farmers, there were two options. If the laborer had farm tools and a draft animal or two, tenancy was the best arrangement. In exchange for an annual rent, the tenant had complete control over what he produced and how he did it. For those with nothing, sharecropping was the only alternative, a system that left the sharecropper with very little autonomy. Very few, Black or white, escaped the vicious cycle of debt engendered by these systems, especially sharecropping.

Freedmen's Bureau

The Bureau of Refugees, Freedmen, and Abandoned Lands was established in March 1865, with the intent of dealing with a massive humanitarian crisis across the South. Commonly known as the Freedmen's Bureau, the agency was tasked with issuing rations and providing medical relief to Black and white refugees, overseeing new labor contracts between planters and freed slaves, administering justice, and working with benevolent societies to establish schools for freed slaves and, incidentally, for poor whites.[39]

In 1866 Congress passed a bill renewing the charter of the Freedmen's Bureau, but President Johnson vetoed it. He thought the bureau's work would hamper the freedmen's transition to independence by giving too much assistance. In what amounted to a referendum on how to approach reconciliation, the Radical Republicans won the 1866 midterm elections in a landslide, and nearly all of Johnson's vetoes were undone.

In July 1865 the Freedmen's Bureau commissioner appointed Gilbert L. Eberhart (1830–1907) as superintendent of schools for Georgia, charged with the dozens of schools that were being organized across the state. The monthly reports generated by district agents in the state chronicle the massive problems facing the bureau. The first efforts of the bureau were focused on assisting the huge numbers of African Americans in coastal Georgia, Atlanta, and Augusta. In areas where there were few African Americans, such as northeastern Georgia, the impact of the bureau was more limited. The agent for the Athens district wrote that "there are not free people enough in the counties of Rabun and Habersham to justify establishment of schools at present. What few there are, live many miles apart."[40]

He was mistaken; while the African American population in Rabun County declined to 117 in 1870 and was no doubt widely scattered, the Black population of Habersham County had actually increased to nearly 1,000. Not until the spring of 1867 did General Pope, then head of the Third Military District, name Dahlonega one of six bureau bases in the state, bringing much-needed resources into the area. That summer and fall there was even an agent in Clarkesville, which made it easier for the freedmen to report abuses, of which there were more than a few.[41]

For white southerners, the Freedmen's Bureau quickly became one of the most hated symbols of "Yankee rule." Many poor whites who might have benefited from its services refused the assistance, simply because the same services were being provided to African Americans. The monthly reports from district bureau agents include assessments of "public sentiment toward Colored schools," and the best that could be said was that it was "improving." Otherwise public sentiment was "indifferent" when it was not "antagonistic." One agent noted that the freedmen were "despised by the white people." In August 1867 an African American school ten miles north of Carnesville was burned to the ground by arsonists, and the bureau agent, William Payne, noted in his report that "they intended to kill the teacher." Not surprisingly, the culprits were never identified.[42]

The agent in Athens was especially distraught at white attitudes when he reported in December 1867, "They have no sympathy whatever from the white people, & no encouragement. They have never received a cent from any white person toward the advancement of education of colored children & the supposition is they never will be assisted by the whites in this part of the country." Two months later he added, "If this class of whites are above sending their children to colored school because of the colored's skin, the Bureau can do nothing more than they have done for a few people in this respect."[43]

A bequest from a northern philanthropist supported the establishment of a Freedmen's Bureau school at Dahlonega in 1865, with classes taught by W. J. Wooten. The monthly reports to the bureau, which continued until 19 March 1869, chronicle a school building with no windows, students with no clothes for cold weather, and the demands of work making class attendance difficult. The bureau established night schools in some places, but when the possibility of a night school in Dahlonega was posed, the agent replied, "No as surely all of the adults make work in the mines, they will not attend school after working in the water all day if they must pay for same."[44]

White, Lumpkin, and Dawson Counties were part of the Rome subdistrict, and the first subdistrict report was submitted in March 1868. At that time there were two schools in the three counties. "The one in Dahlonega is bad condition, is 25 by 30 feet, is worth about $150 & owned by the Freedmen, being a donation from a citizen for a colored school house & church. The other is a log house built by the Freedmen in Cleveland, White Co."[45]

The school in Cleveland apparently failed, and, in the May and June 1868 reports, the bureau agent named that town as a possible site for a Freedmen's school and estimated a "possible twenty-five students." He also recommended a school in Dawsonville, where there might be forty students, and at "Nauchoocha," which the agent never learned to spell properly, but where he estimated there might be twenty to thirty students. There is no other mention of a school at Nacoochee in the bureau's "Records of Assistant Commissioners and Superintendents of Education" for Georgia. If a school had been operating at the Bean Creek church, it is likely the agent would have mentioned it, since the routine reports included reports of private schools if they were present.[46]

In 1868 the Freedmen's Bureau agent in Athens wrote that he thought the hostile conditions would persist "until the mock aristocracy of Georgia is abolished." In fact, white conservatives were already beginning to consolidate their power and attempting to regain some of what they had lost in the Civil War. Indeed, one prominent scholar has asserted that, despite the "fraudulent" outcry about disfranchisement of southern whites, "From the beginning of Reconstruction under the new Constitution, Georgia was controlled by conservative whites."[47] The bureau continued to operate but almost never had adequate funding. From the outset there was white hostility to its mission, even in the North, and by 1869 the bureau had been forced to cut much of its staff. The rise of the Ku Klux Klan and the intransigence of white southerners severely hampered it as well, and in 1872 Congress abruptly refused to reauthorize the act that had established the Freedmen's Bureau seven years earlier.

The terrorism perpetuated by the KKK and other white vigilante groups in the South severely weakened the political power of African Americans and the Republican Party. In December 1870 the Democrats regained control of the Georgia legislature, and the Republican governor, Rufus Bulloch, fled the state in November 1871, just as the last federal troops were withdrawn. In January 1872 Democrat James Smith was inaugurated governor, the last step in white "redemption" of the state from the hated Republicans. With the collapse of northern political support of African Americans in the South, the cause of Reconstruction was doomed. Much was accomplished, but nothing could overcome the white South's opposition to anything that might improve the condition of the freed slaves. Even northern philanthropists began to focus more on the poor whites, especially those in Appalachia, and less on the freedmen.

African Americans were elected to public offices across the South, but they were never proportional to their population and were almost always bitterly opposed by conservative whites. In Georgia sixty-nine African Americans were elected to the constitutional convention or state legislature between 1867 and 1872, but a quarter of them were killed, beaten, or otherwise threatened, and martial law was necessary to force the state legislature to seat them. As African Americans were disenfranchised, a few Black state legislators and U.S. representatives continued to be elected for a few more years. In 1901 George Henry White's term in the U.S. House expired; no more African Americans would be elected to Congress until after World War II. In 1907 the last African American in the Georgia legislature, William H. Rogers, resigned; the next Black legislator in Georgia would be Julian Bond in 1966.

The great W. E. B. DuBois captured the dashed hopes for a benevolent Reconstruction well when he wrote, "The slave went free; stood a brief moment in the sun; then moved back again toward slavery."

Yet, against fierce opposition, African Americans were able to establish their own institutions, especially churches and schools, and gained enough independence and dignity to allow them to endure the dreadful oppression that they faced over much of the next hundred years.[48]

The Ku Klux Klan

Along with the racism of white southerners and the apathy of northern whites, terrorism, pure and simple, ended Reconstruction. Violence against African Americans was commonplace after the Civil War but culminated in the notorious Ku Klux Klan. Founded in 1866 in Pulaski, Tennessee, as a purported social club, the Klan soon morphed into a number of loosely organized vigilante groups and launched a reign of terror in the lead-up to the 1868 elections that continued more or less unabated until 1871. Thousands of African Americans and white Republicans were beaten or saw their homes and barns burned down; hundreds were summarily executed.

There was "a great deal of Klan activity in White County," according to Garrison Baker, the county's best known modern historian, but by 1869 some of it revolved around personal vendettas and extralegal punishment for petty crimes. Baker noted the case of an unnamed man in Nacoochee Valley who had been arrested for stealing a hog from a neighboring farm. Free on bond, he was nevertheless "severely" whipped by Klan members and subsequently left the county. Green B. Holcombe was ambushed and nearly killed that same year because a neighbor thought he had torched his barn.[49]

More serious perhaps were the activities of the Mossy Creek chapter of the Klan. Organized in 1868, it consisted of a dozen or so white men in their early twenties, most with no property and many illiterate. Furious about continued federal occupation, their resistance to federal authority made them natural allies of the area's numerous illicit distilleries, who were always fighting to stay ahead of the "revenuers." In the fall of 1870, they terrorized a Black family at Mossy Creek, some of whom were witnesses to the murder of a deputy marshal searching for a bootleg still. Alarmed, Daniel McCollum and his son George are reported to have walked to Atlanta to ask the governor to send in the state militia, which of course never came.[50]

For a variety of reasons, the Klan ceased to exist as a cohesive organization by the end of 1869, although violence against African Americans would never really disappear. In 1870 and 1871 President Grant signed the Enforcement Acts, which gave the president the power to declare martial law, suspend the right of habeas corpus, and prohibit night riding and the wearing of masks. Under those laws thousands were arrested and violence waned, but by then much of the damage had been done.

While the Klan lay dormant, lynching remained a routine occurrence across the South until well after World War II. The last lynching in Georgia occurred at Moore's Ford Bridge in Walton County in July 1946. By some estimates, between 1877 and 1950, as many as four thousand African Americans were lynched, typically by public hanging, often after being brutally tortured. Gruesome photographs of lynchings in Cleveland and in Demorest are part of the Vanishing Georgia Collection, but there were many others elsewhere, mostly in the South, too many memorialized in awful postcard photographs of the lynching mailed to friends and family.

In February 1915 D. W. Griffith released his revisionist interpretation of the Civil War and its aftermath, *Birth of a Nation*, which proved to be wildly popular nationwide. His depiction of the Lost Cause and glorification of the Klan were, along with Leo Frank's lynching in Marietta, Georgia, in August 1915, part of the catalyst for a second iteration of the Klan. On Thanksgiving night in 1915, fifteen men met on top of Stone Mountain and set fire to a giant cross, a bit of symbolism they took from Griffith's film.

The new Klan was founded by William J. Simmons (1880–1945), who styled himself its Imperial Wizard. While he adopted most of the titles and regalia of the original Klan, his goals were very different, mostly revolving around making his Klan profitable. In 1920 he hired two publicists, who quickly helped drive membership into the hundreds of thousands nationwide. With membership at ten dollars a head, it was a lucrative business, especially after Simmons learned how to play on people's fears in the turmoil after World War I. He expanded the Klan's mission to make it rabidly pro-American and not only anti-Black but also anti-immigrant and anti-Catholic. He appeared before one group of Georgia Klan members and, according to one witness, "drew a Colt automatic pistol, a revolver and a cartridge belt from his coat and arranged them on the table before him. Plunging a Bowie knife into the table beside the guns, he issued an invitation: 'Now let the Niggers, Catholics, Jews and all others who disdain my imperial wizardry, come out!'" Simmons retired as imperial wizard in 1922 but remained "Emperor for Life." By 1925 the Klan had over four million members, with the largest memberships in Ohio, Indiana, and Oregon.[51]

In 1926 the heirs of Jesse Monroe sold their grandfather's old farm along Chickamauga Creek in Nacoochee to Simmons, who envisioned a retreat for Klan members. He built an earthen dam on one of the

William J. Simmons, founder of the modern Klan, in KKK costume. The image dates to the early 1920s, perhaps taken in White County. The costume is typical of the fantastical dress intended to frighten and intimidate. Baudouin, *Ku Klux Klan*, 16.

creek's small tributaries to create the present Sky Lake, and when it washed away in a flood, Simmons had it rebuilt in concrete. Local lore maintains that he had a "small alcove" built beneath the spillway and that it was used for Klan meetings. While that could be true, it seems a rather strange choice for a meeting place at a time when the Klan was marching by the thousands down Pennsylvania Avenue in Washington, D.C.

Simmons's successor, Hiram Wesley Evans, was eventually ousted over charges of corruption and embezzlement, as well as murdering his mistress. Klan membership declined precipitously after that, but it never ceased to exist. There was a resurgence after World War II that provoked Georgia to pass an anti-mask law in 1951, which was a powerful weapon in the battle against the Klan during the Civil Rights struggle.[52]

The Segregated South

The contested presidential election in 1876 led to the withdrawal of the last federal troops from the South, leaving the southern states with more or less free rein in their treatment of African Americans. Interracial marriage had been banned by law in Georgia in 1865, and, when the legislature authorized the state's first public school system in 1872, separate facilities for the races were mandated. Georgia wrote a new constitution in 1877, and among its provisions was a poll tax, which was to be a bulwark of the Jim Crow South.

All voters were subject to the poll tax, including any unpaid taxes from previous years. This had the intended effect of disenfranchising poor Blacks and poor whites, although white people could avoid the tax by proving that their father or grandfather had voted in an election prior to 1867, a test that African Americans could never pass. Establishment of the statewide, whites-only Democratic primary in 1900, literacy tests in 1908, and ongoing intimidation of African American voters ensured that African American participation in elections would virtually disappear in many parts of the state, especially in rural areas.

Aside from lynchings, major hazards for African Americans, especially men, throughout the late nineteenth century and the first half of the twentieth, were debt peonage and convict leasing by the state. Sharecroppers could easily slip into a cycle of debt from which they could not escape, and, although debtors' prisons had been abolished, state law still allowed people to be contracted to private citizens to pay off those debts. There was no contract oversight, and prisoners were often beaten and suffered from malnutrition and excessive work. It was slavery by another name.

Jim Crow

Jim Crow was a derisive term for an African American man, but it came to mean any of a myriad of late nineteenth- and early twentieth-century state and local laws that created the segregated South. Disenfranchisement accomplished, states set to work segregating the races through a tangle of laws that eventually segregated virtually every aspect of daily life. In the 1890s the old racially integrated patterns of living began to disappear. In the early twentieth century, fewer white people had live-in African American servants, and segregated housing was written into deeds and zoning codes all over the nation, but especially in the South. Separate restrooms, water fountains, and waiting rooms for African Americans were routine, as were segregated travel accommodations, restaurants, and almost everything else in the public sphere, including mental hospitals and cemeteries.

The great African American educator Booker T. Washington's famous "Atlanta Compromise" proclaimed the inevitable in 1895: "In all things that are purely social, we can be as separate as the fingers, yet one as the hand in all things essential to mutual progress."[53] W. E. B. DuBois and other African Americans excoriated Washington for his stance, but the following year the Supreme Court gave its approval in the landmark case *Plessy v. Ferguson*, and "separate but equal" became the law of the land.[54]

The Great Migration

Thomas E. Watson's attempt to create a political coalition of poor white and poor Black farmers in Georgia in the early 1890s collapsed in the face of Democratic violence and corruption in the elections in 1892. His subsequent transformation into one of the South's most virulent racists is a fitting symbol for a corresponding harshness that pervaded race relations for most of the first half of the twentieth century. In addition to lynching, African Americans were harassed by "white cappers," armed nightriders who used intimidation and outright violence to drive them away. In 1904 white cappers drove so many African Americans from Franklin County, just forty miles southeast of Sautee Nacoochee, that farmers had difficulty in finding any labor, in spite of high wages.[55]

The 1906 election was dominated by an intense race-baiting gubernatorial campaign, inflamed by false reports in the Atlanta papers of assaults on white women. The bitter rhetoric set the stage for an awful race riot that erupted in Atlanta in September 1906, the worst event of that kind in the city's history. For three days African Americans were hunted down by gangs of whites, who attacked them on the streets, pulled them off the streetcars and beat them, and ransacked and burned Black businesses and neighborhoods. By some accounts up to forty African Americans were killed and scores more wounded. The event was reported internationally and badly tarnished the city's carefully cultivated image as a paragon of the "New South."[56]

Whitecapping in Banks and Habersham Counties in 1907 drove some African Americans out, and in 1912 an African American church in Dawson County was burned and tenants driven off their farms. The worst occurred that same year in Forsyth County, forty or so miles southwest of Sautee Nacoochee. Typically baseless charges of African American assaults on white women resulted in the hanging of two teenaged African Americans and an orgy of white violence against the county's Black citizens. Storekeepers would not sell to African Americans, and whites who hired African Americans saw property worth $90,000 put to the torch. Dozens of African American homes were burned and their occupants driven away. Ultimately, the entire African American population, some 1,150 people, was driven from the county. Some were able to sell their property at a loss, but many farms were simply confiscated as abandoned property. No African Americans would live in Forsyth County again until the 1980s.[57]

Over 90 percent of the nation's African American population lived in the South in 1900, but by 1910 many of them had had enough and were choosing to leave. The *Tifton Gazette* and many white Southerners were not blind to their reasons for leaving, even if they were a little late in speaking out: "They have allowed Negroes to be lynched, five at a time, on nothing stronger than suspicion; they have allowed whole sections to be depopulated of them. . . . They have allowed them to be white-capped and to be whipped, and their homes burned, with only the weakest and most spasmodic efforts to apprehend or punish those guilty—when any effort is made at all."[58]

During World War I a major labor shortage quickly developed on farms and in cities. It became so acute that threats and intimidation were used by white landowners and business owners in a futile attempt to stem the tide. African Americans continued to leave nonetheless, fifty thousand leaving Georgia in 1916 alone. The depression in the South's farm economy under the onslaught of the boll weevil, which was first seen in Georgia in 1915, and collapsing cotton prices after World War I exacerbated the plight of African Americans and gave impetus to the migration. And so, between 1910 and 1930, as many as 1.6 million African Americans left the South, mostly for northern cities, especially New York, Detroit, and Chicago.[59]

The loss of brainpower and labor was a major setback to many African American institutions. Even the church that Martin Luther King Jr. would later pastor in Atlanta, Ebenezer Baptist, saw its congregation halved between 1914 and 1921, and its pastor then, who was Dr. King's grandfather, himself considered leaving.[60] The exodus to the North continued during and after World War II; the large war industries in Oakland, Los Angeles, and San Diego attracted many migrants during that period as well. By 1960, 3.5 million more African Americans had left the South, and by 1970 only 53 percent of the nation's African American population still lived there. Not until the 1990s did the outmigration trend begin to reverse, as African Americans left the Rust Belt for the Sun Belt.

The federal census population schedules clearly show the effect of all this turmoil on the African American communities in White County, which continued to be centered around Cleveland, Mossy Creek, and Nacoochee. In 1890 there were 662 African Americans counted in White County, nearly 11 percent of the population. By 1900 fewer than 600 African Americans had stayed in the county and by 1910 fewer than 400, a little less than 8 percent of the population. The numbers continued to decline slowly after that, and by 1970 there were just 369 African Americans left in White County, nearly all in the Cleveland and Helen census divisions, since the Black community at Mossy Creek had virtually disappeared.[61]

Civil Rights

In May 1954 the U.S. Supreme Court issued its landmark decision in the case of *Brown v. Board of Education*, finding that "separate but equal" inevitably produced inferior facilities, something African Americans had known for a long time.[62] The following year Rosa Parks's refusal to give up her bus seat to a white person precipitated the yearlong Montgomery bus boycott and the beginning of the modern civil rights movement. Furious white resistance included murder, arson, and bombs, which bookend the 1960s with the riots and looting that occurred after Dr. King's assassination in 1968.

Most of all, the white South delayed as long as it could and however it could to comply with the string of court orders, congressional mandates, and even National Guard intervention that eventually forced desegregation of public facilities in some places. The "white only" and "colored only" signs gradually disappeared, but it took passage of the Civil Rights Act in 1964 and the Voting Rights Act in 1965 to fully empower African American voters. Like many places in the South, the desegregation of schools in White County was difficult, and much delayed. In 1964 the county school system was finally integrated, although not without the vocal racism that always accompanied any progress by African Americans. Even then Black students faced continuing hostility, and school buses were not desegregated until 1989.[63]

Bean Creek

Most people who pass through Sautee Nacoochee in the twenty-first century miss the Bean Creek community. Although it is off the beaten path, this community of African Americans dates at least to the Civil War, with roots in antebellum slavery. The community is located south and west of Bean Creek, which the present Bean Creek Road crosses a mile or so north of the Bean Creek church. The branch that runs in front of the church is actually called Ben Creek, or Bristol Branch, as Lumsden designated it on his 1948 map of Nacoochee. Bean Creek Road is now designated as such from its intersection with SR 255, but, historically, its terminus was a half mile west along what was then a portion of Rabun Road. The routes of all these roads were likely more or less set long before white resettlement of the valleys.

Most of the timber in that area had been cut before the Civil War, and mining activity had almost certainly left the landscape in poor condition, but that was generally the only kind of land that whites were willing to allow African Americans to live on and occasionally buy. At the same time, the community became increasingly isolated, as through traffic used the road along the west side of Sautee Valley, especially after the Sautee Post Office was established in 1873. Much of the northern reaches of Bean Creek Road remains rough gravel even today.

The earliest mention of the name is in the 1883 deed by which the Bean Creek Missionary Baptist Church's trustees purchased the land for their church, but use of that name to identify both the creek and the community is surely much older than that. In a recent interview published in *Foxfire*, Ms. Lena Dorsey (1935–2017) recounted the origin of the community's name: "I heard that the name 'Bean Creek' came from a story where this man was riding a mule. He had a bag of beans and the mule, you know, it stumbled and fell, and the beans busted, and beans came down stream. I heard that's how Bean Creek got its name."[64]

GRAHAM

As with many such communities, documentary evidence for the Bean Creek community is sparse. The deed for James R. Wyly's sale of Land Lot 22 to Moses Harshaw in 1837 references "Graham Mill Creek." That might have referred to the small intermittent branch entering Bean Creek at the site of the present Sautee Watershed impoundment southwest of Grimes Nose, but it seems more likely that Bean Creek was known as Graham Mill Creek during the earliest days of resettlement of the valleys.[65]

Moffat's map of Nacoochee in 1837 does not depict Bean Creek, and, although his map is not to scale, it appears that somewhere east of the present church building, he depicts what is apparently meant to be the homestead of someone with the surname Welch. There is no first name, but he may have meant David or Joseph Welch, both of whom appear in the 1830 federal census of Habersham County. David Welch (1804–52) was born in Ireland and was married to a woman named Rachel, with whom he had at least two children. By 1840 they had moved to Augusta, where he died. Joseph Welch, who was born in the 1790s, may have been David's older brother and may be the "J. N. Welch" listed in the population schedules of the 1840 federal census of Habersham County. After that the census recorded no one named Welch, or Welsh, in Habersham or White Counties before the Civil War.

Neither Joseph nor David Welch are known to have owned slaves, but it is certainly possible, if not likely, that they did. African American Welches were a part of the Bean Creek community from an early date. Charles (born about 1835), Anthony (birth date unknown), Henry (birth date unknown), and David Welch (born about 1827), who were almost certainly related, were among the first African Americans to register to vote in White County in 1867. Henry and Anthony left White County in the 1870s, but Charles remained, marrying a woman named Millie and settling near Cleveland. David Welch married a widow named Harriet, and they had several children.[66]

Two other African American Welch families also lived in White County in the 1870s and 1880s, both headed by siblings or cousins of David Welch. Johnny Welch (born about 1823) and his wife, Milly, and their two children were living at Blue Creek; Gilbert (born about 1832) and Angeline Welch (born about 1855) were living with their three children at Town Creek. In 1900 Charles and Millie Welch were still living at Cleveland, but by 1910 there were no African American Welches anywhere in the county. No Welch burials have been documented in the Bean Creek Cemetery. Whether there was a connection between these African Americans and the white Welches in the county in the 1820s and 1830s remains uncertain but seems likely.

The Bean Creek community may have grown up around housing for slaves working the mines along Bean Creek, perhaps including those of Moses Harshaw and his sons, but there is no direct evidence for that supposition. Quarters for the Harshaw house servants were likely near the main house in Land Lot 44, and others were perhaps near mining operations where Bean Creek Road crosses Bean Creek in Land Lot 22. Both land lots were owned by Moses Harshaw and encompass much of the Bean Creek community.[67]

Except for the Dean Mine, all the mining in and around Nacoochee was placer mining. Presumably, the enslaved laborers did all the digging and separated the gold from the ore, probably with the use of a stamp mill, but no site-specific information has yet been found. Harshaw used slave labor in mining his gold lots, but exactly where the eleven mines he is purported to have worked were located is uncertain. Nevertheless, near where Bean Creek Road crosses the creek, as late as 1940, there was still a quarry in operation in the northwest corner of Land Lot 22. It has since been filled and turned to agriculture. Harshaw's son-in-law, David R. Lowery, had thirteen slaves in three houses in White County, but he paid taxes on several parcels of land in the early 1860s, so it is not certain where those houses were located. It is likely that they were in the forty acres he paid taxes on in Land Lot 44. Lowery was born in North Carolina around 1817 and, with his wife, Ann, may not have come to Georgia until the 1850s. He appears to have moved to Webster County in southwest Georgia in the 1860s, but the African American Lowreys at Bean Creek may well descend from some of his slaves.

Because gold had been found along Bean Creek, there was considerable land speculation in and around what became the Bean Creek community. Before his death in 1859, Moses Harshaw partitioned Land Lots 22 and 23 into forty-acre "gold lots," and other landowners partitioned their lots as well. It is unclear how much mining was done on these lots, but it was probably enough to ruin the land for agriculture. After the Civil War, the new owners of the old Harshaw house had to fill in the shallow pit mines that had been dug over much of their property before it was fit for planting.[68]

WILLIAMS

As noted earlier, there is uncertainty about where George Walton Williams housed the hundred slaves that he said he brought from the Sea Islands plantations abandoned to the federal troops in 1862, but it could have been near Bean Creek. The White County tax digests document that George and his brother Edwin were taxed on jointly owned property encompassing over 4,500 acres in White County, among which were more than 100 acres in Land Lots 22, 42, and 43.

George Williams reported that the slaves he had brought from the Sea Islands returned there after the war. One wonders then how to account for the fact that the increase in White County's African American population between 1860 and 1870 was far above that of surrounding counties. Hall County had twice as many slaves as White and Lumpkin Counties, but the size of its African American population barely budged in the 1860s, even as the white population made modest gains of 3 percent. In Lumpkin County the African American population actually declined during the period, while the white population increased by 13 percent. In contrast, in Habersham County the African American population grew by 14 percent in the 1860s, which was more than twice the rate of growth of the white population. White County's white population grew by a third in the 1860s, while its Black population more than doubled, increasing an astounding 105 percent. There were 274 African Americans in the county in 1860; ten years later there were 564 "colored" residents enumerated in White County. The reason for this growth remains uncertain, but more than a few of Williams's slaves may not have returned to the Low Country.[69]

BEAN CREEK INSTITUTIONS

Perhaps more than anything, African Americans wanted their own institutions, which naturally required a community. It is not surprising that the first African American churches and schools appeared where there had been numerous African American slaves and, after the war, enough people to start a church and a school. Once either one of those institutions was established, it gave an important anchor to the community as well as a magnet for further growth, as illustrated by the African American communities at Cleveland, Mossy Creek, and Nacoochee.

Bean Creek Church

As noted in chapter 6, no certain date for the establishment of Bean Creek Missionary Baptist church has yet been documented. While certain members of the community believe the date to have been either 1862 or 1865, some of the enslaved African Americans may have begun holding prayer meetings in an old log cabin across the road from the present church before the Civil War. Formal organizations such as Baptist churches probably had to wait until after the war.[70]

Mary Ann Nicely (*left*), born a slave at Nacoochee in 1851, and her son, Ed, and his wife, Lessie. Where the photograph was taken is uncertain, but it could have been at the old log building where the congregation first held services. Photo by Andy Allen, *Reflections*, July 2006.

The Baptists did not have the formal requirements that made the establishment of a Presbyterian or Methodist Episcopal church, Black or white, more difficult. Baptists had fewer restraints, but the problems that the African American Baptists at Dahlonega had in withdrawing their letters of membership—which was the traditional way that Baptist membership was recorded—in order to establish their own Baptist church illustrates one of many points of negotiation that had to occur between African Americans and white Americans in the postwar era. The congregation at Bean Creek could have gone through a similar exercise, but the white Baptist congregation at Nacoochee, if that is indeed the origin of the Bean Creek church, might have been more accommodating than their counterparts in Dahlonega.[71]

Bean Creek School

While the Bean Creek church is thought to have been established about 1865, it was several years before a school could be set up, since few freed slaves had the resources, literacy, or education to support a school immediately after the war, and there is little evidence that the wealthy white families at Nacoochee jumped into the breach. In 1868 Martin R. Archer, the Freedmen's Bureau agent at Dahlonega, expanded his efforts to establish schools for the freed African Americans in Dawson and White Counties. He was "astonished" at the poverty of both Black and some of the white citizens of those counties, writing that "I was not aware of the extent of this destitution until the effort was made to organize schools."[72]

The voluminous records of the Freedmen's Bureau provide a plethora of possibilities for documentary research well beyond the goals of this study. The collections are being digitized, but at present indices are primarily helpful for those researching family names. The genesis of the Bean Creek school might be found in those records, but a search of the records of the Education Division of the Bureau of Refugees, Freedmen, and Abandoned Lands, 1865–71, turned up no evidence of a school at Bean Creek in 1867 or 1868, which was the last mention of Nacoochee in those records.[73]

By 1880 at least twenty-one African American families and 125 individuals were in the Nacoochee District, and a school was almost certainly in existence, probably established through the county's poorly funded public school system. When James H. Nichols sold an acre in the southwest corner of Land Lot 43 for ten dollars to the trustees of the Bean Creek Missionary Baptist Church in October 1883, the deed stipulated that the land could be used only for religious and educational purposes. It is likely that the church was already being used as a school, a common arrangement in those days. A separate school building was constructed later, but exactly when has not been documented.

Bean Creek Stores

Walter Lumsden's 1948 map of Nacoochee suggests that the store closest to the Bean Creek community was at the intersection of Rabun Road and SR 255. By 1960 state highway maps show that it had closed, although two other stores were depicted, one of which was at the heart of the Bean Creek community. Located across the road from and south of the church, the store may not have come into existence until well after World War II, since it is not depicted on the State Highway Board maps before 1960. The second store was on the north side of Rabun Road about halfway between SR 255 and Bean Creek Road. Who owned and operated any of these stores and the extent of their inventory have not been documented.

THE PEOPLE

Freed African Americans, who were not named in the 1860 census, often took their former master's surname after the war. As a result, most of the white surnames found in the 1860 slave schedules appear as African American surnames in the 1870 census, including Williams, Richardson, Starr, Trammell, Brown, Wyly, and Dorsey. Perhaps not surprisingly, there were no African Americans named Harshaw in White County, although there were several Black Harshaw families over the Blue Ridge in Towns County as well as all over western North Carolina. In most cases, however, it is very difficult to match an unnamed slave in the 1860 census with named African Americans in the 1870 census.

In 1870 nearly 20 percent of the county's African Americans lived at Nacoochee. Since the census taker typically went door to door, it is often possible to have a snapshot of not only individual families but also their neighbors, who were often their relatives as well. A closer look at the census schedules also tells us something about how the lives of African Americans might have been integrated with those of white Americans in the twenty or thirty years after the Civil War.

In 1870 a few individual African Americans were listed as part of white households, including siblings Brit, Jesse, and Fanny Jones, who were servants in Edwin Williams's household and probably lived in separate quarters on the property. All in their late teens, they would have provided critical help in the household. A few African American families also lived in relative isolation, scattered among white families around the county. More often, however, there were two or three or more African American households near one another. Along Dukes Creek the 1870 census found pioneers John L. and Sarah Richardson's household surrounded by no fewer than seven African American households, including thirty individuals, almost certainly living in the Richardsons' slave dwellings that had been enumerated in 1860. Three families were named Welch, while the others were Sutton, Kirkpatrick, and Houston. The only African American Richardsons in White County in 1870 were Shadrack Richardson and his wife, Emily, who lived at Tesnatee with their eight children. In the population schedules for the 1870 census, he listed his occupation as "digging for gold."

Larger communities of African American households also developed at Cleveland and Mossy Creek, but the largest African American community in White County was on Bean Creek. As noted earlier, it may have grown up around Moses Harshaw's quarters for his miners in Land Lot 22, but there may have been Lowery slave houses in Land Lot 44 that were part of the community as well. In addition, present research suggests that George Walton Williams might have housed his hundred slaves on the 120 acres he and his brother Edwin owned in Land Lots 23, 43, and 44 during the Civil War. These too could have contributed to the genesis of an African American community at Bean Creek even before the Civil War. Wherever they came from, by 1870 at least ten families and sixty-seven people were enumerated as part of the Bean Creek community.[74]

PROPERTY

Only 13 percent of African Americans in Georgia owned real estate in the late nineteenth century, about a third of that of white property owners. While African Americans at Sautee Nacoochee owned their church, they had little else in terms of real estate. The White County tax digests in the late 1870s and 1880s, for example, list sixty-eight African American polls (i.e., adult males eligible to vote), who collectively owned a total of 514 acres in the county. In contrast, there were 709 white polls owning 138,725 acres of land, which averages 159 acres per white poll versus fewer than 8 acres for each Black poll.

Most white landowners would not sell to African Americans, since to allow them any independence would necessarily increase the cost of labor. As a result, very few African Americans acquired property in White County or anywhere else in the Piedmont in the nineteenth century. Only with the agricultural economy booming after 1900, along with the tightening labor market brought on by the onset of the Great Migration after 1910, were many African Americans able to buy their own property.

Trammell

At Nacoochee Alfred Trammell is the only African American recorded as owning his own farm in 1900; everyone else rented. Born in July 1838, he and his wife, Sarah Jane, were married in 1860 and had two children, only one of whom was alive in 1900. Neither Alfred nor Sarah Jane could read or write. The loss of *White County Deed Book B* and *Deed Book C* has left a large gap in the records, and exactly where his property was located has not been determined. However, Trammell Drive and the four Trammell households depicted in Walter Lumsden's 1948 map on Bean Creek Road north of the church probably approximate its location.

Brown

By 1910 five more African American families owned their own homes at Nacoochee. They included Thomas J. Brown, who was born about 1860, and his wife, Marie. He was the son of Joseph Brown and Mary Brown and probably a former slave of Daniel Brown's son James H.

Brown. After the war Joseph Brown made his living prospecting for gold. Whether gold allowed Thomas Brown to acquire property is not known.

Dorsey

Emma Dorsey also bought a house before 1910. She was born at Nacoochee about 1870 to Edward (1846–1928) and Matilda Dorsey (born 1854). They were pioneers in the Bean Creek community after the Civil War, and they and generations of their descendants are buried in the Bean Creek Cemetery. It is likely that Edward, and possibly Matilda as well, were slaves of Andrew J. Dorsey (1790–1864) and his wife, Nancy Smith Dorsey (1786–1874). North Carolina natives, they are buried at Mossy Creek United Methodist church cemetery. At the age of ten, as the 1870 census population schedule recorded, Emma Dorsey's occupation was "working on farm." She had ten children, but their father has not been surely identified. Censuses conflict as to whether she learned to read but indicate she could not write. She spent her life washing and cooking, presumably for white families. Listed as a widow in the 1920 census, Emma Dorsey died in 1944 and may lie in an unmarked grave in the Bean Creek Cemetery.

Jones

Annie Jones, whose maiden name has not been found, was born about 1854 and married Briton Jones, who was around the same age, in 1870, the same year he was also enumerated in the household of Edwin P. Williams. They had seven children, one of whom was Columbus Jones, born in May 1883. They were sharecroppers in the Chattahoochee District of White County in 1900, but Briton Jones died before 1910. Given her employment as a laundress and his in "odd jobs," according to the census, it is uncertain how they were able to purchase their home. Annie Jones apparently died before 1920, and by then Columbus Jones was in DeKalb County, working as a blacksmith.

Nicely

Henrietta Nicely was born in 1855, the daughter of Henry and Jennie Jarrett, who may have been slaves of Devereaux Jarrett. She married Jim Nicely, who may have been a son of Anica Nicely; all three are probably buried in Bean Creek Cemetery. The Nicely family was a large one, and all of their familial relationships have not been established, but, by the time the 1910 census was taken, Henrietta was widowed and living with her twelve-year-old nephew. She was "truck farming," but what she raised for the market is uncertain. Henrietta Nicely died in September 1925.

Other Families

By 1920 at least ten families in the Bean Creek community owned their own homes; by 1930 there were a dozen. All had surnames familiar in the history of Sautee Nacoochee: Lowery, Dorsey, Jarrett, Jones, Austin, Anderson, White, Brown, Trammell, and Richardson. There were a number of others as well who do not show up in the census. Continuing searches of White County public records, including tax digests, estate records, and records of deeds and mortgages, would certainly reveal additional significant information about the evolution of the African American community at Bean Creek.

Part of William B. Lumsden's map of Sautee Nacoochee in 1948, with the Bean Creek community at the center. Note the dashed line separating the "colored settlement" from the "white settlement." Courtesy of Sautee Nacoochee Community Association.

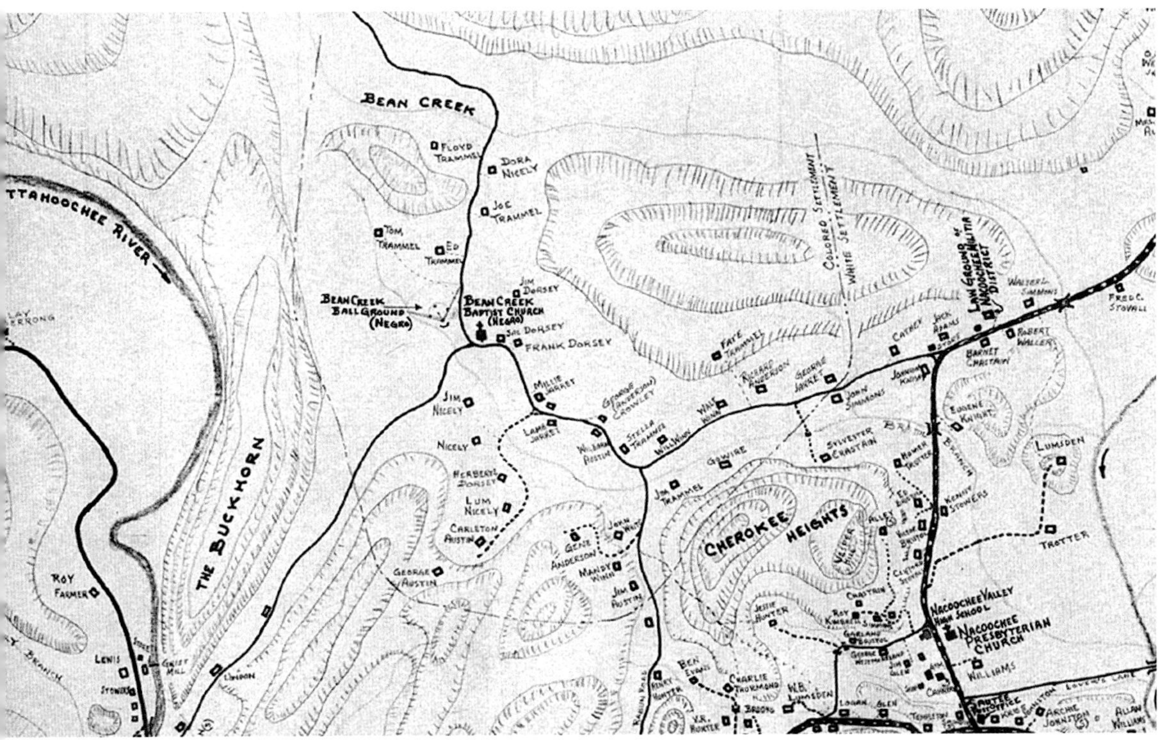

EMPLOYMENT

After the Civil War, most African Americans in northeast Georgia were left to work out the terms of their freedom more or less on their own. A few local white residents did what they could, but many more were hostile to their interests or simply did not care. Generally without land or much of anything else, ex-slaves had only their labor to sell, unless they were fortunate enough to have learned a skill or trade. Some wanted to do anything but farm and left for Atlanta, Chattanooga, and elsewhere, where they hoped another alternative would present itself.

For those who stayed, there were not a great many choices, at least until automobile travel became common in the second quarter of the twentieth century. As one resident remembered, "Wasn't nothing else to do but to work on somebody's farm. They worked till they got through, wasn't no set time."[75] The federal census population schedules confirm the fact that the vast majority of African Americans, and whites as well, relied on agricultural work to make a living, mostly working as sharecroppers. In the 1870–80 censuses of Nacoochee, African American occupations were variously given as "farm hand," "farm labor," or, if they owned or rented their own place, "farmer."

A very few of the women were working as servants, washing and ironing and cooking in private homes. Of the remaining women, most were "keeping house," according to the census, which entailed the hard grind of laundry and cooking in the days before modern appliances, as well as seeing that the family's chickens, mule, and other farm animals were cared for. Most families had a garden and depended on hunting and fishing to augment the family's diet.[76]

The only early exceptions to these patterns were Joseph Brown (born about 1825), "digging gold" according to the 1870 census, and his son Thomas Brown (born about 1860), a "gold digger" in the 1880 census. Beginning in the 1890s, a revival in gold mining gave employment opportunities to more African Americans at Nacoochee, with the 1900 census enumerating nine men employed in pit mines such as those dug all around Bean Creek. The work did not last, as most of the mines quit operating during World War I. There was a brief revival in the 1930s, with three or four African Americans finding work as drag operators at one or more of the gold mines.[77]

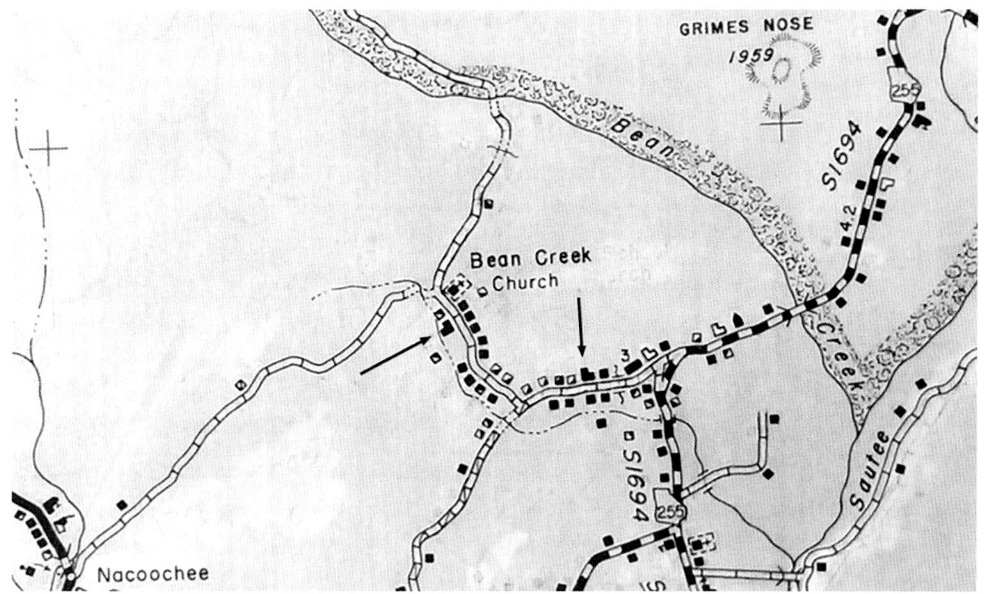

Detail from the Georgia Transportation Board map of White County, 1960, showing the Bean Creek community. Solid squares are occupied residences; open squares denote unoccupied structures. Arrows have been added by the author to indicate the stores on Bean Creek Road. Georgia Department of Transportation, "Archive Maps."

Construction of the Gainesville and Northwestern Railroad to Helen in 1912 and 1913 provided a number of jobs to African Americans at Nacoochee. A dozen African Americans were employed by the railroad by 1920, working as graders, spike drivers, drillers, and section hands. There were still a few African American section hands in 1930, but the railroad shut down the following year, eliminating even those few jobs.

Three or four African Americans at Nacoochee found work in the lumber business in the 1910s and as many as nine in the 1920s, either as loggers or common laborers in the sawmills and lumberyards at Helen. Like so many others, those jobs largely disappeared during the Depression. A few African Americans got away from farm labor, including Charlotte and Minerva Brown, who both worked as nurses at Nacoochee in 1900. Alexander Wells and Lam Jones operated blacksmith shops, the latter's shop probably at Bean Creek. Columbus Jones was working as a carpenter in the first decade of the twentieth century, and James Nicely was a well digger. As hotels, boarding houses, and restaurants opened to serve a nascent tourist industry in the first quarter of the twentieth century, a few jobs, mostly as cooks, were available for African Americans, and more African Americans were employed as cooks by white families as well.

After World War I, African Americans found occasional work, too, on some of the poultry farms that were established in the 1920s. In the 1930s some of the New Deal programs also gave work to a few African Americans at Nacoochee. One of those programs may have been responsible for bringing an African American doctor to Bean Creek. Her name was Alice Brown, a widow born in 1880, but little else is known about her beyond the 1940 census notation that she was in private practice. She may have been the same Alice Brown who died in Atlanta in 1964.

After World War II African Americans continued to leave rural areas, many of them moving to places like Atlanta, where a thriving Black community and better job opportunities continued to draw people to the city. The Black population in White County continued to decline, and in the twenty-first century there were more Hispanics in the county than African Americans.

8

Agriculture and Commerce

In the first quarter of the nineteenth century, northeastern Georgia was resettled by farmers, mostly white (both rich and poor), and until World War II the economy revolved around agriculture and its ancillary industries such as gristmills, sawmills, blacksmithing, and distilleries. All of those were necessary components of any agricultural economy, as were hunting, fishing, and home industries such as weaving. By the 1840s the valleys of northeastern White County had been thoroughly resettled, with the vast majority of the residents engaged in agriculture. In his *Statistics of the State of Georgia* in 1849, Rev. George White described Nacoochee Valley as having "three stores, one hotel, one church, and several mechanics' shops. The valley is about eight miles long and about half a mile wide. It is one among the most beautiful valleys in the world. The land is productive, rewarding the farmer with liberal crops of corn, wheat, &c. more than 1,200,000 dollars worth of gold has been found in this valley."[1]

After the Civil War, railroad construction vastly reduced the isolation of Nacoochee, as lines were completed through Cornelia, seventeen miles southeast of Sautee Nacoochee, in 1871 and through Clarkesville, less than ten miles away, in 1883. Although the railroad did not reach Nacoochee until 1913, it had long since made available a host of consumer goods that supplied a growing number of country stores.

Farming in the Foothills

Because the land and climate did not lend themselves to the culture of cotton, the agricultural economy in the hills, valleys, and mountains of northeastern Georgia differed in both scale and kind from that of much of the rest of the state. Nevertheless, agricultural historian James C. Bonner found that "Nacoochee Valley . . . took on some of the characteristics of a Middle Georgia plantation community in the two decades before the Civil War." Indeed, Edward Williams, Daniel Brown, Moses Harshaw, and John L. Richardson came to Habersham County in the early 1820s to establish farms, and it was not by accident that they claimed the richest lands. Many of their neighbors were subsistence farmers with the simple goal of feeding their family and livestock and somehow raising cash to pay taxes and buy what little was needed that could not be bartered. Some raised small amounts of tobacco for barter; but cowhides, deer and beaver skins, and small furs could be exchanged with local merchants such as Charles L. Williams for salt, coffee, flour, ammunition, and other necessities.[2]

The schedules from the federal agricultural censuses of 1850–80 provide a plethora of data on farms of the period. However, the available microfilm copies of the Habersham County schedules from 1850 are illegible in places, and data on the farms in and around Sautee Nacoochee has not been located. Those farms can be found, however, in the agricultural censuses 1860–80, showing farm size, produce and livestock, and farm machinery owned, among other things.

Farms were not large in the hilly terrain of northeastern Georgia. In 1860 the federal agricultural census of White County tallied 274 farms with more than three improved acres, which generally included land under cultivation or in pasture. Half of those had fewer than fifty acres, while another third had fewer than a hundred acres. Only forty-three farmers in the county worked between a hundred and five hundred acres; there were none of the thousand-or-more-acre plantations found in other parts of the state.

Although Edwin Williams is among those named in the 1860 agricultural census, the various columns after his name were inexplicably left blank, perhaps because all of his land was farmed by tenants, who were reported separately. His brother Charles L. Williams, however, appears as the largest farmer at Nacoochee in 1860, with three hundred acres of improved land and another five hundred acres of unimproved land. The farm was valued at $7,000 plus $570 in farming implements and machinery, far more than any of his neighbors. Alonzo Harshaw

A view of Henry Conley's farm, in the valley where Helen would develop in the early twentieth century. Similar scenes would have been characteristic of much of Sautee Nacoochee in the nineteenth century. Brooks and Greear, *Images of America*, 9.

had only half as much improved acreage in his father's farm in Sautee, but he claimed another 650 acres of unimproved land.

John L. Richardson had a hundred acres of improved land and eight hundred acres in unimproved land, all valued at $15,000 because of the gold mines on the property. The farms of Jehu Trammell, Henry Conley, and Prior Pitner were also among those with more than a hundred acres of improved land and farms valued at $4,000, $3,000, and $1,900, respectively. There were several residents in the valleys with much more modest farms, including the Littlejohns on the south side of the river near Dukes Creek, who had only thirty-five improved acres. Only about half of Georgia's farmers owned the land they farmed; after the Civil War, even fewer would own their land. The landless were among those who worked for their more prosperous neighbors as sharecroppers or tenants.

White County began losing population in the 1890s, as textile mills in Gainesville and elsewhere lured people away. There was a corresponding decrease in the number of farms, which fell from 1,008 in 1900 to 914 in 1910 and 974 in 1920. Average farm size, which had been as high as 255 acres in 1860, fell to 117 acres in 1910 and 98 acres in 1920. Included in that was improved acreage that had averaged 55 acres in 1860, just over 30 acres in 1910, and 29 acres in 1920.

LIVESTOCK

In 1860 the agricultural census schedules show that nearly every family had at least one horse, and wealthy farmers like John L. Richardson, Elijah Starr, and Charles L. Williams had five or six. Most had a mule or two as well, and there were a few working oxen in the valleys. Everyone kept at least one "milch cow," and many had between five and six. Nearly every farm maintained a herd of sheep, typically ten to fifteen, and there were likely goats as well, although the agricultural census schedules did not count them. And every farm kept "swine," typically fifteen to twenty, since salted pork was a mainstay in nearly everyone's diet.

As with so many things, Charles L. Williams maintained far more livestock than any of his neighbors. In 1860 he had thirty-eight mules, but it is not clear how they were worked. There were fifty-eight cattle, in addition to his eighteen milk cows, and sixty swine, which would probably have produced a surplus of salt pork that could be sold from his store. Until the last quarter of the nineteenth century, cattle and hogs were often free-ranging, feeding seasonally on the plentiful hardwood mast in the forests and causing significant environmental damage along the way. Hogs were particularly destructive with their constant rooting, and some small plant species were rooted to extinction in many locales.[3]

The Civil War took a severe economic toll, and not just on those who lost slaves. Between 1860 and 1870, the county's population grew by 50 percent, but its livestock was significantly diminished. The number of oxen remained unchanged, but the number of mules rose from 186 to 205, reflecting a growing dependence on mules in agriculture. Sheep increased from 1,950 to 2,341 over the decade, but in contrast the number of horses actually fell from 630 recorded in 1860 to 486 in 1870, and the number of milk cows went from 816 in 1860 to 784 ten years later. Even the number of swine fell from 5,607 in 1860 to 4,177 in 1870 and did not surpass 5,600 again until the 1880s. By 1880 the number of horses, cattle, and milk cows had surpassed antebellum levels, but the number of oxen continued to decline as they were replaced by mules. There were fewer sheep as well, as commercial fabrics became more widely available and affordable. All these trends would continue into the early twentieth century.

PRODUCE

European settlers had brought to the Americas the wheat, rye, oats, and barley that had been their traditional crops but quickly found that these grains did not always grow well, especially in the South. Very soon they adopted the culture of "Indian corn" or maize, which provided food for people and fodder for animals. Ground into meal, corn was eaten by nearly everyone nearly every day. Even after the kernels were removed, the husks could be used for brooms and stuffing for mattresses, and the cobs could be added to fodder, although they did not have much nutritional value. Cobs also could provide excellent bedding for animals and were used for fuel and for making charcoal. There were even recipes for corncob jelly.

When George Walton Williams was a boy, he got "gold fever" and urged his father to dig for gold. Edward Williams's answer? "Now, George, you see the corn before you, plow four furrows between each row. This field is a sure gold mine—one that has never failed me. We will make corn to sell to those men who spend all their time hunting for gold." Edward Williams apparently had a reputation as one of the state's largest producers of corn. In 1852 he was even awarded a silver pitcher for the "best acre of upland corn" at one of the agricultural fairs he regularly attended and at which he often exhibited. It was not just corn either; the following year he received another silver pitcher for "best method of reclaiming land."[4]

Corn was typically interplanted with squash or pumpkin, which shaded the ground and helped to keep down weeds and slow evaporative water loss from the soil, and beans or gourds, which could use the corn stalks to support their vining habit. Beans were most often planted with corn since they had the advantage of fixing nitrogen in the soil. Cow peas, which were not vines, also fixed nitrogen in the soil and were another staple that could be dried for year-round use.

Nearly every farm produced 200 or 300 bushels of corn, but Dr. Starr produced 600 bushels in 1860, Jehu Trammell 700 bushels, and Alonzo Harshaw 750 bushels. By far the largest producers of corn in the county were Henry Conley, who produced 1,000 bushels in 1860; John L. Richardson, who produced 1,100 bushels; and Charles L. Williams, who produced 2,500 bushels. After the Civil War, Capt. James H. Nichols could also produce 2,500 bushels of corn on 125 acres. Almost no one grew barley, but many farmers grew some wheat, typically producing fewer than 25 bushels. The exceptions were Charles L. Williams, who produced 100 bushels in 1860, and Dr. Starr, who produced 104 bushels. Two-thirds of Nacoochee farmers grew some rye, but in smaller quantities than wheat, although Williams produced 100 bushels in 1860. Rye was generally used to make whiskey. Perhaps a third of Nacoochee farmers grew oats, typically less than 30 bushels, but Williams again was the exception, with 100 bushels of oats in 1860.

White County produced 655 pounds of rice in 1860, but not at Nacoochee. Most farmers in the valleys produced a few bushels of peas and beans, and most of them grew a few bushels of Irish potatoes as well. Nearly everyone produced sweet potatoes, typically ten times the amount of Irish potatoes and part of what made Georgia the largest producer of sweet potatoes in the country. The top sweet-potato producers at Nacoochee in 1860 were Alonzo Harshaw with 85 bushels and Jehu Trammell with 100 bushels. A few people kept "market gardens,"

A view of the north side of Nacoochee Valley, a spectacular image of the valley in its agricultural heyday. Brooks and Greear, *Images of America*, 12.

producing vegetables for sale, but the value of these was generally less than twenty-five dollars per year. Unusually, Henry Conley made fifty dollars from his market garden in 1860. A few people produced butter, but almost no cheese. There was some production of molasses (Alonzo Harshaw produced twenty-one gallons in 1860), but honey was more widely produced.

Tobacco and cotton were grown primarily as cash crops, and both were grown in White County. The county grew over four thousand pounds of tobacco in 1860, but none of that production appears to have been in the valleys. In 1860 there were 68,000 farms in Georgia, producing seven hundred thousand bales of cotton. White County had 274 farms, which produced only one hundred bales of cotton, and none of that at Nacoochee. Only Rabun, Towns, Union, and Pickens Counties produced less.

Production of corn and sweet potatoes, two of the most important food crops, fell sharply in the 1860s, even as improved acreage rose from 15,000 in 1860 to 17,700 in 1870 and 20,600 in 1880. Corn production fell from 117,185 bushels in 1860 to 80,811 in 1870, but by 1880 White County farmers were raising nearly 150,000 bushels of corn. As more land was devoted to cotton and other crops, only 139,456 bushels of corn were produced in White County in 1910. There were 19,900 bushels of sweet potatoes harvested in 1860, but that too fell to 8,899 in 1870 and was only at 12,200 in 1880. In 1910 production rose to 19,205. These declines in productivity in the 1860s continued into the early 1870s and were a statewide phenomenon precipitated by the war and by the Panic of 1873 and the ensuing economic depression that lasted for much of the decade.

Even after the economy recovered in the late 1870s, the farm economy continued to suffer from stagnant and falling prices until the 1890s, when the cotton economy began swelling to its peak in the first decade of the twentieth century. From only 150 bales of ginned cotton in 1900, White County produced 327 bales in 1909 and then nearly doubled that to 629 bales in 1913. It was peak King Cotton.

Domestic Manufactures

Another component of most agricultural economies was domestic, or "homemade," manufactures, the value of which was tallied in each federal agricultural census from 1850 to 1870. The category encompassed anything made, whether for home use or for sale, that was not reported in the produce schedules. Before the Civil War, wool spun into thread was used to weave fabric for clothing, including socks, shirts, and jeans, and for rugs and other household articles. Some farmers even tanned small amounts of leather to make shoes for their children. In his *Statistics of the State of Georgia* in 1849, Rev. George White noted of Habersham County, "The females of this county are remarkably skillful in weaving jeans. Beautiful saddlecloths are also made, and sent to Clarkesville, where they meet with ready sale." It should be noted that these "jeans" were not the common blue cotton denims that had been manufactured in Italy as early as the eighteenth century and Levi Strauss made famous in the 1870s but rather made from a coarser, woolen material.[5]

The agricultural fairs typically included exhibits to showcase a variety of homemade goods, mostly by women. In addition to canned goods, the South Central Agricultural Society presented prizes in "Domestic Manufactures," which included the best cotton thread, best domestic fringe, best rag carpeting, best woolen hearth rugs, best cotton counterpane or quilt, best woolen and cotton coverlet, best cotton socks, best woolen jeans, best woolen socks, best linen diaper, and best carpet and binding.[6]

Cotton could be woven, of course, but the raw material was often not readily available in the mountains. After the war the weaving of wool began to decline, as commercial cotton fabric became more affordable. The widespread use of fabric flour and feed sacks led almost immediately to their being reused to make clothing or quilts. In Habersham County the value of homemade items or "family goods," as the 1840 census put it, rose from $25,528 that year to $31,275 in 1850.[7] Ten years later the combined value of those goods in Habersham County and the newly created White County had more than doubled to $66,900. The poverty of the 1860s reduced the value of homemade products by nearly half, and, as the railroads expanded and made cheap commercial goods more readily available, fewer and fewer products were produced at home.

Scientific Agriculture

As early as the 1830s, some Georgians were recognizing that cheap land that resulted from the state's land lotteries had damaged agricultural and social conditions. One critic at the time thought the lotteries the "most unjust, unequal, immoral" way to distribute public lands. It "begat a careless, slovenly, skimming habit of farming" that was akin to killing cattle only for the hide. By 1850 soil exhaustion and erosion could not be ignored. In 1851, barely twenty-five years after the county's founding, a Troup County planter in the western Georgia Piedmont wrote that "we are awfully bad off up here, having nearly worn out one of the prettiest and most pleasant counties in the world." He, like many others in the period, bemoaned the appearance of "some of our large plantations," when he looked out on "the waving broom sedge, the barren hillsides, and the terrible big gullies."[8]

In addition timber was widely used as fuel and fencing material; by 1850 it was in short supply in some of the older counties. There were even recommendations to build wood-framed rather than log slave houses as a way to conserve timber. Not only was soil fertility depleted and timber resources wasted, poor farming practices ruined many of the state's waterways as well. Charles Lyell, the greatest geologist of his day, wrote from Columbus in 1847, "I am assured that a large proportion of fish, formerly so abundant in the Chattahoochie have now been stifled by the mud."[9]

In northeastern Georgia the situation was not so dire, perhaps, but wasteful farming practices still reduced yields and exhausted the soil. With the increasing use of Peruvian guano after 1846 and of chemical fertilizers after 1850, some of the limitations of worn-out land could be overcome, although most of Georgia's small farmers could ill afford that expense. In addition there was a growing awareness of the damage the South's devotion to cotton was doing to the region's economy. While small farmers remained more or less self-sufficient, by the Civil War the South was already importing hay from Maine, potatoes from Nova Scotia, apples from Massachusetts, butter and cheese from New York, pork from Ohio, and beef from Illinois. That situation only worsened after the war. Southern agricultural leaders urged diversification, recommending a greater emphasis on livestock, poultry, orchards, vineyards, vegetables, and fodder, but soaring cotton prices blunted the push for diversification and left the South dependent on cotton until the boll weevil made the point moot after World War I.[10] With the thriving apple industry in North Georgia today, it is difficult to imagine that it was not always so, but in 1860 the value of the state's orchard products was only $176,000 while that of its cotton crop was at $30 million. In White County barely $3,000 in orchard products was reported.

Jarvis Van Buren (1801–85) was a New York–born architect and contractor who came to Clarkesville in 1838 and built the Grace Episcopal church (1839–41), Woodlands (1847–51), and several other residences and other buildings in the area over the next few years. A man of many abilities, he bought ten acres and a family of slaves in 1840 and set to work establishing a nursery. He was especially interested in apples and roamed the orchards of the Southern Appalachians, gathering what became a large collection of seedling apples, many of which he painted in meticulous detail.[11] He was a prolific author as well and published numerous articles in the *Southern Cultivator*, which was published in Augusta from 1843 to 1872, and other influential agricultural journals promoting Southern orchards and vineyards. His work was especially important for attempting to preserve native varieties:

> Many of these [apples] were originated by the Cherokee and Creek Indians, who, it appears, were entirely ignorant of the process of propagating by grafting, but depended upon the sowing of seeds, which were collected in their intercourse with the whites. When the Indians left the country, their lands were occupied by our citizens, and since the enthusiasm for cultivating fruit has become awakened within the past ten years, these desirable varieties have been made public. Amongst our best winter apples are the Equinetely, Tillaquah, or Big Fruit, Chestoa, or Rabbit's Head, Elarkee, and Cullawhee, all of Indian origin—the latter the largest apple known.[12]

In 1853 Van Buren was among the organizers of the Horticultural Society of Georgia, later renamed the Pomological Society of Georgia. At the society's exhibition on 31 July 1860, he exhibited twenty-one varieties of apples and twenty-three varieties of pears. How many of his apple and pear trees might have been planted at Nacoochee is not known, but it seems likely that Edward Williams and others at Nacoochee would have bought from Van Buren's famous nursery in the 1850s. Despite Van Buren's efforts, poor transportation hindered the development of a commercial apple industry, and not until 1895 did Harry R. Straight of Cornelia establish the state's first commercial orchard. Others followed, including the Rabun County orchard of Col. John Porter Fort, who won prizes at the national apple show in Spokane, Washington, in 1908 and 1909. A four-hundred-acre orchard was also established at Yonah in 1907.

Agricultural societies were established and agricultural fairs held in the late 1840s and 1850s across the state, and Edward Williams was apparently a strong supporter. The Southern Central Agricultural Society held its first Agricultural Fair and Internal Improvement Jubilee at Stone Mountain in August 1846; the fair continued there until 1850, when it was moved to Atlanta. In 1851 the fair was held in Macon, Georgia, and, as the Georgia State Fair, has been held continuously since that time except for four years during the Civil War. Edward Williams attended nearly all the agricultural fairs and probably also subscribed to the *South Countryman*, the *Southern Cultivator*, and some of the other agricultural journals of the period. At the state

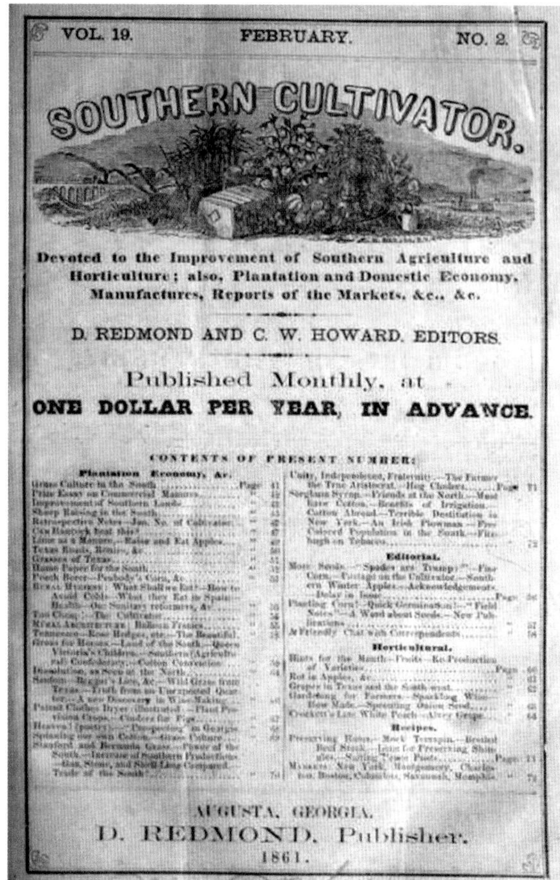

VOL. 19. FEBRUARY. NO. 2.

SOUTHERN CULTIVATOR.

Devoted to the Improvement of Southern Agriculture and Horticulture; also, Plantation and Domestic Economy, Manufactures, Reports of the Markets, &c., &c.

D. REDMOND AND C. W. HOWARD, EDITORS.

Published Monthly, at

ONE DOLLAR PER YEAR, IN ADVANCE.

CONTENTS OF PRESENT NUMBER:

Plantation Economy, &c.

Editorial.

Horticultural.

Recipes.

AUGUSTA, GEORGIA.

D. REDMOND, Publisher.

1861.

The *Southern Cultivator*, one of the South's most influential agricultural journals in the mid-nineteenth century. Biodiversity Heritage Library, Smithsonian Libraries and Archives, Washington, D.C.

fair in 1852, he was awarded a twenty-five-dollar silver cup for "best acre of upland corn." According to his biographer, he was one of the state's "large corn growers," but precisely how large that might have been remains unknown since he cannot be located in the 1850 agricultural census.[13]

George Walton Williams credits his father with the introduction of new grasses into the valley, including herd's grass, timothy, and clover. Timothy is a common perennial grass that grows over most of Europe and became popular for fodder and hay in England in the mid-eighteenth century and in the United States in the early nineteenth century. Clover too was introduced into North America from Europe. Thomas Jefferson touted the benefits of clover in restoring and maintaining soil fertility. Williams most likely brought seed with him in the 1820s and so was among the first to plant timothy and clover at Nacoochee.[14]

Much has been written about the "cheese dairy" Edward Williams built on Tray Mountain in the 1840s, beginning with Charles Lanman's account in 1849. Vermont-born Joseph E. Hubbard (1823–85) is credited with inspiring Williams's interest in cheese during a chance meeting as he traveled through Nacoochee in the late 1840s. Hubbard must have taught Williams well: at the state fair in 1849, and again in 1850, Williams received the first premium, or prize, of five dollars for "a specimen of cheese, of superior quality." That and another entry from Chattooga County elicited extra praise in the 1849 fair's final report: "In fact these samples of Georgia made cheese loudly proclaim that we need no longer be tributary to any other section for a superior article of cheese." The report also included Edward Williams's own description of the operation of his Nacoochee dairy. Besides noting the fans powered by an overshot waterwheel that ran continuously to keep flies off the cheese, he stated that "often no coloring matter is needed in the manufacture of our cheese, and . . . the cattle have no food, save the herd's grass, and native grasses of the valley."[15]

Operated by slave labor, the dairy was at first located part way up Tray Mountain, but by the 1850s Williams had moved it to the valley, perhaps near Williams Mill at Sautee Creek since he continued to use water to power fans that ran continuously over the cheese. The dairy did not survive Williams's death. Although one George Waring was operating a small dairy on the Soque River in 1860, it must not have survived for very long either. In 1862 Charles W. Howard, one of the state's best-known agriculturalists, stated that no cheese was being made anywhere in Georgia. He criticized the state's farmers and planters for their dependency on northern dairy products, which had disappeared from the southern market as soon as the war broke out. The price of butter rose to fifty cents a pound, when it could be had at all.[16]

Gentleman Farmers

The continued preservation of the rural character of Sautee Nacoochee is due in no small part to the succession of large landowners who were not dependent on agriculture for their livelihood. While they enjoyed an agricultural lifestyle, and generally expected the farm to pay for itself, they were independently wealthy, either through inheritance or professions, and did not depend on income from the land to sustain themselves. The era of the gentleman farmer at Sautee Nacoochee began with the purchase of Daniel Brown's old homestead by James H. Nichols (1835–97) for $5,594 in February 1869. Nichols had been a merchant and druggist in Milledgeville in the late 1850s and soon married Kate Latimer. They had one child, Anna Ruby (1860–1947), namesake of the famous waterfalls on Smith Creek in northern White County.[17]

When the Civil War broke out, Nichols enlisted as a captain in the Confederate army. Rising to the rank of colonel, he fought in Lee's Army of Northern Virginia and was paroled at the end of the war. He came home to find that Milledgeville had been ransacked but not burned by Sherman's army. Worse, his wife had been assaulted by Union soldiers and never got over the ordeal. In 1869, at the western end of Nacoochee Valley, Nichols built what remains one of the state's most magnificent private residences. Called West End, it is the focal point of an extraordinary collection of twenty-three outbuildings, fourteen built before 1890, and a large working farm that was valued at $18,000 in 1870. That year he had 300 acres under cultivation or in pasture and another 1,500 acres of woodland and "waste" area, figures that do not do justice to the estate. In an 1892 publication titled *Health Resorts of the South*, George Chapin wrote, "Captain Nichols has gathered around him everything that makes life pleasant, a large farm, well stocked rich fields, trained hounds, and plenty of game, fish ponds, a choice library, billiard room, gas, pure spring water throughout, green house, and fountains."[18] Kate Nichols was eventually institutionalized, and, after Anna Ruby's marriage, Captain Nichols was alone in the house. In 1893 he sold West End and moved to Atlanta. He died on 19 November 1897 while visiting Nacoochee.

The new owner of West End was Calvin Welborn Hunnicutt (1828–1915), a native of Mecklenburg County, North Carolina, who moved to Atlanta in the late 1840s. In 1857 he married Letitia Ann Payne (1833–86). She is said to have had seven children, but only six have been identified: Luther (1858–99), Joseph Edgar (1863–1926), Mary (1861–1933), Letitia (1865–1965), Eddie (1867–1900), and Sarah (1871–1948). As Atlanta's business district went up in flames in November 1864, Hunnicutt lost the small fortune he had made as a wholesale merchant and druggist. "Undaunted," according to his obituary, he was part of the postwar city's astounding recovery. By 1870 he and Albert Belingrath had formed the firm of Hunnicutt and Belingrath, wholesale merchants for a variety of "home furnishings," especially plumbing and gas lighting fixtures. Another fortune made, he built a large house on Spring Street in Atlanta. In 1877 he was appointed to Fulton County's newly authorized Board of Commissioners for Roads and Revenue and served for fourteen years, including eight as commission chair.

After the death of his wife in September 1886, Hunnicutt continued to live in his big house on Spring Street, along with three of his daughters and their husbands and several grandchildren. In 1893 he bought West End from Captain Nichols, intending it as a retreat for his large family, although it is not clear exactly how much it was used. The deaths of his oldest son, Luther, in 1899 and daughter Eddie the following year were a severe blow, and Hunnicutt turned over his business to his son Joseph. In 1903 he sold West End.[19]

West End, built by James H. Nichols in 1869. Photograph by author, 2015.

The new owner was Dr. Lamartine Griffin Hardman (1856–1937), a native of Jackson County, Georgia, and the small town then known as Harmony Grove but renamed Commerce in 1909. The son of a prosperous Baptist minister and doctor, he graduated from the Medical College of Georgia in 1877 and went on to postgraduate study at the University of Pennsylvania and the famous Guy's Hospital in London. Returning home, he established a medical partnership with his brother, and they went on to build a hospital in Harmony Grove that provided a much-needed alternative to the hospitals in Athens and Gainesville.[20]

Hardman was an entrepreneur on a grand scale, and much of his work was focused on his hometown. In 1893 he and his brother organized Harmony Grove Mills, producing cotton sheeting and thread, the first industrial development in a town where cotton mills would dominate the economy through much of the twentieth century. Lamartine Hardman served as its president from 1899 until his death. In addition he opened the Hardman Drug Company, was one of the organizers of the Northeastern Banking Company, and founded the Commerce Telephone Company. Hardman was also a gentleman farmer on a large scale, reportedly one of the state's largest farmers by 1900. In 1902 he was elected to the state legislature, representing Jackson County. As a state senator from 1907 to 1908, he drafted the legislation that would prohibit alcohol sales in the state. After two failed attempts, Hardman was elected governor in 1926 and reelected in 1928.[21]

While West End was never a full-time residence, Hardman enlarged the farm and turned it into a productive, year-round endeavor that, after 1907, included a dairy. After his death in 1937, ownership of the property remained in the Hardman family, but by the 1980s most farming operations had ceased. Beginning in the 1990s, Governor Hardman's descendants worked with the Georgia Trust for Historic Preservation and the Trust for Public Land to develop a plan for the continued preservation of the property. In 2002 ownership was transferred to the Georgia Department of Natural Resources, which now operates it as an historic site, ensuring preservation of one of the state's most important landscapes.

Gristmills and Sawmills

Most industry in antebellum Georgia was related to agriculture, which formed the base of the state's economy until the last half of the twentieth century. There was perhaps no more necessary industry than gristmills, which were sometimes called cornmills. A few could mill flour, but that was a more laborious and expensive operation. As a practical matter, most mills in the South ground corn to make cornmeal, a staple in the diet of all Southerners, Black or white. From the beginnings of agriculture, grain had to be ground to make bread. By the end of the first millennium CE, water-powered mills were operating in Asia Minor and Europe. By the Middle Ages, there were thousands of them in England, generally just a few miles apart and serving a community of a few hundred people. Millers typically made their living by grinding their neighbors' grain and taking what was generally known as a "miller's toll" in the form of a percentage of the meal or flour that was ground.

Waterwheels were of two principal types. A horizontally mounted paddle wheel created a "tub mill," in which the power of the turning wheel was transferred directly to the grinding stones.[22] Much more efficient were vertically mounted wheels, with the moving water directed either over or under the wheel, and a system of gears that multiplied the power to turn the grinding stone, which caused it to turn much more rapidly than the wheel in a tub mill. Water-powered mills depended on the force of falling water, so they were typically located where there were natural falls or shoals and where a change in elevation could be enhanced by the construction of a mill pond that would retain enough water to operate the mill. As a result, certain locations were the site of multiple mills, sometimes simply renamed when there was a new owner or they had to be completely rebuilt due to flood damage or to accommodate new technology.

Antebellum sawmills were also water powered and were often part of the same mill, where, with a shift in the mill's power train, the miller could saw lumber during off-seasons. Wooden gears and framing were inexpensive for these "sash saws," which used a reciprocating, up-and-down motion to do their work, but there was often not enough water power to saw large timbers. Until the advent of affordable steam-powered and portable sawmills after the Civil War, sills, corner posts, and other large house-framing members continued to be worked by hand.[23]

Some of the Cherokees no doubt operated gristmills, but if any of those survived the resettlement of the valleys, they have not been identified. By 1840 twenty-two gristmills and nine sawmills were in operation in Habersham County, about half of each in what would become White County. In 1850 thirty gristmills, twenty sawmills, and a single flour mill were in Habersham County. By 1860 the combined value of flour and meal produced in Georgia totaled more than $4.5 million, more than any other manufactured product in the state. Seventeen mills were in operation in White County in 1870, but several more were built as the agricultural economy began to boom in the late nineteenth century. Many of the mills ceased operation as that same economy collapsed after World War I, but a few continued into the 1960s. Nora Mill is one of the few active mills that remain in the state.

The dam built by John Martin in 1893 at what is now Nora Mill, shortly after being reconstructed to replace Daniel Brown's dam from the 1820s. The mill race had not been completed. T. Lumsden, *Nacoochee Valley*.

NORA MILL

The site of Nora Mill is one of the best mill locations in White County. Located on the Chattahoochee River in Land Lot 57, this grist- and sawmill is believed by some to have been built by Daniel Brown in 1824. It is reported, probably accurately, to have been the first dam on the river. That date may be accurate, but, according to Habersham County land records, Brown did not acquire the property until after the spring of 1825. His son James Harwell Brown (1812–57) continued operation of the mill after his father's death, but it appears that it may have ceased operations after the younger Brown's death. Captain Nichols acquired the property in 1869, when it was referred to as "J. H. Brown Old Mill place" in some of the quitclaim deeds associated with the sale, but George Walton Williams makes no mention of a mill in his narrative tour of Nacoochee in 1874.[24]

In 1876 Nichols sold the mill site to John Martin (born February 1857), a Scotsman who came to White County prospecting for gold. In addition to the large house that Martin built across the road from the mill site, he built a new mill using a waterwheel constructed by Vince Sims, who had just built his own mill on Dukes Creek, north of what is now SR 356. In 1893 Martin rebuilt Brown's old mill dam, which was rebuilt again in the early twenty-first century. In 1905 Martin sold the mill to Lamartine Hardman and moved away. Hardman named the mill after his late sister, Nora. He installed a generator that provided electric lighting for his farm as well as for the Crescent Hill Baptist church. It is not clear when the mill ceased operations, but in 1998 a group of investors bought it from the Hardman family, and, with the help of Ron Fain, a descendant of some of the county's earliest settlers, it was put back into operation.

WILLIAMS MILL

At the opposite end of Nacoochee Valley, another prime mill site presented itself, not on the river but rather on Sautee Creek. A mill may have first been built on the site by James R. Wyly, the contractor who built the Unicoi Turnpike and whose hotel was just east of the creek. Long known as Williams Mill, it was acquired by Edwin Williams after Wyly's death in 1855. In 1874 Edwin's brother George noted of the location, "Here we find a grist mill, saw mill, and blacksmith shop, all in full blast. The old Confederate gun factory has been converted into a work shop. It was in this building Col. E. Williams turned out during the war 'pikes' by the thousand for Governor Brown's Georgia Militia."[25] Lumsden notes that there had also been a tanyard at the site, another one of Edwin Williams's many enterprises. In 1989 Dr. Thomas Lumsden wrote that "the heart pine mudsill of the dam" was still in place about 150 yards upstream from the SR 17 bridge, as was the foundation of the mill itself just downstream of the dam on the east side of the creek.[26]

STOVALL MILL

On Chickamauga Creek, just upstream from the present SR 255 bridge, an exposed gneiss shoal offered another mill site, and Rev. Wilkes Timothy Leonard (1804–62) built a dam and gristmill there in the early 1830s, but it is unclear how long it operated.[27] After the Civil War, the mill site was acquired by Frederick Dover (ca. 1827–before 1910), one of a large family of siblings and cousins who came to Habersham County in the 1820s. A farmer with nine slaves in 1860, the war cost him most of his personal wealth, but he was able to build a grist-, shingle-, and sawmill complex, along with a cabinet shop, on Chickamauga Creek in Land Lot 20. He also built a covered bridge over the creek, which was washed away in the early 1890s but replaced by the existing covered bridge in 1895. Dover may have died as early as the 1890s, and it is not clear how the mill was operated after that. In 1917 his heirs sold the property to William Isaac Stovall (1863–1945), who replaced the original overshot waterwheel with a turbine in the 1930s before ceasing operations in 1938. The mill and dam were destroyed in a flood in 1964.

OTHER LOCAL MILLS

As noted in the preceding chapter, an 1837 deed from James R. Wyly conveying Land Lot 22 to Moses Harshaw mentions Graham Mill Creek, which may have been an earlier name for Bean Creek. In 1823 Archibald Graham purchased Land Lot 22 and apparently operated a mill there. He died before 1830, when James and Thomas Graham, who were probably his sons, were enumerated in the Habersham County census. None of the family can be located in any other antebellum census of Habersham or White Counties, and no other details of the family have come to light.

Mauldin Mill stood south of Unicoi Road on the Habersham County side of the eponymous creek until after World War II. Little is known of the mill's history, including the identity of its builder. In 1850 the federal census enumerated three Mauldins who were heads of households in Habersham County: Francis Mauldin (1789–1860), Alexander Mauldin (1792–1868), and Jeremiah Mauldin (1795–before 1880). Born in South Carolina and probably brothers or cousins, they appear to have been prosperous farmers. All three can be identified in the antebellum federal census, but Habersham County land records have not been searched for any property they might have owned. Francis and his family moved to Towns County in the 1850s, while Alexander moved his family to Rabun County before 1860. Jeremiah; his wife, Keziah; and their children remained in Habersham County, and it may be that it was he who operated the mill. Moffat's map notes three mills: the mill now known as Nora Mill, Williams Mill, and a third mill, which was apparently located on Dukes Creek in Land Lot 91. In the 1850 census, the first to show occupations, no millers were shown in the Nacoochee District, although mills were certainly there. The Richardson-Lumsden House (1830) is believed to have been built using lumber from a sash sawmill on Dukes Creek that was operated by John Greer Lumsden (1789–1845), whose son Thomas Reid Lumsden (1821–1914) married Sophronia, daughter of Rev. John L. Richardson. No other documentation for that mill has been located.[28]

While most of the farmers in and around Sautee Nacoochee would have used one of the gristmills noted earlier, there were other mill sites in northern White County, some used by a succession of owners. One of those sites was at the junction of Dover and Dukes Creeks in Land Lot 64, about a mile west of SR 75 Alt, as the crow flies. John Lee Pardue (1839–1919) may have established the first gristmill at that location as early as the late 1850s, but it was destroyed in a flood.

Buckner Vincent Sims (1849–1928) built another gristmill at that location in 1875, around the same time that he built a waterwheel for John Martin's rehabilitation of Daniel Brown's old mill on the Chattahoochee River. Sims also operated a grain thrasher, a syrup mill, and a woodshop there until the mid-1920s. In 1929 his son Richard B.

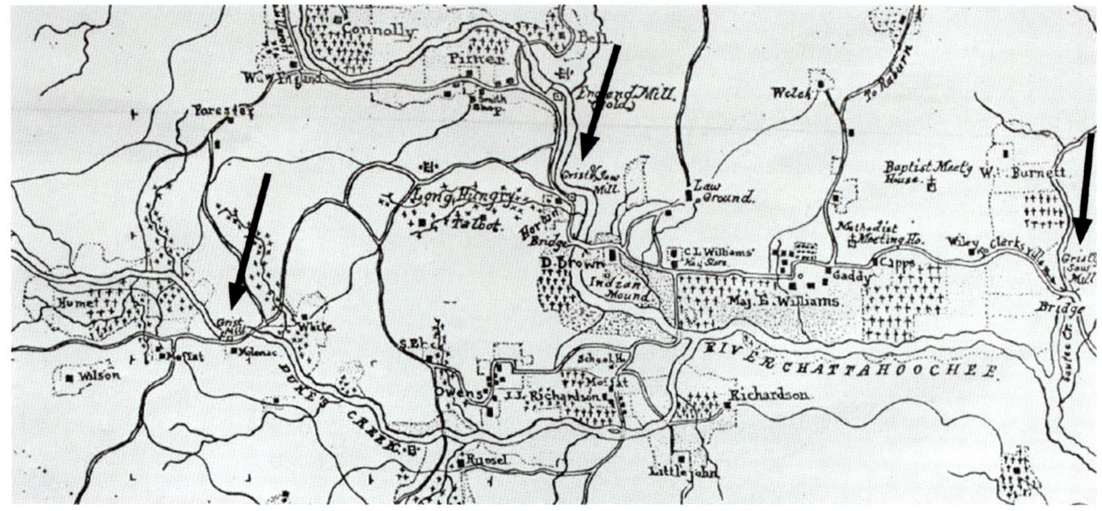

Detail from Moffat's map of Nacoochee in 1837, annotated to locate the three mills depicted on the map. The site of Stovall's mill on Chickamauga Creek is not encompassed by this map. Most of the roads shown on this map can be correlated with roads shown on the 1937 USFS map, but many of them are no longer in existence in the twenty-first century. Courtesy of Sautee Nacoochee Community Association.

Sims (1888–1978) built a new grist- and syrup mill and continued to operate it until 1963.[29] Finally, when Captain Nichols bought part of the estate of Jehu Trammell (1793–1869) in 1870, the deed references the mill Trammell operated on Smith Creek in Land Lot 27, just a few hundred yards upstream from the Chattahoochee River. It is likely that Trammell's mill remained active throughout the antebellum period.[30]

Distilleries

Along with gristmills and sawmills, distilleries were ubiquitous in most agricultural societies, and the antebellum South was no exception. While more prosperous people in rural areas could afford imported Madeira wines, one of the few wines that actually benefited from a rough sea voyage, most people could not. Typically, numerous small stills were operated by farmers and millers producing alcohol for local sale and consumption. Historians have long held that stills were necessary because poor transportation made it impossible to get large quantities of unprocessed grain to market. In reality subsistence farmers had little produce to spare, and distilled grains offered an easy way to raise cash to pay taxes and to buy sugar, coffee, gunpowder, salt, and the few other necessities that could not be produced on the farm. Taxes aside, many people just enjoyed their whiskey and brandy. The rising campaign in support of liquor prohibition in the early nineteenth century was one of the issues that caused a schism in the Baptist Church even before the great split over slavery in 1845.[31]

In the early colonial period, imported beer, wine, and rum were the drinks of choice, but their increasing cost helped drive an increase in domestic production of hard ciders, brandy, and whiskey. In 1769, for example, a five-hundred-acre farm near Augusta was advertised for sale, along with an eight-acre peach orchard, from which "a great quantity of peach brandy" could be made.[32] Southerners used a wide variety of fruits for fermenting. Martha Washington kept recipes for making wines from lemons, gooseberries, blackberries, and elderberries, but efforts toward domestic grape production routinely failed. Southerners preferred apples and peaches, with the latter often grown because they bore fruit several years earlier than apples. In the late eighteenth century, new apple varieties were developed that were sweeter and bore fruit much faster than older varieties. In addition alembic stills became cheaper and more widely available and helped bring about a great expansion in both cidering and brandy distilling.

After the Revolution British restrictions on trade drove the cost of rum out of reach of most U.S. consumers, precipitating the rise of domestic whiskey production. Corn could, of course, be fermented, but the result was a raw, fiery drink. In Scotland distillers preferred barley, but it did not grow well in the South, and rye quickly became the grain of choice. Producing a third more alcohol than wheat, rye also produced a much smoother whiskey than corn. As a result, many farmers, including those in and around Sautee Nacoochee, planted small amounts of rye, almost all of it used for whiskey making. It is interesting to note, perhaps, that domestic brandy and whiskey were never aged and so were virtually colorless. By the late nineteenth century, whiskey aged in oak barrels, which imparts its distinctive golden hue, was the most common conception of the drink.[33] In 1840 the federal census of manufactures for Habersham County counted thirteen distilleries, producing 4,635 gallons of distilled or fermented liquors. Neighboring Lumpkin County produced 6,280 gallons while Franklin County, which had the lowest population, produced 13,000 gallons of spirits in 1840.

George Walton Williams was proud that his father abstained from alcohol, and he himself was inclined toward prohibition as well. In

1870 he wrote, "I am glad to say, that among all the laborers I have not seen a man under the influence of liquor. This speaks well for these honest-hearted people. Liquor is not allowed to be sold in Nacoochee Valley and I hope never will be."[34] None of the memoirs of the valleys mention stills, although they were probably present, even if they were just small stills set up from time to time to distill whatever grain or fruit was available. Since distilleries remained entirely legal, they were not hidden away and often were housed in permanent, specially built structures, always close to a spring or branch. Such a structure must have stood on the small branch that flows into Sautee Creek from the north along the line between Land Lots 18 and 19. A "stillhouse branch" was noted at that location when Moses Harshaw bought part of Land Lot 19 in 1834.[35]

In Great Britain attempts to tax whiskey always resulted in a sharp rise in illicit production of alcohol, and the same was true in the United States. The country's first constitutional crisis in the form of the so-called Whiskey Rebellion in western Pennsylvania in 1794–95 was the result of a new excise tax on whiskey producers, a tax that was soon abolished. Until the Civil War, excise taxes were levied only on commercial producers. The farmer with his pot still, producing small amounts for local consumption, remained untaxed and legal. That changed in 1862, when the federal government announced a twenty-cent excise tax on each gallon of domestically distilled spirits, which increased to two dollars a gallon by the end of the Civil War. For the average southern farmer, that rate was intolerable, since whiskey had sold for as little as twenty-five cents a gallon in 1860.

Along with military occupation, the agents of the Bureau of Internal Revenue came to Georgia and set to work attempting to collect the whiskey tax. What followed was North Georgia's "moonshine war," a decade of fierce resistance and conflict that included the murder of revenue agents and distillers alike. Since the smell of the distilling process was impossible to disguise, whiskey making was transformed from a legal, open operation before the war to a series of illegal stills hidden deep in mountain coves. For much of the next century, federal revenue agents battled illicit liquor production—and were largely unsuccessful in their efforts.[36] So began the era of southern moonshiners, so called for their habit of running their stills at night, when they were less likely to be discovered. These illicit distillers, however, called themselves "blockaders," doing a necessary service that was only technically illegal, much like the Confederate blockade runners during the Civil War.[37]

After decades of effort, the temperance movement forced the prohibition of alcohol in Georgia in 1908 and across the nation in 1919. Along with that came a boom in moonshine whiskey, especially in the mountains of northeastern Georgia. With automobiles making long-distance travel practical, moonshine from White County and elsewhere was easy to sell in Gainesville and Atlanta. Through the first half of the twentieth century, the county was "in the upper ranks" of illicit liquor production, and, especially during the logging boom, "whiskey and moonshine flowed freely."[38] National prohibition was repealed in 1933, but the demand for moonshine and the easy profits that one could make, provided one was not caught, kept business strong. Better methods of detecting illicit stills and stiffer penalties finally put an end to the era of the moonshiners in the 1970s.

Crafts and Trades

Farming was a way of life for most people in White County until after World War II. Even so there were always individuals who learned and practiced specialized trades, many of them passed from father to son. Carpenters and cabinetmakers were enumerated in the antebellum censuses of what is now northern White County; similarly, masons and other mechanics of the building trade have always lived and worked in and around Nacoochee and continue to do so. There were several blacksmiths working in the area in the nineteenth century, as well as a cooper, a wagon maker, a gunsmith, a shoemaker, and a weaver. At least one woman at Nacoochee was always able to find work as a seamstress, and a "taylor" was even recorded living next door to Charles L. Williams in the population schedule for the 1850 census. Most of these tradespeople farmed too, and may have worked at their trade only as farmwork slowed down.

After the Civil War, some of these occupations disappeared, and the character of others changed considerably. Cheap, mass-produced hardware, shoes, clothes, and guns, for example, drove the tinsmiths, shoemakers, weavers, hatters, gunsmiths, and other small craftspeople out of the market and reduced blacksmiths to shoeing horses for a living. At least two wheelwrights continued to make and repair wagon wheels in the 1880s, and a shoemaker continued to ply his trade. By 1900 there were only two blacksmiths, and no one in the Nacoochee District was employed in any of the traditional crafts and trades, except for the building trades. Those, too, changed, as the advent of "balloon-framed" construction made carpentry much simpler and joiners redundant.[39]

Labor

Although most were farmers or farm laborers, many men gave an occupation of "day labor" or, in the early twentieth century, "odd jobs" in the population schedules for the federal census. They were mostly unmarried and young and usually but not always African American. A few unfortunate families were headed by men such as these who were only seasonally employed. In addition to employment in asbestos and gold mining, a few men were employed by the railroads by 1870 as graders, spike drivers, and drillers. It was a job that would have required days away from home until the lines were brought closer to Nacoochee, but it continued to be a source of local employment as late as the 1930s. Even before the giant sawmill at Helen opened in 1913, small local saw- and gristmills employed a few locals and continued operating until most closed in the 1930s.

Many people, Black and white, found work in providing services for Nacoochee's gentry. Gov. Lamartine Hardman's expansion of his farm operations, which included the establishment of a dairy, gave employment to several Nacoochee residents and their families before World War II. In the 1930s dozens of unemployed men and women went to work for the Civilian Conservation Corps, the Works Progress Administration, and other federal programs created by the New Deal during the Great Depression. Roadbuilding projects employed several men as mechanics and laborers, and a number of residents of Nacoochee must have participated in Wauchope's archaeological survey of northern Georgia in 1938 (see chapter 2, page 41).

Women had decidedly fewer employment opportunities in the nineteenth century and then mainly if they were single or widowed;

once they were married, work outside the home was avoided if possible. Teaching school was often an option for women, and there were two nurses at Nacoochee in 1900. Until more job opportunities opened for women after World War II, women continued to provide the most mundane services. Census data suggest that there were always one or two women at Nacoochee who were seamstresses, making and repairing clothes for the community. A few cooked for private families, boarding houses, or hotels, and, as always, women provided most of the labor in running a household.

After the Civil War, more than a few girls and young women, mostly white, were employed as "domestic servants." Beginning in 1850, the decennial census recorded individual occupations; when the 1870 census was taken, for example, three sisters from one family and four from another were all employed as servants. In the first quarter of the twentieth century, the occupation was often described as "maid," but the duties did not necessarily change. Finally, there were always several women who made a living doing "washing and ironing," as the job was often listed in the census, often doing the laundry of others in their own home. The widespread use of mechanical washing machines in the early twentieth century, and then electric washers in the 1920s, finally destroyed the demand for their labor.

Professional Services

In the earliest years of white resettlement of Nacoochee, professional services were apparently not found outside the county seat of Clarkesville, where there were two doctors and four lawyers in 1829. In 1846 the *Augusta Sentinel-Chronicle* reported of Nacoochee, "There are four or five families who have lived in this valley all in sight of the river for twenty-three years, who have never had a case of fever and ague—and no resident physician in the valley. The many [doctors] who have settled here from time to time, have been very unceremoniously starved out."[40] By 1850, however, Dr. Elijah F. Starr had married Edward Williams's daughter and set up practice in the valley, and the federal census continued showing a doctor in the Nacoochee District through 1940.

Retail Trade

Self-sufficiency was the primary goal of most nineteenth-century farmers, but no one could live without a few necessities that could not be produced on the farm. Salt was the primary necessity, since it made possible the preservation of meat, as well as cabbage and other pickled vegetables. Gunpowder and ammunition had to be acquired as well, since most farmers depended on at least some hunting to supplement what they raised on the farm. Then, as now, coffee was deemed an essential by many, and there was always a demand for sugar, finer fabrics than coarse homespun, needles and thread, and farm implements, as well as clothing and shoes. George Walton Williams began his career as a merchant on Wall Street in Athens, Georgia, in 1839 and later wrote, "We did not deal in fancy stocks, but drove an honest trade in exchanging sugar and molasses, for eggs, chickens and butter!"[41]

As the network of railroads expanded in the decades before the Civil War, a much larger array of merchandise became available, including large quantities of salt pork and other staples that were not being raised locally. After the Civil War, general stores dispensed kerosene for lighting, along with lamps, chimneys, and wicks. In the early twentieth century, many of them began selling gasoline as well. They also often housed a post office and naturally became a place to meet and talk with neighbors or just while away some time. They were the secular community center on weekdays, as the local church was the religious center on Sunday.

Well before the Gainesville & Northwestern Railroad was completed to Helen in 1913, the people of White County benefited from the railroads. In 1872 the Atlanta & Richmond Air-Line Railway was complete from Atlanta to Cornelia, seventeen miles south of Sautee Nacoochee; ten years later the Blue Ridge and Atlantic Railroad was built from Cornelia through Clarkesville, just eleven miles away. In addition mail-order catalogs from Montgomery Ward (1872) and Sears Roebuck (1886) increased the scope of merchandise available to even the most rural Georgians.

With cash scarce in many areas through much of the nineteenth century, much trade was by barter, with farmers bringing surplus butter, eggs, vegetables, hides, and furs to exchange for the manufactured goods they needed. Especially after the Civil War, merchants' willingness to extend credit was a critical link in the economic survival of many farmers as they awaited the next harvest of corn and cotton. At the same time, bad luck or improvidence trapped many sharecroppers and tenant farmers, white and Black, in a vicious cycle of increasing debt that left them only a step removed from slavery.

The department and specialty stores that were made possible by the proliferation of manufactured goods and the rise of commercial advertising in the late nineteenth century were a far cry from the country store. After World War I better roads and the increasing use of automobiles made it possible for rural Georgians to participate in the consumer revolution, and one by one the country stores began to close. In more rural places such as White County, where none of the roads were paved until after World War II, the general stores remained a necessity for a little longer than in more developed parts of the state.

WILLIAMS STORE

The first stores in Habersham County were in Clarkesville, the county seat ten or twelve miles east of Nacoochee. In 1829 Sherwood's *Gazetteer* noted four stores in that town, but, if the Williams family tradition is correct, there was also a store at Nacoochee by that time as well. Located at the head of the intersection of the Unicoi Turnpike and what was then the main road from Nacoochee to the south and southwest, the Nacoochee store was operated by Charles Lathrop Williams (1811–86), the eldest son of Edward and Mary Brown Williams. According to his younger brother, George, Charles began "trade at the early age of twelve," which would have been about 1823. No other details of this early venture have been documented, but it is likely that the young Williams was a clerk in a store established by his father or, perhaps more likely, his grandfather, Daniel Brown, who had operated a store in Burke County prior to his migration to Georgia. Whatever its origins, the Williams store was the heart of historic Nacoochee from the 1820s until after the Civil War.

Lumsden and others state that the Nacoochee Post Office was established in Charles Williams's store in 1826, and that Williams served as postmaster for thirty years. To have been appointed postmaster while still a minor seems unlikely, but he did serve in the position until 1856. According to the National Register nomination for Nacoochee,

Williams built what is now known as the Dyer House in 1828, while still a teenager and several years before his marriage in 1836. In the meantime he made a fortune during the gold rush buying and trading prospectors' gold for merchandise. The small store, which stood in front of his house, was probably contemporaneous with the house, and when the store was torn down around 1900, a significant amount of gold was retrieved from the dirt under the building.[42] The Williams store was a family affair, and at least two of his sons worked in the store, which was still in operation in 1880. The business appears to have gone far beyond selling dry goods, however. The largest slaveholder in the valley, Williams was also a major agricultural producer and had the largest herd of cattle in the valley. Given the number of milk cows he owned (eighteen in 1860), he probably operated a small dairy as well. He also owned thirty-eight mules in 1860, far more than he could have used on his farm, suggesting that he may have rented or sold them.[43]

Williams prospered after the Civil War and continued to operate his store on the Unicoi Turnpike. In the 1870 census he was not simply a

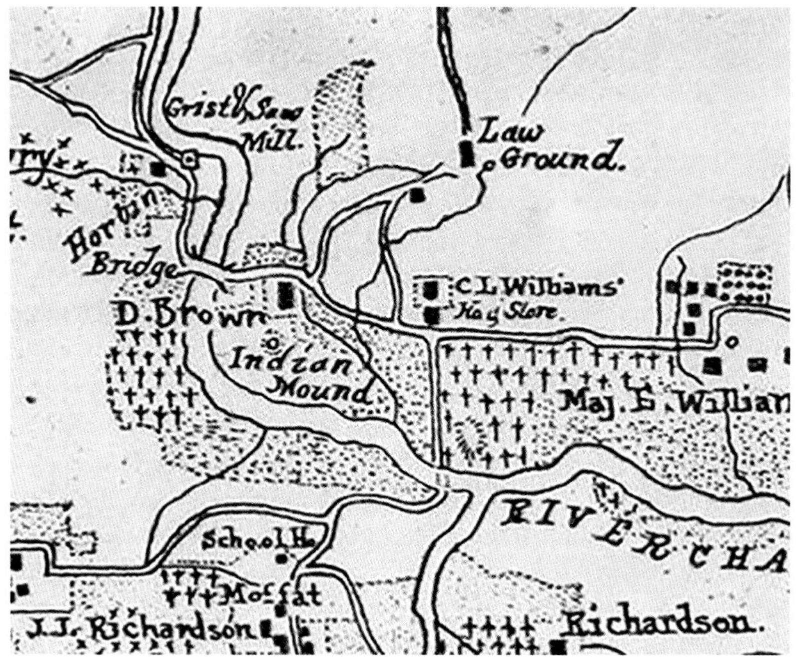

Detail from Moffat's map of Nacoochee, locating Charles L. Williams's home and store at the westernmost of two important historic intersections in Nacoochee Valley in the nineteenth century. Courtesy of Sautee Nacoochee Community Association.

"merchant" but rather a "dry goods merchant." The railroad would not be completed into Cornelia until 1872, but the array of merchandise that would have been offered was far greater than it had been even in 1860. His son Charles Lee Williams (1849–1930) clerked in the store for a few years, but by 1880 he and his wife, Clara Felicia Jamison (1857–1942), were in Washington, D.C., where both spent their working lives in the federal bureaucracy before retiring to Nacoochee, where they are buried.

Charles L. Williams Sr. continued to farm, of course, and operate his dry-goods store for the rest of his life. His son Thomas C. Williams (1838–96) apparently took over the business after his father's death in 1886, but there is no census or other documentation to confirm that. After Thomas died rather prematurely, their neighbor John H. Alley succeeded him as Nacoochee postmaster, followed by his wife, Katie, who was postmaster when the post office was moved to Hardman's brick stores in 1913. The old Williams store probably closed around that time and was soon torn down. For generations the center of "Nacoochee proper," as his brother put it, had been Charles Williams's store. With removal of the post office to a site closer to the railroad in 1913, a new Nacoochee was born.

SAUTEE STORE AND POST OFFICE

It is not clear if the present Sautee Store (ca. 1872) was the first store at or near that location where the valleys intersect. It was an ancient intersection of trails and later roads, and not by accident did Edwin Williams build his house up the hill opposite it in 1845. Perhaps Charles Williams's store farther up the valley was able to serve the smaller, antebellum population, but, as the economy recovered and the population grew after the war, the cramped old-fashioned stores typical of the second quarter of the nineteenth century proved inadequate to the explosion of merchandise available, as railroads and mail-order catalogs connected people to a burgeoning consumer economy.

Completion of a railroad from Atlanta through Cornelia in 1872 was surely part of what precipitated the construction of a second general store in Nacoochee Valley, as was perhaps a bit of sibling rivalry. Charles Williams's sons, Thomas C. Williams (1838–96) and James H. Williams (1845–1909), both worked in their father's store, but, as he came of age after the Civil War, James took advantage of the situation and opened his own store at the opposite end of the valley. With increasing consumer demand, they may not have even been real competitors. The building was also the site of the Sautee Post Office from the time the office was established in 1913 until 1959, when its services were combined with the old Nacoochee Post Office to create the present Sautee Nacoochee Post Office. The store was rehabilitated in the 1970s and reopened by Astrid Fried as a gift shop specializing in items from her native Norway. The current owners have retained mementos of the past in and around the ancient building, making it a popular attraction for locals and visitors.

HARDMAN STORES

Construction of the railroad through Nacoochee included construction of a depot in 1912, near where the Unicoi Road crossed the river. About the same time, Lamartine Hardman constructed a brick duplex commercial structure just southeast of the depot. Hardman's first tenant was Parks Lester Hood (1880–1971), a younger brother of William Leonard Hood. In 1902 he married Birdie Bell Oakes (1884–1911), and they had a son, William. After her death he married Mary Charlotte "Lottie" Jenkins (1892–1983), and they had two sons and two daughters. After the railroad closed in the early 1930s, Hood relocated to a store across from his house on the Unicoi Turnpike, three quarters of a mile west of its intersection with SR 255. For many years his store had the only gasoline filling station in the valleys.

The other tenant in Hardman's stores was Thomas B. Henderson (1872–1957). The son of John Henry and Mary Henderson and a native of White County, the younger Henderson married Zora Mae Conley (1869–1922) in 1895. They had at least five children: Mary, Thomas Jr., John C., Henry H., and Dolly A. Before 1900 Henderson began operating a store at Aerial, a couple miles southeast of Sautee, and continued there until moving to Nacoochee. He was also the depot agent for the Gainesville and Northwestern Railroad until it closed in the early 1930s. Henderson's store remained in business into the 1940s.

The Nacoochee railroad depot in the foreground (*right*) and Hardman's store in the background (*center*), circa 1930. Courtesy of Sautee Nacoochee Community Association.

SAUTEE VALLEY STORES

In Sautee Valley there were at least six stores documented by the State Highway Board maps from 1940 to 1960, and four of these are documented in the *Sautee Valley Historic District National Register: Nomination Form*. There were at least two more on Bean Creek Road, but the chronology of all these enterprises is not well documented. In 1878 Edward Perry West (1857–1937) acquired Moses Harshaw's old house in Sautee Valley, and he may have been responsible for building the small store and post office that still stood along the road in front of the house until around 2005, when it was torn down. William Isaac Stovall (1863–1945) bought the property in 1893, around the same time that James E. Hood (born 1856) was made postmaster of the new Lynch Post Office, which operated out of the Stovall store until the post office was moved a mile or so north in 1905.[44]

Hood's Store

James E. Hood was the first postmaster when the Lynch Post Office was established in 1893 and, as noted earlier, may also have operated the store in front of the Harshaw-Stovall House. There were Hoods at Nacoochee before the Civil War, but it is unclear how James might be related to them. No details of his early life have been documented,

Undated photograph of Leonard and Bud Hood leaving their store in Sautee Valley. T. Lumsden, *Nacoochee Valley*.

but in 1900 the federal census enumerated him simply as a farmer at Nacoochee. William Leonard Hood (1878–1933), who is probably buried in an unmarked grave at the Nacoochee Methodist church, operated a store that sat on the south side of SR 255, where the road went through a sharp dogleg turn near the Stovall House. The dogleg was eliminated, perhaps along with the store, when the highway was realigned in 1950.

In 1902 Hood married Ethel Lyon (1888–1972), and by 1910 they and their two children, as well as his widowed mother, were living at Sautee, where they farmed some acreage next door to his brother Parks. At least by 1910, however, the census shows his occupation as "merchant," although it is not certain he had built his own store at that point. Hood, who sometimes used his middle name, and his wife, Ethel, operated the store through the 1920s. After his death in 1933, she continued with it until after World War II.[45]

West's Store

Walter Lumsden's map of Sautee Nacoochee in 1948 depicts the store that Orville B. West (1871–1964) operated in front of his house at that time, and both structures remain contributing structures in the Sautee National Register Historic District. Orville West was the son of farmers James B. and Jane West of Towns County, Georgia, but by 1900 was a merchant in the small crossroads community of Batesville in Habersham County. Much like his contemporary Leonard Hood, however, West had a farm at Nacoochee in 1910. Unlike Hood, however, the census does not document his occupation as a merchant but rather consistently enumerates him as a farmer. According to the National Register nomination for Sautee Valley, West built his house in 1921, along with the presumably contemporary gable-fronted store building just west of his house, both of which are still standing on the southeast side of SR 255, about a half mile northeast of the Harshaw-Stovall House.[46]

Sautee Store, circa 1900. T. Lumsden, *Nacoochee Valley*.

OTHER STORES

The 1880 census documents three other merchants in the Nacoochee District, in addition to the two stores in Nacoochee Valley, two of them descendants of the North Carolina pioneers in the early 1820s. James H. Westmoreland (1846–1933) was one of those listed as a "dealer in general merchandise" in the 1880 census schedule. That same year he married Julia Ann Craig (1857–1920), with whom he had eleven children. The business must have failed or he simply quit it, and he returned to farming. He is buried at Cleveland.

Virgil Robertson (1840–1934) was enumerated as a "farmer & dealer in general merchandise" in the 1880 census population schedules for Nacoochee. He had been born in Alabama, the son of Benjamin and Alif Robertson, but by 1860 the family was in Habersham County, where he married Huldah Logan (born 1844). As with Westmoreland, it is not clear how long Robertson was a merchant in White County, but, after his first wife's death, he remarried in 1886 and moved to Young Harris. In the early 1900s he and his family moved to California, where he died and is buried. Joseph A. Richardson was also a dealer in general merchandise in 1880. He was a son of Joseph and Mary Richardson and perhaps a nephew of John L. Richardson. Like Westmoreland, he apparently did not continue as a merchant but returned to farming. He died before 1920 and is buried at Nacoochee.

There were several other stores in the valleys, but they are poorly documented. A storehouse was apparently in existence when William Isaac Stovall (1863–1945) bought Moses Harshaw's old house in 1893, but there is no indication that Stovall himself ever operated a store. In addition to the Wests' and Hoods' stores, the highway maps document two store buildings near the intersection of Rabun Road and SR 255. One of those was associated with the house of William S. Allen and was used as the Lynch Post Office after he was appointed postmaster in 1905. The highway maps also depict two commercial buildings in the Bean Creek community that were still open in 1960. No other documentation for those stores has been located. Finally, a store was operating at Nora Mill by 1940 and may have been open for decades before that. It, too, sold gasoline after World War II. A store was also located on the south side of SR 17, just southeast of its intersection with SR 255, before World War II, but it closed in the 1950s.

Piggly Wiggly opened the first self-service grocery store in 1916. In the 1920s chain groceries such as A&P were common in the larger cities and towns. After World War II chain grocery stores opened in smaller towns like Cleveland and Clarkesville, slowly driving the smaller grocers out of business. A few small old-fashioned grocery stores that were not self-service survived until the 1960s, and those that survived into the late twentieth century found, like Sautee Store, a new niche market.

9

Mining, Manufacturing, and Timber

Agriculture remained the occupation of the vast majority of people in Habersham and White Counties until after World War I, but the exploitation of natural resources began at an early date. Most famously, the discovery of gold on Dukes Creek in 1828 precipitated a major gold rush, and steatite, iron, and asbestos have also been mined within a few miles of Nacoochee. Railroads made industrial logging operations possible, and, between about 1890 and 1930, much of the forests of northeastern Georgia was clearcut. Logging brought a railroad through Nacoochee and gave rise to Helen, but the more or less exhaustive exploitation of natural resources supported a local nonfarm economy only until the timber reserves were depleted. Conservation of natural resources became a byword of the early twentieth century, and the work of the U.S. Forest Service and New Deal programs such as the Civilian Conservation Corps began the restoration of a landscape that again supported the entire community.

Mining

People have been quarrying the rock and minerals in the hills around Nacoochee for thousands of years, beginning with soapstone, or steatite, which continued to be quarried into the historic period. Quartz was sought out for making points and knives since chert deposits were few and far between in northeastern Georgia. After 1829 gold mining remained a major component of the local economy through most of the rest of the nineteenth century. Local deposits of iron ore were also being quarried by 1830, and foundries on the Soque River two miles southwest of Clarkesville and at Mossy Creek in southern White County were in production, with at least some of that iron finding its way to local blacksmiths. Asbestos deposits on Sal Mountain, three miles south of Nacoochee, were mined and refined from 1892 to 1920 and were the nation's largest source of asbestos for part of that time.

QUARTZ, SOAPSTONE, AND GNEISS

As noted in chapter 2, the early people in northeastern Georgia used locally available quartz for fashioning their points and knives, but they also used chert from other regions, especially northwestern Georgia, where it is commonly found. These people identified accessible sources and sometimes even kept those places in the cycle of their seasonal migrations. Not so readily available as local quartz, but still present, soapstone has been a valuable material since the prehistoric period, especially when made into bowls and cooking slabs. According to the Georgia Geological Survey, "*soapstone* is properly applied to an impure form of steatite which contains varying amounts of chlorite, tremolite, pyroxene, magnetite, pyrite, quartz, and carbonates of calcium and magnesium."[1]

Historically, all the soapstone produced in Georgia was mined and used locally, typically with the miner being the end user as well. Because of its thermal qualities, soapstone was used for fireplaces, hearths, chimneys, and furnace linings, but it was also used for gravestones, door steps, foundations, and well linings. By the twentieth century, marble and granite had largely replaced soapstone for funerary monuments, and brick and concrete had replaced it in building construction.

In 1914 the Georgia Geological Survey reported significant soapstone deposits in Elbert and Stephens Counties and an especially pure deposit on the White-Lumpkin county line due west of Cleveland. None of these were ever mined commercially. The geological survey identified a smaller deposit five miles northwest of Cleveland and two miles from Asbestos Station, noting that it was "exploited to a limited extent for local use." A similar deposit was noted a mile south of Helen and a few miles west of Nacoochee. "It is exposed near the crest of a small ridge, where it has been quarried for local use, such as in the construction of chimneys, etc."[2] Other local stone was quarried for other

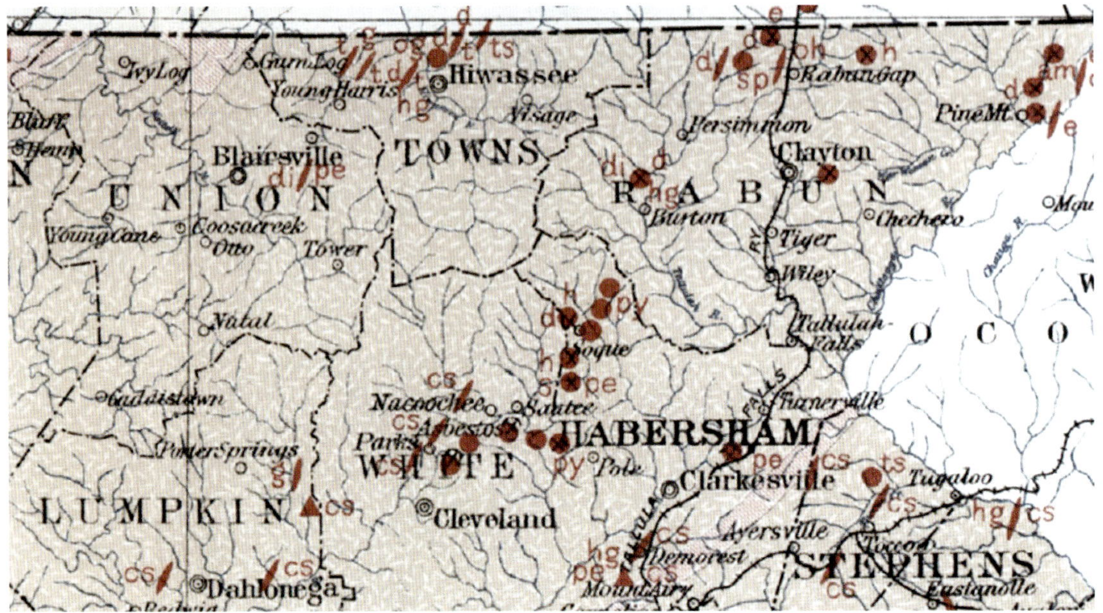

Part of a 1914 Georgia Geological Survey map, showing deposits of asbestos, talc, soapstone, and other minerals in northeastern Georgia. The red circles denote asbestos deposits, red triangles denote soapstone deposits, and the red elliptical shapes denote rock outcroppings. Hopkins, *Map Showing Distribution*.

uses. Dr. Lumsden notes the millstone quarry on Chimney Mountain five or six miles north of Sautee Nacoochee. There rough blanks were cut from a series of gneiss outcrops known locally as Stair Step Ledges and hauled to the mill for final shaping and finishing.[3]

IRON

The earliest iron objects are nine iron beads found in an Egyptian tomb, dated about 3200 BCE. That iron was meteoric in origin; it was another thousand years before people in Anatolia began mining and smelting iron ore, marking the beginning of the Iron Age. The iron industry in Europe, Asia, and Africa was well developed by the time the American colonies were established, and in 1646 the first ironworks in British America began operating at Saugus, Massachusetts. By the second quarter of the eighteenth century, six blast furnaces, nineteen hammer forges, and numerous bloomeries were in the colonies. In the South pig iron was being exported from Virginia and Maryland to Britain in the 1720s, but the first belt of iron-bearing ore to be mined and smelted in the Deep South was in South Carolina near Camden, where furnaces began operating before the Revolution. That belt trended southwest into Georgia, but the poor quality of its ore always limited production. The state's main belt of metamorphic, iron-bearing ore was discovered in northwest Georgia after the Cherokee were dispossessed, but a second, much smaller belt was being mined in the 1820s. Composed of Brevard schist, the belt runs from Rabun County in a southwesterly direction across Habersham County and southern White County, through metropolitan Atlanta, to Heard County and on into Alabama.[4]

Habersham County has one of the oldest industrial sites in Georgia, on the Soque River some ten miles southeast of Sautee Nacoochee. A bloomery (i.e., a place where pig iron was produced from the raw ore) was opened there as early as 1830 by Jacob Stroup (1771–1846), who came from a long line of German iron masters. He built several ironworks in York and Union Counties in South Carolina before 1830, when he sold his King's Mountain Iron Works and moved to Habersham County. He built ironworks on the falls of the Soque River but, in 1837, sold the furnace and moved to Bartow County, where deposits of higher-quality iron ore ensured a better future for the iron industry.

The Soque ironworks operated for only a few years after that, although as late as 1840 it was producing 129 tons of cast iron while consuming 488 tons of charcoal fuel and employing thirty-six men. By 1850 iron production at Soque ceased as the entire industry shifted into northwest Georgia. Although there was a brief attempt to revive the old Soque ironworks during the Civil War, they were abandoned after that. In 1882 the site was sold to a group of Atlanta investors who built a textile mill, the forerunner of Habersham Mills, the county's largest employer until it closed in 1999.[5]

By the early 1850s, a second ironworks was operating in Habersham County at Mossy Creek, some ten miles, as the crow flies, south of Nacoochee, where there was ample water power to drive the bellows and other components of the operation. In 1854 New Jersey–born Horatio Hennion (1827–1905) came to inspect the property, and, in spite of finding the foundry in poor condition and operating at a loss, he bought it from Andrew Gilmer for "five or six hundred dollars," repaired the dam, and brought the foundry back into production. In 1855 he married Gilmer's sister, Margaret Jane Service (1837–1923) of Union

County in upstate South Carolina, where there were ironworks that Hennion also at one time considered as an investment.[6]

Hennion bought sixty acres of ore land eight miles from Mossy Creek, but the ore was filled with slate and difficult to mine. Nevertheless, he continued operating his ironworks until the Civil War, when the negative response from his neighbors to his strong Unionist views made him fear for his life. In 1862 he sold the property and moved to western North Carolina. Hennion's memoir records the new owner as "Tom Lumson," but it is likely that the buyer was in fact Thomas R. Lumsden (1821–1914), a wealthy planter in Talbot County, who could have bought the ironworks as an investment, although there is no record of the Mossy Creek ironworks operating after 1862.[7] How much of the cast iron that came out of the Soque and Mossy Creek ironworks was put to use by local blacksmiths is uncertain, but at least some of it must have been. Ultimately, the relatively poor quality of the ore and the expense of overland transportation of iron in wagons was a factor in making local iron production unprofitable.

ASBESTOS

Just south of Nacoochee is Sal Mountain. With its peak at 2,241 feet above sea level, it was created by the same geological processes that created Lynch Mountain and Mount Yonah, to the northeast and southwest respectively. One of the nation's first—and at one time its largest—asbestos mines was on Sal Mountain. Not until the late twentieth century was it realized that asbestos was hazardous to human health. Before that time it was widely used, principally because of its unique qualities of being both incombustible and a poor heat conductor, along with physical properties that allowed it to be woven.

The ancient Greeks used asbestos for lampwicks and wove it into fabric; Charlemagne is said to have had a woven-asbestos tablecloth, but there was no commercial production until the nineteenth century, when deposits were identified in Cyprus, Siberia, the Ural Mountains, and the Italian Alps. By the 1850s asbestos was being used in the manufacture of paper, gloves, purses, ribbons, and girdles; by the late nineteenth century, a wide range of commercial applications had been identified. While many of those early uses were only experimental, asbestos continued to be used in a wide variety of ways until its hazards were adjudicated in the late twentieth century.

In the meantime asbestos was used for woven, fireproof rope and asbestos fabrics that were used in automobile tires, brake liners, theater safety curtains, carpets, upholstery, and protective gear for firefighters. Virtually no building was constructed in the first three quarters of the twentieth century without some asbestos material. Most heating systems used asbestos fabrics for insulation of furnaces and heating ducts, and it was widely used in components of electrical systems. It was also combined with cement to make roofing tiles, siding, shingles, and floor tiles. Since the 1980s the abatement of asbestos hazards in existing building has been an annual multibillion-dollar industry.

Asbestos is almost always found in association with corundum, a crystalline form of aluminum oxide that has long been used as an abrasive. There are three types of asbestos, distinguished by the orientation of the asbestos fibers: cross fiber, slip fiber, and mass fiber. The asbestos on Sal Mountain is a mass-fiber amphibole asbestos known as anthophyllite. By 1890 there was considerable prospecting for gold in the so-called corundum belt, which extends from Canada through the Piedmont into Alabama and which naturally led to the discovery of asbestos as well. Most of it proved to be of a relatively poor grade and unprofitable to mine, since demand was only just beginning to grow. The exception was the asbestos deposits on Sal Mountain, contained in six separate masses spread over eighty acres on the northeastern flank of the mountain.[8]

In 1892, not long after discovery of the deposits, Thomas W. Hix organized the Sal Mountain Asbestos Company, which operated continuously until it went out of business during World War I. Hix was born at Rockland, a small town in Knox County on the coast of Maine, in 1837 or 1838. Beyond that nothing is certain of his life, since he has only been surely documented by the 1900 federal census when he was boarding at the Alley House at Nacoochee. Otherwise he is lost in the sprawling Hix family, which had settled on the coast of Maine in the early eighteenth century. Hix's Sal Mountain operation must have gotten off to a quick start. The *Cleveland Progress* reported on 30 June 1893 that the company had "an exhibit covering 100 square feet of floor" at the great Columbian Exposition in Chicago, "showing the different

The buildings of the Sal Mountain Asbestos Company, circa 1914. Hopkins, *Map Showing Distribution*, plate 7A.

stages of its manufacture into paints, bricks, packing, crucibles, feltings, roofings, linings, etc."[9]

All mining was done by open cuts and surface quarries. Sometimes dynamite was used, but usually rock was simply broken up manually with sledgehammers and then hauled in wheelbarrows to the drying shed. Once dry, the Sal Mountain rock was put through an "American jaw crusher," a cast-iron, steam-driven machine that broke the ore into smaller pieces. Passed through a rotary crusher after that, the ore proceeded through a "Raymond pulverizer," made by a Chicago company that is still in business today. The pulverizer was sometimes called a fiberizer since it turned the ore into asbestos fiber, with residue blown off into a "dust room." Unlike most other types of asbestos, little of the Sal Mountain ore was wasted in the process. The fiberized material was packed into bags and shipped by wagon to the railroad at Clarkesville and on from there to manufacturers in the North and Midwest.[10]

At its most productive in the 1890s, the Sal Mountain mine was shipping twenty-five tons a day, making it the largest asbestos mine in the country. Many of the numerous "miners" shown in the 1900 census of Nacoochee were no doubt mining asbestos and not gold. By 1910, however, competition from higher-grade and more valuable chrysotile asbestos discovered in Vermont, Wyoming, and Arizona left Georgia as the only producer of amphibole asbestos in the country. The Sal Mountain Asbestos Company also owned nineteen acres a half mile from Asbestos Station on the Gainesville and Northwestern Railroad, which was completed to that point in 1912. The following year the Sal Mountain milling equipment was moved to Gainesville and the raw ore was taken there by railroad to be milled. By then all but one of the six nodes of asbestos on the mountain were depleted, and in 1918, with the United States' entry into World War I, the mine closed. The mine reopened after the war, but never at the same scale. By World War II all production had ceased entirely.[11]

Another significant asbestos deposit was found along the Tallulah Falls Railway five miles northeast of Clarkesville and a half mile south of Hollywood. Some of that deposit may have been mined as early as 1907, but real mining began in 1913, when a nineteen-acre tract was purchased by a group of unidentified investors who formed the Asbestos

The asbestos-mining operation at the Sal Mountain Asbestos Company site near Asbestos Station. Mount Yonah is visible in the background. Hopkins, *Map Showing Distribution*, plate 14A.

Mining and Manufacturing Company. By the end of the year, fiberized asbestos was being shipped from the mine. The company had intended "to manufacture a part into cement and later into pipe coverings" and other products, but it is not clear if that ever happened. As with the Sal Mountain mine, the Hollywood asbestos mine could not compete with the better-quality asbestos that was being produced elsewhere, and the factory was shuttered after the United States entered World War I in 1918. The 1914 Geological Survey report also noted asbestos deposits on the Calhoun mine property on the southwest flank of Sal Mountain and at a site near the county line five miles northeast of Sautee. Ten miles southeast of Sautee and not far from the Chattahoochee River, a third deposit was noted in "3 outcrops of talcose-asbestos rocks." None of these deposits were ever commercially mined.[12]

GOLD

The Spanish *entradas* that began with Hernando de Soto in 1539 were part of an age-old quest for gold. Many later historians have argued that the Spanish found it in the mountains of northern Georgia, but, as seen in chapter 3, there is no sure documentation for that. There were local reports of log structures discovered along Dukes Creek during mining operations in the 1830s, and many have assumed they were of Spanish origin, dating to the 1560s. The idea of sixteenth-century Spanish mines is not entirely implausible, but whatever physical evidence there was for them has long since been destroyed or lost, and no documentary evidence has been located.[13]

When gold was finally discovered in Georgia, it was found in one of three types of deposits, described by the Georgia Geological Survey as "(1) vein deposits; (2) placer deposits—consisting of beds of auriferous stream gravel, both ancient and modern, which include as a subclass, gulch and hillside deposits consisting of soil and decomposed rock brought from higher levels by rain-wash and the action of gravity and having usually a more or less irregularly occurring sub-stratum of angular or slightly sub-angular gravel; or (3) auriferous saprolites or decomposed rock in place."[14] If a rich vein of gold-bearing quartz was found, deep trenching or tunneling were sometimes needed to extract the ore, but many of the early mines were placer mines, so-called after the Spanish homonym, *placer*, which means shoal, alluvial, or sand deposits along a stream. These deposits consisted of gold eroded from bedrock, sometimes nearby but often some distance away, and washed out with other heavier materials to create the deposit. Many of the deposits around Nacoochee, especially on Bean Creek, are gold-bearing saprolites, which are essentially quartz veins that are disintegrating.

While early prospectors and modern tourists might pan for gold, once a deposit was found and deemed worth mining, a miner could, with a minimum investment, process much more material using a rocker box, also known as a "cradle," or a sluice box to separate out the gold. Gold-bearing ore had to be broken down before it could be processed, and stamp mills were built for that purpose. The early mills used water power to raise and lower large iron or sometimes iron-covered wooden

Henry Sandham, *The Cradle*, a miner working a placer mine in 1883. CPH.3C00734, Prints and Photographs Division, Library of Congress, Washington, D.C.

blocks to "stamp" (i.e., crush) the ore so the gold could be washed out. One historian of gold mining described these mills:

> These primitive mills were constructed of wood, with iron shoes and die-plates. A few of these old-fashioned mills may still be seen in operation in Georgia in the Nacoochee valley, seemingly serving the purpose of the tributers and petty quartz miners, and it is stated that they are operated at a fair profit. They are cheaply constructed, a 10-stamp mill with water-wheel and building complete costing about $150. The amalgamation is done on a copper plate of the width of the battery and about one foot long.[15]

In 1851 miners in California began hydraulic mining, which used high-pressure streams of water to blast away entire hillsides of potential ore that could then be sluiced for gold. The technique destroyed everything in its path, including waterways downstream that became choked by debris and silt. While certainly effective and immensely profitable for the miners, hydraulic mining caused massive flood damage in some areas and ruined the land for generations. As a result, hydraulic mining was banned in the United States in 1884.[16]

Carolina Gold

In his *Notes on the State of Virginia* (1782), Jefferson wrote that he knew of only a single instance of gold being found in Virginia, "interspersed in small specks through a lump of ore" found below the falls of the Rappahanock River.[17] In 1799 Conrad Reed, the twelve-year-old son of John Reed, who had come to this country as a Hessian mercenary during the Revolution, picked up a large metallic rock in a streambed on the family's farm in Cabarrus County, North Carolina, some twenty-five miles east of Charlotte. Not knowing what it was, the Reeds used the rock as a door stop for several years until it was discovered to be gold. By 1804 Reed was mining and would grow rich off the nation's first gold mine. Gold was soon discovered in a belt stretching from McDowell, Burke, and Rutherford Counties, west of Cabarrus County, and Montgomery and Randolph Counties on the east, southward to Lancaster and Union Counties in South Carolina. By 1824, 2,500 ounces of North Carolina gold had been deposited at the Philadelphia Mint.[18]

In 1825 Denison Olmsted, a professor of chemistry and mineralogy at the University of North Carolina, published a paper titled "On the

Gold Mines of North Carolina." It attracted considerable attention, especially because Olmsted was able to characterize something of the geology of the fields:

> In almost any part of this region, gold may be found, in greater or less abundance, at or near the surface of the ground. Its true bed, however, is a thin stratum of gravel enclosed in a dense mud, usually of a pale blue, but sometimes of a yellow colour. On ground that is elevated and exposed to be washed by rains, this stratum frequently appears at the surface; and in low grounds, where the alluvial earth has been accumulated by the same agent, it is found to the depth of eight feet; where no cause operates to alter its original depth, it lies about three feet below the surface.[19]

Later that same year, the first gold-bearing quartz vein in the United States was found in Montgomery County, sixty miles east of Charlotte, and successive discoveries soon showed a gold belt stretching northward into Virginia and to the southwest through Georgia and into Alabama.[20]

Georgia Gold

There are several gold belts in northern Georgia, but the Dahlonega Belt is the most substantial. Varying from two to six miles wide, the belt runs from Alabama through Haralson and Paulding Counties; the northwest corner of Cobb and the southeast corner of Bartow Counties; across Cherokee County through Dawson, Lumpkin, White, Habersham, and Rabun Counties; and on into Macon County, North Carolina.[21]

The particulars of the earliest discovery of gold in Georgia have long been argued but never definitively answered. By some accounts a Cherokee boy playing in the Chestatee River in 1815 found the first gold nugget, which his mother promptly hid. While that tale may be apocryphal, the Cherokee certainly knew there was gold on their land but rightly did not advertise the fact. There are reports of gold being discovered in McDuffie County, a few miles west of Augusta, in 1823 and at Villa Rica in Carroll County in western Georgia in 1826. Both locations have documented deposits of gold, but no historical documentation to support those early discovery dates has yet been found.[22]

Multiple accounts place the discovery of gold on Dukes Creek, where some of Nacoochee's richest mines would be located. Thomas Bowen

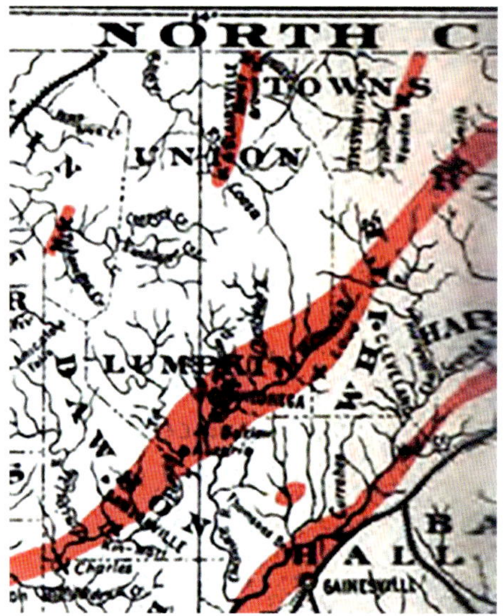

Detail from Georgia Geological Survey map, 1896, showing the gold belts in North Georgia. One belt passes across Dukes Creek, the Chattahoochee River just west of Nacoochee, and the upper reaches of Bean Creek. McCallie, *Map Showing the Distribution*.

(1788–1887) said in an interview in his old age that he made the first discovery when a storm uprooted trees along Dukes Creek, and he found gold among the roots. Another widely repeated story, and one with some credibility, is that an unidentified slave of Maj. Francis Logan (1802–78), returning from the gold fields in Rutherford County, North Carolina, recognized similarities in the rock and soils along "Nacoochee River," now known as Dukes Creek. Performing assay work in an iron skillet, he confirmed that it was gold. Logan was a native of Rutherford County, North Carolina, and probably knew Major Williams and others in the valley, but he eventually settled in the western part of the county, where there was another belt of potentially lucrative gold deposits.[23]

The earliest historical documentation for gold in Georgia is a report from 1 August 1829 in the *Georgia Journal*, a newspaper in Milledgeville, the state capital: "GOLD.—A Gentleman of the first respectability in Habersham County, writes us thus under date of 22d July: 'Two gold mines have just been discovered in this country, and preparations are making to bring these hidden treasures of the earth to use.' So it appears that what we long anticipated has come to pass at last, namely that the gold region of North and South Carolina would be found to extend into Georgia."[24] A report published in 1854 on the "metallic wealth" of the United States also notes that the first discoveries of gold in Georgia were made in the summer of 1829. The same report stated that $212,000 of Georgia gold was in the first shipment to the Philadelphia Mint in 1830. The gold rush was on.[25]

Whoever made the first discoveries, and there were probably multiple, nearly simultaneous findings in what are now Lumpkin and White Counties, it seems likely that they were actually made months earlier than that summer. Prospectors were generally loath to disclose any information about the results of their work, at least until they had staked a legal claim. Local historians have said much the same thing and posit 1828 as the date that gold was first discovered.[26]

At the time no one knew the extent of the gold deposits. The *Georgia Journal* reported that "as many as six or seven thousand persons were, soon after [the discovery], engaged in washing for gold in that region."

For the early residents of Nacoochee, the experience would have been unsettling at the least. The reporter knew what would follow, probably from experience in the Carolina gold rush, and wrote, "And another anticipation of ours will hereafter come to pass, namely, that it will be a sad day for Georgia, when the precious metals are found in any great abundance in her soil. The best thing that the Legislature could do, would be to prohibit under severe penalties the working of any gold or silver mines in the State. Those who live to see the result will be convinced to their sorrow that this advice is not founded on a slight or partial consideration of the subject."[27]

Of course, the legislature did not outlaw gold mining, and within a few months Governor Gilmer wrote, "Many thousands of idle, profligate people [have] flocked in from every point of the compass, with pent up vicious propensities, when loosed from the restraints of law and public opinion, [which] made them like the evil one in his worst mood."[28] Benjamin Parks offered what might be the quintessential description of the period when he told a reporter for the *Atlanta Constitution* in 1894: "They came afoot, on horse back and in wagons, acting more like crazy men than anything else."[29]

The months following the first reports of gold would have been difficult for the Williamses and Browns and the other pioneers at Nacoochee, since a large proportion of the prospective miners would have passed through on their way west. Nevertheless, the Williams family prospered by continuing to grow grain and operating one of the few early stores near the gold fields. Charles L. Williams bought much of the gold produced around Nacoochee, acting as an agent for banks in Charleston. In 1860 he stated that between two and three million pennyweights of gold, or as much as three or four Troy tons, had passed through his hands over the preceding thirty years. In the end merchants often found more profit in the gold rush than the miners.[30]

Within a short time, the frenzy had moved to richer deposits farther west in what is now Lumpkin County. In the spring of 1830, Governor Gilmer wrote John McPherson Berrien, the U.S. attorney general, "I am in doubt as to what ought to be done with the gold-diggers. They, with their various attendants, foragers, and suppliers, make up between six and ten thousand persons. They occupy the country between the Chestatee and Etowah rivers, near the mountains, gold being found in the greatest quantity, deposited in the small streams which flow into

Hydraulic mining in White County. Brooks and Greear, *Images of America*, 20.

those rivers."[31] In 1835 Congress created branches of the U.S. mint in New Orleans, Dahlonega, and Charlotte. A building was constructed and machinery installed in 1837, and the Dahlonega mint began production in February 1838. Before it closed in 1861, more than $6 million in coin were minted from Georgia gold.

Most of the easily accessible gold was mined by the early 1840s, and by 1845 mining "was falling off rapidly." By the time of the enormous gold strikes in California in 1849, mining in northern Georgia was becoming the province of large companies willing to make the investments necessary to dig and process gold-bearing ore. Yet gold prospecting and mining, much like the present-day state lotteries perhaps, always retained an allure for some. As one historian noted in 1854, "The Georgia gold excitement did not last very long; but there, as in the other Southern States where this metal was found, gold-washing continued to be followed by the majority of the washers, not as a regular pursuit, but as one to be taken up and dropped again, as circumstances directed."[32] When the gold mining industry was revived after the Civil War, most of the mines were acquired by groups of investors who organized corporations for the specific purpose of mining in a given area. Investors were typically northerners, but, between 1865 and 1899, no fewer than sixteen British companies registered in London for the development of land and gold mines in Georgia, including mines at Nacoochee.[33]

A water-powered stamp mill on Dukes Creek. Brooks and Greear, *Images of America*, 15.

Some of the antebellum mines were worked using slave labor, but after the Civil War, in a parallel to sharecropping, individuals worked as "tributers," mining for a percentage of the gold that was recovered. A state geologist described them at Nacoochee in 1896: "There are quite a number of petty operators, some washing gravel in sluice boxes, others mining rich, narrow seams in the saprolite and 'beating' the ore in wooden stamp-mills, as, for instance, at the Thompson mine near the Yonah Land Company's property, where the mining operations were formerly carried on by a mother and son, the latter digging the quartz and carrying it on his back to the mill, while his mother attended to the beating."[34] While a substantial amount of gold was taken out of the mines in northern Georgia in the nineteenth century, most deposits were tapped out in the early twentieth, and the cost of extraction soon exceeded the value of the ore that could be produced, which put an end to commercial mining.

White County Gold

There was significant mining activity around Loudsville, nine or ten miles southwest of Sautee Nacoochee, near where Town Creek joins the Tesnatee River, but the most active mines in White County were around Nacoochee. The valley was renowned for the large gold nuggets, some reportedly as large as two pounds, that have sometimes been discovered in the Chattahoochee River and its tributaries and even in the alluvial soil of the valley floors, all washed out of the narrow gold belt that slices from northeast to southwest across northern White County, just to the north and west of Sautee Nacoochee.[35]

Countless holes were dug by prospectors, as they searched for the limits of the gold belt. Ultimately just a few areas proved to have significant deposits. Around Nacoochee one of these areas was along the Chattahoochee River on the north side of Hamby Mountain, which lies between the river and Dukes Creek and apparently took its name from one Edom Hamby, a pioneer and neighbor of the Conleys and Englands in 1830.[36] Other areas with significant deposits were on the south slopes of Hamby Mountain and in the upper watersheds of Dukes and Bean Creeks.

More than one reporter noted the possibilities for mining if the tremendous "over-burden" of alluvial soil, in some places six or eight feet deep, could be removed from the Nacoochee Valley floor, which would be an archaeological and environmental nightmare. While the valley itself was not dredged, a barge was used to dredge the Chattahoochee River from Sautee Creek to Dukes Creek about 1890. Winched upstream, the barge was maneuvered into Dukes Creek, where it was subsequently left to rot. Pit ponds from that dredging operation are still visible on the east side of SR 75 at Dukes Creek.[37] All the earliest mining was placer mining, which typically required a minimum investment in equipment, but, in the early 1830s, the first vein of gold-bearing quartz was discovered by Benjamin Reynolds (1792–1855). He used slave labor to expose the ore by making a cut fifty feet deep and seven hundred feet long. The "Reynolds vein" continued to attract mining interest into the twentieth century.[38]

There would be many more of these scars in the landscape, which were noted by travelers as early as 1843: "From thence the road winds to the southern sections of the valley, where the inhabitants are defacing the beauty of the scene, in the successful opening of gold mines, and where the renowned Mount Yonah lifts its lofty head."[39] George White's *Statistics of the State of Georgia*, published in 1849, reported the gold mines then active in what is now White County:

> Loud's vein has been a rich mine; not now in operation. Has been excavated to the depth of 135 feet. Gordon's, near Loudsville, is considered rich. Lewis's, one mile from Loudsville, would be valuable were water convenient. Holt's, two miles from Loudsville, is thought to be rich. Richardson's mines, on Duke's creek, in Nacoochee valley, have yielded 150,000 pennyweights of gold. They are still worked. Forty hands employed. Deposit mine. White & McGee's mines. Vein and deposit. Have yielded 66,000 pennyweights of gold in eight years. Gordon & Lumsden's mines, on Duke's creek. Vein and deposit. Produced in 10 years 100,000 pennyweights of gold. Williams's mine, on the Chattahoochee, has been in operation about 20 months, and paid fair wages. Littlejohn's mine, on Duke's creek, is an excellent vein. Has been worked two years, and has yielded 30,000 pennyweights. Harshaw's mine, on Sautee creek, has yielded largely.[40]

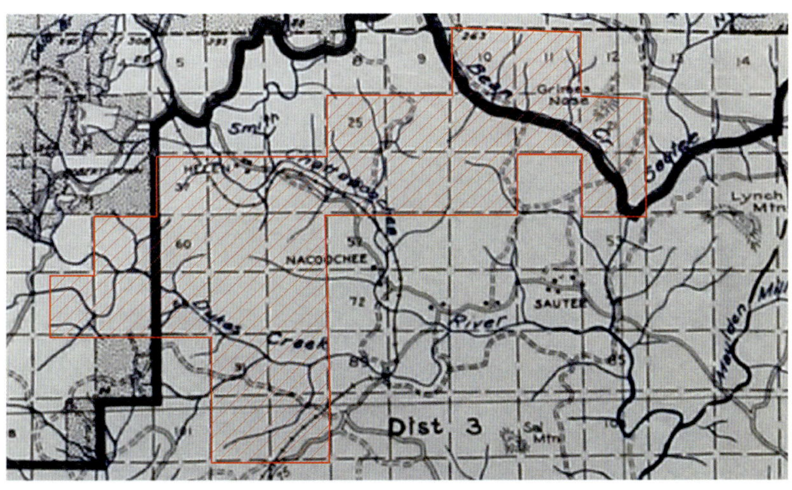

Detail from U.S. Forest Service map, 1937, annotated by the present author to delineate the land lots on which there was significant mining activity in the nineteenth century. Georgia Secretary of State, Office of Surveyor General, Historical Map File, 1937, Georgia Archives, Morrow, Georgia.

Most of these mines remained active for decades, although often under different owners. The outline of mining operations around Nacoochee that follows is not exhaustive but is meant to provide a sense of what areas were repeatedly mined.

Dean Mines

In 1851 miners in California developed the process of hydraulic mining, which used a high-pressure water cannon that could wash away entire hillsides in a matter of days. About 1858, Josiah Robinson Dean Sr. (1796–1871) brought that technique to Nacoochee when he began hydraulic mining on Land Lot 38. Dean was born in Massachusetts, where he married Betsey Wheaton Chase (1793–1852) and with whom he had at least four children. By 1860 Josiah Dean and three of his sons were in White County, where the federal census enumerated the elder Dean, a widower and "miner" boarding in the household of Charles Williams at Nacoochee.[41]

To support their hydraulic mining operation, Josiah Dean Sr. and his two sons, John Jr. and Simon Dean, joined T. Collett Leventhorpe (1815–89) and two Boston investors, M. C. Smith and C. D. Butler, to form the Nacoochee Hydraulic Mining Company, which was incorporated by the Georgia legislature in December 1857, with a legal address at Charles Williams's store at Nacoochee. In 1858 Leventhorpe joined the Deans at Nacoochee to begin operations, and his wife later visited as well and stayed with the Williams family.[42]

The company constructed a series of ditches and troughs to carry water from the base of Dukes Creek Falls, in Land Lot 33, three miles away. Built to avoid trestles, it wound its way down the mountain in a course nearly eight miles long. Known as the Hamby Ditch, it was a local marvel of engineering and provided enough water for the miners to make massive cuts into the earth on the north side of Hamby Mountain. "With its branches, which spread out on the ridges, thousands of acres of land can be irrigated and washed," George Williams euphemistically observed.[43]

After the Civil War, hydraulic mining continued and another flume, known as the Yonah Ditch, was built along the south side of Dukes Creek to improve the efficiency of mining on that side of Hamby Mountain. Other smaller "water ditches" were built elsewhere, including one serving the Bean Creek mines. Hydraulic mining inflicted massive environmental damage; even today, the effects of this man-made erosion of the landscape are still visible around Nacoochee. Josiah Dean Sr. died in 1871, but his oldest son, Josiah Jr., continued mining and farming until he died in the early 1880s. The Dean mines were valuable property and soon attracted foreign investors.

Martin Mines

In 1882 a British army officer organized a group of investors to form Georgia Gold Mines. They intended to buy mining property in White County but could not raise the capital, and the company was dissolved in 1884. Around the same time Nacoochee Placer Gold Mines was organized for a similar purpose, but that venture too failed for want of investors. In December 1885 William Crammond Martin (1861–1911) of Dundee, Scotland, acting as agent for a group of London merchants, accountants, and bankers, bought the "Dean Mines at Nacoochee" from the widow of Josiah R. Dean Jr. for £2,000 and £48,000 in company shares. Martin was probably a younger brother or cousin of John Martin (1857– after 1920), also of Dundee, but that has not been proven. He is not known to have been related to the Cherokee judge Martin, who claimed and then gave up one of the reserves at Nacoochee after the War of 1812.[44]

As a young man, John Martin went to work for the *Dundee Courier* but soon took an interest in the nascent field of electric lighting and

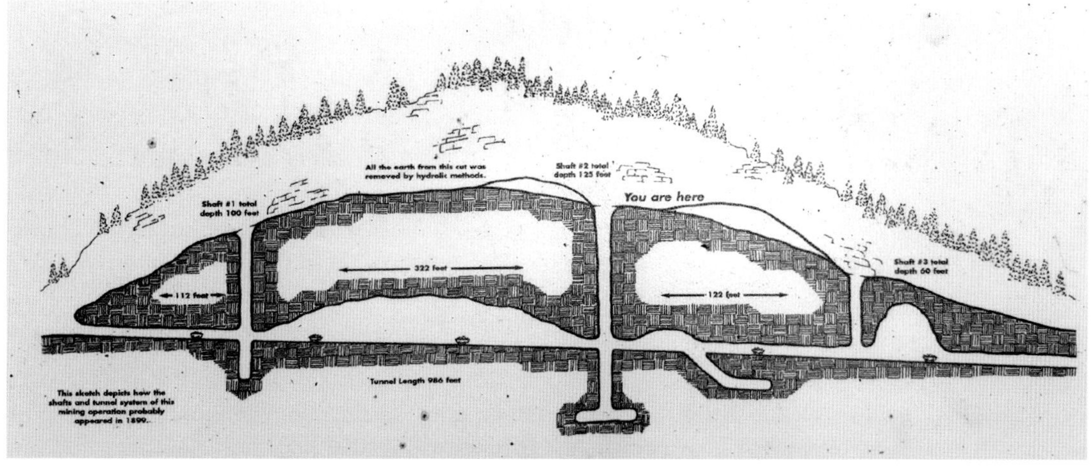

Cross-section of the Martin mine on Hamby Mountain. Courtesy of Smithgall Woods State Park.

moved to London. He first came to Georgia in 1885 and again in 1887, probably on behalf of London investors. In 1889 he moved permanently to the state. Shortly after his arrival, he reconstructed Nora Mill and built his residence across the road from the mill; both structures remain standing today. Martin also went on to become one of the county's largest property owners, as he acquired some twenty thousand acres of land, primarily for the rights to its timber and water.[45]

Richardson Mines

On the south side of Hamby Mountain and the watershed of Dukes Creek, John L. Richardson had the good fortune of having settled on what turned out to be some of Nacoochee's best gold lots and was made a rich man because of it. His daughter, Sophronia Antoinette Richardson (1827–50), married Thomas Reid Lumsden (1821–1914) in 1845. They had two children, John Greer Lumsden (1846–1936) and Jesse Richardson Lumsden (1848–1933), before her untimely death at the age of twenty-three.

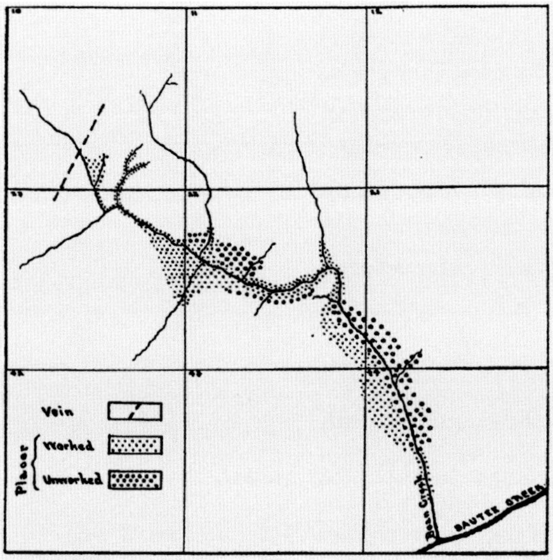

After the Civil War, the Richardson mine was reopened by Thomas and Sophronia's sons, Jesse and John. On 17 April 1880, the *Cleveland Advertiser* reported the brothers' "White County Bonanza": "The Messrs. Lumsden, Jesse and John, have found and are now operating on the richest mine perhaps ever discovered in northeastern Georgia. They have with two hands taken out near three thousand pennyweights of gold nuggets in less than three weeks, making over one hundred dwts. per day to the hand. Two of the pieces weighing over three hundred dwts a piece. The prospect for future operations is equally as good. The mine is on the John L. Richardson property. Who can beat this?"[46]

Plat map of Jesse and John Lumsden's gold lots along Bean Creek, 1896. Yeates, McCallie, and King, *Preliminary Report*, 38, fig. 1.

Calhoun Mines

Begun around 1830 as part of then vice president John C. Calhoun's considerable investments in gold mining, the Calhoun mines eventually incorporated the Richardson mines and went through a succession of owners after Calhoun's death in 1850. In 1893 the company was reorganized as the Yonah Mountain Land and Mining Company. They consolidated three older mines and ultimately owned or leased 4,800 contiguous acres along Dukes Creek. "Very productive" placer deposits were mined on all the creek's branches, with the largest on Richardson Branch. The seven-mile-long Yonah Ditch that began at Dukes Creek Falls produced a sixty-five-foot head of water that powered a hydraulic elevator at the lower end of the valley, which graded the ore.[47]

Jones Mines

On Bean Creek the first discoveries were in Land Lot 10 in 1830, and the lot was famously productive after that. It was one of the two lots designated as "school lots" in the 1820 lottery but must have been sold shortly after gold was discovered. Whoever owned it, the lot was mined, "more or less continually," after that, and much of the work done after the Civil War used hydraulic mining and some tunneling.[48] After the war the lot was acquired by Charles Colcock Jones Jr. (1831–92) and became known as the Jones mines. After his death Albert G. Rice Sr. of Atlanta redeveloped the mining operation. A milling plant

A landscape destroyed by mining, with Yonah in the distance. Brooks and Greear, *Images of America*, 24.

was built along Bean Creek Road, near its first crossing of the creek in the southwest corner of Lot 10. The plant included a water-powered fifteen-stamp mill and a Wilfley concentrating table, which used vibration to sort the ore. Much of the hillside northeast of the milling plant had been subjected to hydraulic mining.[49]

Jarrett-Childs Mines

Devereaux Jarrett is thought to have been the original owner of these mines on Bean Creek, which he worked with slave labor. Land Lot 23, immediately to the south of Lot 10, was the center of operations for the Childs Mines, which was developed after the Civil War by Asaph King Childs (1820–1902), a wealthy merchant in Athens, Georgia. Here, too, hydraulic mining was used: "The material from the cut was flooded through a long sluice line to a stamp mill situated on Bean Creek near the Jones mine. A very large amount of material was treated in this manner."[50] The milling plant was on Bean Creek near the Jones mine and included forty-five stamps, operated by water power supplied from a ditch fed by streams higher up the mountain. There was also a series of washing and sizing plate screens that helped ensure that as much usable ore as possible was retrieved. Mining operations apparently ceased in 1888. After Childs's death in 1902, his son-in-law David C. Barrow, chancellor of the University of Georgia from 1906 to 1925, inherited the property, but the mine is not known to have been operated after that.[51]

Lumsden Mines

About 1890 Jesse Richardson Lumsden (1848–1933) built a house in Sautee Valley and mined several land lots on Bean Creek. The overburden of topsoil was several feet thick, which inhibited much serious mining along the lower reaches of Bean Creek, but the rest of the creek was worked.

Manufacturing

Through much of the nineteenth century, there was little manufacturing in northeastern Georgia. There were blacksmiths making horseshoes, hinges, and other articles; a few tanners turning hides into usable leather; and a few leatherworkers making harnesses, shoes, and other articles. The occasional cooper continued to work making and repairing barrels, and there was even a wagon and carriage maker in business in Clarkesville before 1840. Although the carriage maker remained in business at least until 1860, White's *Statistics* did not include it in his overview of manufacturing in Habersham County in 1849: "There is no cotton factory in the county; 3 gold-mills, 4 jug factories, 8 or 9 distilleries, 20 saw-mills, 30 grist mills, 1 flour-mill." In 1850 the federal census found only twenty-nine people employed in manufacturing in all of Habersham County, which then included White County. In 1864 Edwin Williams's blacksmiths produced "Joe Brown pikes" at a shop on Sautee Creek, one of a number of such futile efforts to provide materiel for the dying Confederacy. His brother later remembered a "Confederate gun factory," but that seems unlikely, given the technical challenges to the manufacture of firearms. In any case any such manufacturing did not survive the war.[52]

WILLIAMS'S TANNERY

The federal census reports five tanneries in White County in 1860, with a product worth of $3,350. One of those was apparently being operated by Edwin Williams, probably using slave labor, at his mill site on Sautee Creek. Nothing has been found to document how large it might have been or how long it might have operated, but it was apparently the only tannery in the Nacoochee District before the Civil War.[53]

The business of tanning hides to make leather is an ancient one, and the very nature of the tanning process made tanneries obnoxious neighbors, due mainly to the stench from bits of rotting flesh left on the raw hides and from the urine and dung that was sometimes used in preparing them. The tanning process itself takes its name from tannin, an astringent organic compound found in many plants, especially oak and hemlock.[54] With a ready supply of water, all that was needed for a tannery was some sort of trough in which to soak the hides and a scraper for removing hair and other unwanted material. One of the favorite containers could be hollowed out of eight- to ten-foot-long sections of yellow poplar, two or three feet in diameter, and split in half. A mixture of hardwood ashes and water created a lye solution, in which the hides or skins were soaked for several days to loosen hair for removal.

For the actual tanning, bark was stripped for tannin, with chestnut oak (*Quercus prinus*) the favorite source of some in northeastern Georgia. The bark was cut into small pieces and soaked in water for several days to create tannic acid. After the hides had been thoroughly cleaned and all the hair removed, they would then be soaked in the bark solution for weeks or months, which alters the protein structure of the hide and turns it into leather, durable and less prone to decomposition and ready to be finished in a variety of ways.[55] The tanning process was hard on the environment, especially so with large tanneries, which could denude entire landscapes for bark and pollute rivers with their effluvia. At Nacoochee Williams might have used bark waste from his sawmill, but salt, tannic acid, decomposing flesh and hair, and all the other waste wound up in Sautee Creek, one of many ways the natural environment was thoughtlessly degraded in the nineteenth century.

GLEN'S WOOL FACTORY

In his narrative tour of Nacoochee Valley in 1870, George W. Williams noted the "wool factory of James Glenn, Esq."[56] James Glen (1836–1927) was the son of John Glen (1792–1880), who was born in Renfrewshire, Scotland, but was in New York when he married Eliza Baldwin (1804–70) about 1826. Their only known children were James and his sister, Mary, who was born about 1829. The Glens were in Henry County, Georgia, a few miles southeast of Atlanta, by 1850, when the federal census listed his employment as "manufacturer." What he might have manufactured has not been documented.[57]

In 1853 John and Eliza Glen moved to Habersham County, where they acquired thirty acres in Land Lot 88, making them neighbors of the Richardsons, the Littlejohns, and others who lived along the lower reaches of Dukes Creek. In 1860 their son James married Susan Littlejohn (1831–1909), the daughter of pioneers Abraham and Sarah Littlejohn. They had at least four daughters and two sons. John and James Glen were both enumerated as millers in the 1860 census of White County.[58] The Glens built a new mill on Dukes Creek, blasting a short channel in a section of the creek. Fourteen feet wide, one hundred feet long, and as much as thirty feet deep, the channel funneled an extraordinary flow of water that was able to drive an undershot waterwheel with enough power to operate the rollers used to press wool into felt, with the final product being felt hats.[59]

Because of the simplicity of its manufacturing, felt was being produced thousands of years ago, long before the development of thread and woven fabrics. Felt can be made from many types of animal fur, but the physical characteristics of sheep's wool make it ideal for the "wet felting" process that created the finished felt material.[60] As the rollers pressed and repressed the wool, which was mixed with water, the directional scales and kinks found on each hair created a dense matrix of felt. To make a hat, the material was formed into a loose cone, by hand, and then dipped in hot water, which caused the fibers to shrink, making a denser material. Then using steam and pressure, the felt was molded on wooden forms to shape the crown. With the brim trimmed, the hat might be treated with a stiffener before being sanded to create a uniform texture. It is impossible to estimate how many hats the factory might have turned out or for how long it operated. Although George W. Williams pointed out the Glen "wool factory" in 1870, it may have already ceased to operate by that time. The federal census that year found John Glen "infirm," while his son James had returned to farming.

MOSSY CREEK POTTERY

The final component of the early manufacturing economy in White County was pottery. Glazed, but otherwise unadorned, stoneware jugs were found in virtually every household, used for storing whiskey and syrup and for making butter churns and jars for fruits, vegetables, and meat. Much of the stoneware used at Nacoochee would have been made at Mossy Creek. While Georgia's ubiquitous red clay makes excellent brick, other clays were preferred for the bowls, pitchers, churns, and crocks necessary on any farm. Southern White County had especially good deposits of stoneware clay, which John Burrison described as "a fine-grained, relatively pure clay, which fires to a light gray or tan . . . and tough enough to withstand rough usage on the farm." Pottery was first produced at Mossy Creek in the 1820s, when the Davidson, Craven, and Dorsey families migrated there from North Carolina, bringing their folk pottery traditions with them. With more than ninety potters over the course of time, the Mossy Creek potters were the genesis of a pottery tradition that has, in some instances, continued for several generations.[61]

In 1840 the federal census manufacturing schedules counted six potteries or "jug manufactories" in the entire state; three of those were at Mossy Creek, where five people were employed. By 1849 there was a fourth, and by that time, too, Christopher Columbus Meaders (1808–87) had moved to Mossy Creek and opened a dry-goods store. In 1893 his son John Milton Meaders (1850–1942), who had been peddling local stoneware in the mountains, established his own pottery. In the late twentieth century, his great-grandson Lanier Meaders (1917–98) would be one of the best-known folk potters in the country.

Timber

The first European settlers in northern Georgia in the eighteenth and early nineteenth centuries found the same vast temperate forests that had covered the landscape since the last ice age. The forests were dominated by old-growth chestnut, tulip poplar, hemlock, and several species of oak and pine, along with many others, and had a rich understory of smaller trees, shrubs, and plants. There were few open glades or meadows, although lightning strikes could start fires that would burn thousands of acres, and the Indians regularly burned rich valleys like Nacoochee to keep the land open for agriculture and to improve hunting. The Indians also made use of the forests in other ways, but the huge stands of timber were mainly seen as shelter for wildlife and plants that helped sustain their way of life. Timber might be selectively cut, laboriously with stone tools and fire, but only for certain special uses.

European colonists brought iron axes and saws, which greatly expanded the practical limits of cutting and using timber and made it possible to construct tens of thousands of log cabins in the eighteenth and nineteenth centuries. Once the logs had been cut, leaving a clearing in the forest for the cabin, building the cabin itself could be accomplished by two men in just a matter of days. In the rush to create farmland for the tide of new settlers in the eighteenth and early nineteenth centuries, great swaths of virgin timber were cleared. One of the most common ways to quickly clear a field for agriculture was to girdle the standing timber—that is, to remove a strip of bark around the circumference of the tree, causing it to die. Corn could be planted amid the trees as soon as the leaves were gone, and, as the dead trees fell, the wood could be collected for fuel. An enormous amount of timber was simply consumed as firewood, since a typical family might use ten cords of firewood annually, or the wood of ten trees twenty-two inches in diameter.[62]

Technological advancements made logging much more efficient, with better steel for axes and crosscut saws widely available in the mid-nineteenth century. While one might wonder how long it must take to cut down a tree without a chainsaw, much less clearcut an entire forest, two experienced sawyers with sharp axes and a saw could take down a thirty-inch-diameter tree in as little as fifteen minutes. Water-powered mills were built all over northeastern Georgia from an early date, mostly for grinding corn, but some were built so that the waterwheel could power not only the rotation of grinding stones but also a reciprocating sash saw. White County had seven sawmills in 1860, but some of them may not have operated fulltime. Most of the timber was brought in one log at a time by a local farmer, who needed the lumber for his own use.

INDUSTRIAL LOGGING

In the decade or so before the Civil War, steam-powered mills were built in larger cities and across South Georgia, where the virgin forests of long-leaf pine and cypress were being cut. Logging and sawmills were big business in the 1850s, with more than 16 percent of Georgia's workforce employed in logging by 1860. In the 1860s and 1870s, Georgia was the nation's largest lumber producer and in 1880 was still producing more lumber than any other southern state. The state was also second only to North Carolina in the number of sawmills, and many of those were portable, steam-powered mills. Weighing several tons, they still could be moved in a couple of days, and they greatly increased the logger's range. Forests were almost always clear-cut (i.e., all the trees in an area were removed), which remains the most common kind of logging today because it is the most efficient way to remove large quantities of timber and because it is generally safer for the loggers themselves.

At the same time, clear-cutting does enormous environmental damage, although this was little recognized before the last quarter of the nineteenth century. After a tree is downed and limbed, the logs are dragged away, or "skidded," from the site, which destroys everything in its path and creates points of erosion that could last for decades. Everything was simply "torn to pieces," one logger noted. "You hardly ever left a tree of any size standing and all the little [ones] was tore down." Worse still, the limbs and other debris were simply left behind, inevitably to dry out and, sooner or later, catch fire and burn. Severe erosion followed, washing away the rich but thin layer of topsoil and silting up streams.[63]

In the first half of the nineteenth century, the forests of New York and New England and the forests of long-leaf pine in eastern North Carolina were the main sources of U.S. lumber. By the time of the Civil War, the majority of that timber had been cut, and the lumber companies moved on to the forests of Michigan and Wisconsin. Within twenty years the end of those forests was in sight, and the timber barons turned their attention to the South. For four decades the Southern Appalachians would be the source of much of the nation's lumber, as the industrial loggers did their work and moved on.

Railroads

Before railroads commercial sawmills generally depended on waterways on which to transport logs from the source to the mill, somewhere downstream, and so most were located along larger waterways. The virgin forests adjacent to larger waterways thus were among the first to be destroyed. Railroads made possible the era of industrial logging, which flourished in the Southern Appalachians from about 1880 until after World War I. The railroads created a huge demand for forest products, especially timber for crossties. Each mile of track required around 2,500 crossties, depending on the species of wood used, and those had to be replaced every five to ten years. At the same time, railroads offered a solution to the age-old problem of transporting large quantities of logs and gaining access to new supplies of timber, which ensured great profits for large commercial sawmilling operations along railroad corridors.[64]

Land and Timber Companies

The era of industrial railroad logging in the Southern Appalachians began with the Scottish-Carolina Timber and Land Company, which was organized in Glasgow in 1884 by a group of Scottish investors. The company's U.S. agent, Alexander Alan Arthur (1846–1912), brokered title to or options on many tens of thousands of acres in East Tennessee and western North Carolina, beginning in 1885. Dozens of other companies were formed to exploit the region's vast stands of virgin timber, some for logging and sawmill operations and others for simple speculation in real estate and timber rights. Southern entrepreneurs there were, to be sure, but most of them depended on northern and foreign investors to stay afloat.[65]

A typical temporary logging railroad; when logging was done, the rails would be salvaged, but the rest would be abandoned with the desolated landscape. Brooks and Greear, *Images of America*, 45.

In 1901 brothers Edwin Andrew Gennett (1874–1941) and Nathaniel Chapman Gennett (1878–1948) bought a bankrupt sawmill on the Chattooga River in Rabun County, along with timberland there and across the river in Pickens County, South Carolina. It was the first of what would eventually be twenty-five mills owned by the Gennett Lumber Company. Sons of a poor Italian immigrant who had made a fortune with a wholesale and retail grocery in Nashville, the Gennett brothers would make their own fortune in timber, much of it in northern Georgia. In addition, their timberland would form a significant part of the U.S. national forest when it was established after World War I.[66]

Another timber company that had a major impact in northern Georgia was the Pfister & Vogel Leather Company of Milwaukee, Wisconsin, organized in 1853 by the German immigrants Frederick Vogel and his cousin Guido Pfister. The company rapidly grew into the world's largest producer of tanned leather and consumed enormous amounts of tannin, a critical ingredient in the tanning process that is best obtained from oak trees. In the early 1900s, with Frederick Vogel's sons August (1862–1930) and Frederick Jr. (1851–1936) running the company, the Pfister and Vogel Company acquired sixty-five thousand acres of timberland in Union and Lumpkin Counties for as little as two dollars an acre. They built a sawmill on the site of today's Vogel State Park, which operated until it was closed in the mid-1920s, after all the readily available timber had been logged.[67]

Robert Pinckney Tucker (1865–1920), scion of a family of Low Country planters ruined by the Civil War, played a major role in the deforestation of the mountains. In the late 1890s, he is said to have borrowed $5,000 and begun buying and selling tens of thousands of acres of land for lumber and mineral rights and as hunting preserves for wealthy northerners.[68] The scale of his business is suggested by his reply to a New York client in 1901: "There is a good opportunity for both of us to make some money. I now have in my books some of the finest hunting preserves in the United States. Most of them are smaller than 20,000 acres. . . . There are other properties that can be purchased, and I could deliver, in this locality, anywhere from 10,000 to 50,000 acres of as fine hunting lands as can possibly be had."[69]

In early 1904 Tucker paid John Martin, who had made a fortune mining gold on Hamby Mountain, $26,000 for twenty thousand acres in northern White County that he had acquired over the preceding thirty years. Tucker bought an additional sixteen thousand acres, most of it in White County but some in neighboring Union and Towns Counties. In 1907 Tucker decided to sell his timberland in Georgia, but the initial contract was entangled in the affairs of a bankrupt buyer and eventually fell through. Not until November 1908 was Tucker put in touch with J. Hunter Byrd of Saint Louis, Missouri, but events moved quickly after that. In January 1909 Byrd went to Charleston and closed the purchase of Tucker's thirty-six thousand acres for $235,000, four times what Tucker had paid for the property just a few years earlier.[70]

BYRD-MATTHEWS LUMBER COMPANY

Joseph Hunter Byrd (1880–1966) was a descendant of one of the first families of Cape Girardeau County, Missouri. He was the son of Abraham Ruddell Byrd (1851–1922), who had made a fortune in flour mills before retiring to Texas as a rancher in 1902, leaving his nephew Ruddell Monroe McCombs (1870–1953) to run the mill. In 1904 Hunter Byrd and his younger brother, Abraham Byrd Jr. (1882–1971), moved to Saint Louis, where they worked for a few years at John Mitchell's Alsop Process Company. Around 1908 the Byrd brothers and their father formed A. R. Byrd & Sons Investments.[71]

By 1909 the Byrd brothers had the resources to buy timberland in Georgia, but not to actually establish a sawmill and commence logging operations. For that they turned to their father and to Charles D. Matthews (1843–1917) of Sikeston, Missouri, who had grown wealthy through his mercantile business, grain trading, real estate, and banking. In the spring of 1909, a deal was struck organizing the Byrd-Matthews Investment Company, the parent company for the Byrd-Matthews Lumber Company. The lumber company's sawmill would have a steam-powered electric generator, a planing mill, and a drying kiln. By the time it was in full operation in the spring of 1914, it was one of the nation's largest sawmills. "Harvest" of the vast forests of mostly virgin timber in Rabun, Union, Towns, and northern White Counties was already underway in 1913 and would continue more or less continuously for the next fifteen years.[72] The Byrd-Matthews Company also organized the Gainesville and Northwestern Railroad, which was incorporated in Georgia in February 1912 as a subsidiary of the

Byrd-Matthews Investment Company. Surveying for the line had begun as early as 1910, and by November 1911 clearing of the right-of-way had begun. Depots were also constructed, including the one at Nacoochee, and trains began running in 1913.[73]

HELEN

Many consider John Elvin Mitchell (1871–1938) the founder of Helen. Born in Tullahoma, Tennessee, he had only a fourth-grade education but became a notable inventor with hundreds of patents to his name. In the 1890s he established a plant in Saint Louis that manufactured a variety of mill machinery, and he perfected and acquired patents on the "Alsop process" to bleach flour, an invention that significantly boosted the profits of the flour mills of the Byrds and other mills that used the process.[74]

Every sawmill had its associated mill town of worker housing, stores, a hotel, a bank, and perhaps a church. Without the town the mill could not be operated, but it is not clear how much town planning typically went into their development. Most of them were thrown up rapidly and only slowly improved, until they were abandoned when the timber ran out and the sawmill shut down. Exactly why John Mitchell got involved with the development of Helen is not clear; perhaps it is just an example of his wide-ranging interests. In October 1911 he bought 225 acres on the east side of the river and another 135 acres on the west side of the river in the small valley upstream from Nora Mill. The larger parcel he sold to Byrd-Matthews Lumber Company as a site for its enormous new sawmill, while on the smaller parcel Mitchell laid out streets and platted lots along the Unicoi Road across the river from the mill. He even established a bank "for the benefit of mountaineers and backwoodsmen." He named the town Helen, in honor of the only daughter of Ruddell McCombs, the Byrds' cousin, who was by then president of the company.[75]

The Byrds and Matthews rarely visited, and on-site management was provided by Robert O. Byars (1853–1916), a Saint Louis cabinetmaker by trade. His death in June 1916 must have been a serious blow to the company, especially since by that time the Byrds owed Matthews over $2 million in notes and interest. In March 1917 the mill closed, as Matthews took control of all of the company's assets and searched for a way to reopen the mill. Later that year the Morse Brothers Land & Lumber Company, with home offices in Rochester, New York, took over the mill's operation and brought it back into production.[76]

When the Morse Brothers company was organized, it took control of all the mill assets and 17,000 acres in the Tucker tract that had not been logged. It also folded in 18,000 acres of timber that it already owned on Blood Mountain, adjoining some of the old Byrd-Matthews holdings. Separately, Matthews sold the 18,500 acres cut over by Byrd-Matthews. The Morse Brothers company continued operating its sawmill at Helen until May 1931, when collapsing demand forced its closure. By then nearly 56,000 acres of all but the most inaccessible areas of northern White County had been clear-cut and the timber turned into millions of board feet of lumber.[77]

The decline in the timber harvest ultimately made the railroad unsustainable, and service to Nacoochee and Helen was discontinued in 1928, dealing another blow to an economy already diminished by the onset of the Great Depression. The sawmill at Helen was an economic lifeline for the local economy, which had suffered from the steady population drain brought on by the opening of textile mills at Gainesville and elsewhere and by the timber companies' acquisitions, which displaced many small farmers.

In 1910 the population of White County reached its lowest point of 5,110; but by 1920 it had increased to 6,106. Much of that growth was the result of the logging industry and especially the sawmill at Helen. In 1910 only 250 people were in the Chattahoochee District, which included the Helen Valley; in 1920 there were 864. By 1930, however, the railroad was gone and the sawmill was barely operating, and half the district's population had left. The population of Nacoochee, meanwhile, continued a steady decline that began in the 1890s, falling from 998 in 1900, to 765 in 1920, and to 720 in 1940.

PROFIT AND LOSS

In 1900 one hundred thousand acres of mostly virgin forests were still in White County, most of it in the northern part of the county, but by the 1930s much of that had been cut down.[78] What was left was a badly compromised landscape and natural environment. Bear, deer, turkey, and most other species of wild game disappeared, as did the trout and other aquatic life that were smothered by the silt from eroded hillsides. While White County might never have suffered the almost total environmental destruction wreaked by the copper smelters around Copper Hill, Tennessee, the county's landscape was, historian Baker lamented, "scarred and decimated," especially in the northern part of the county. "Much of the area looked like a barren wasteland." One local resident, returning to the scenes of her youth, wrote that "I saw the stripping of the trees from the mountains. In the valleys were huge piles of virgin timber waiting to be picked up . . . and all around me was devastation. I did not like what I saw."[79]

Logging the great forests made millions for the timber barons, but only a fraction of that wealth was returned to the local community. Worse, logging brought on population decline and severely disrupted the local economy. While there had always been speculation in gold lands, those lands were inherently limited after the gold belt had been mapped. The rampant speculation that accompanied industrial loggers in the late nineteenth and early twentieth centuries was of another

An engine of the Gainesville and Northwestern Railroad, southbound from Nacoochee. A characteristic of most photographs from the late nineteenth and early twentieth centuries is the notable lack of large trees. Brooks and Greear, *Images of America*, 37.

A view of the Byrd-Matthews sawmill, looking southeast. Brooks and Greear, *Images of America*, 34.

sort entirely and forced up land values so much that many small farmers could not survive the higher taxes even if they wanted to stay.[80]

The population of White County, including Nacoochee, began to decline in the 1890s and continued to drop in the first decade of the twentieth century. Along with that, the average farm size declined some 20 percent between 1900 and 1910 in northeastern Georgia, southwestern North Carolina, and southeastern Tennessee, and much of that loss was the result of land acquisition by the timber companies. For those farmers who stayed, the decline in farm size made it more difficult to subsist and severely restricted a family's ability to acquire any land for the next generation.[81] One of the government agents noted, "The emigration tendency in the vicinity of this tract was so strong that the remaining settlers have been unable to maintain schools and churches or keep roads in good condition. This situation has made it easy for a body of land of the size of this tract to be assembled."[82]

A less obvious cost of gold mining and logging was the destruction of a way of life. Fence laws, pushed by the timber and railroad interests, and acquisition of the woodlands by private and corporate timber companies were a severe blow to a traditional way of life that had gone on for centuries. Gifford Pinchot (1865–1946) noted as much when he wrote of those farmers displaced by George Vanderbilt's hundred-thousand-acre hunting preserve near Asheville in the 1890s: "They regarded this country as their country, their common. And it was not surprising, for they needed everything usable in it—pasture, fish, and game—to supplement the very meager living they were able to scratch from the soil."[83]

It was that near-vanished way of life that students at Rabun Gap–Nacoochee School began to document in *Foxfire* in 1966. Huge tracts of land bought by the lumber companies, often at rock-bottom prices, were left in near-ruinous condition, and the local economy, which had once been mostly self-sufficient, was decimated. It would take a long time to recover, but the stage was set for scientific forest conservation. Today Georgia has more commercially managed forest land, 24.4 million acres, than any other state, and much of northern White County is part of the Chattahoochee National Forest.

10

Conservation, Recreation, and Tourism

From the beginning, management of the nation's natural resources has operated between two poles of support. One, represented by Henry David Thoreau, Frederick Law Olmsted, John Muir, the Sierra Club (1892), and the Audubon Society (1905), remains dedicated to preserving the natural world, not as mines and stands of timber but simply for its own sake. The other pole is the vigorous conservation ethic of Theodore Roosevelt and Gifford Pinchot, who believed that forests and all natural resources should be managed and used for the benefit of humans. Which of these viewpoints prevails at any given time is, of course, a function of political and popular support.

By the time the industrial loggers had exhausted the supply of easily accessible timber and abandoned the Southern Appalachians, work was already underway to repair the damage. The task was so large that it took federal involvement through the Civilian Conservation Corps (CCC) and the Works Progress Administration (WPA; renamed the Work Projects Administration in 1939), and especially the ongoing work by the Department of Agriculture's Forest Service, to make much progress. The conservation work supported by federal and state programs made possible the revival of hunting, fishing, hiking, and other kinds of outdoor recreation. Most of all, conservation, along with automobiles and better roads, encouraged what in the late twentieth century became the foundation of the area's economy: tourism. The late twentieth century success of the reimagined Helen and the increasing market for second homes in the mountains also led to tremendous population growth, threatening the very qualities that were the attractions in the first place. Public-private partnerships and individual efforts have nevertheless made great strides in ensuring conservation of the natural and scenic resources of the mountains of northeastern Georgia.

Conservation

For much of the country's early history, land and natural resources seemed in endless supply, and little thought was given to how they were used and exploited. With the country's foundations built on the English common-law conception of the sanctity of private property, law and custom took a laissez-faire approach that allowed private and corporate property owners free rein in how they used their land and resources. The result was too often appalling waste, especially in logging and hunting.

The great forester Gifford Pinchot (1865–1946), the country's first native-born professional forester, summarized the situation in the late nineteenth century, although the attitudes he described had persisted since the nation's founding:

> To waste timber was a virtue and not a crime. There would always be plenty of timber. The lumbermen. . . regarded forest devastation as normal and second growth as a delusion of fools. . . . And as for sustained yield, no such idea had ever entered their heads. The few friends of the forest were spoken of, when they were spoken of at all, as impractical theorists, fanatics, or "denudatics," more or less touched in the head. What talk there was about forest protection was no more to the average American than the buzzing of a mosquito, and just about as irritating.[1]

Frederick Law Olmsted (1822–1903) and others extolled the beauty and wonder of nature and how easily that beauty could be destroyed. In 1864 George Perkins Marsh (1801–82), considered by some the nation's first environmentalist, published *Man and Nature, or Physical Geography as Modified by Human Action*. In addition to describing the extent of changes that humans had already wrought on the natural environment, Marsh intended "to point out the dangers of imprudence and the necessity of caution in all operations which, on a large scale, interfere with the spontaneous arrangements of the organic or the inorganic world; to suggest the possibility and the importance of the restoration of disturbed harmonies and the material improvement of waste and exhausted regions."[2]

That same year, 1864, President Abraham Lincoln signed the Yosemite Grant Act, which conveyed what would later become Yosemite National Park to the State of California "upon the express conditions that the premises shall be held for public use, resort, and recreation; shall be inalienable for all time."[3] It was the first federal action that designated an area for public use and preservation and was a major first step in the creation of a national park system, which began with the creation of Yellowstone National Park in 1872. By the time the National Park Service was established in 1916, there were thirty-four national parks and national monuments.

Warnings that the nation could one day run out of timber gained attention after the Civil War. Dr. Franklin B. Hough (1822–85) was one of the leaders in the nascent campaign to be better stewards of the nation's resources. In 1873 a paper he presented at the annual meeting of the American Association for the Advancement of Science, titled *On the Duty of Governments in the Preservation of Forests*, was especially influential. Two years later the American Forestry Association was organized to lobby for much the same thing.

With this pressure, in 1876 Congress authorized Hough to make a survey of the nation's timber resources. The result was his *Report upon Forestry*, twenty-five thousand copies of which were printed and distributed to public officials in 1878. The report was effective in making the argument for conservation of the nation's resources, particularly its forests, and in 1881 the U.S. Department of Agriculture established its Division of Forestry, with Hough as its first chief. In 1882 Hough recommended that all federal timber lands be withdrawn from sale or lease until there was better regulation of the process. Although there were attempts to abolish the division and fire Hough, the Division of Forestry was made permanent four years later.

Hough's work and increasing public support led to the Forest Reserve Act, which was passed in 1891. The act gave the president the power to "set apart and reserve, in any state or territory having public [i.e., federally owned] land bearing forests, in any part of the public lands, . . . as public reservations." During his term of office, President Benjamin Harrison created fifteen reserves, covering thirteen million acres of public land, and his successors continued to make similar designations.

In 1897 Congress passed the Forest Service Organic Administration Act, which established a statutory basis for management of the forest reserves in order "to improve and protect the forest within the reservation, . . . securing favorable conditions of water flows, and to furnish a continuous supply of timber for the use and necessities of citizens of the United States."[4] That would form the basis for management decisions for the national reserves for most of the twentieth century.

In 1898 Gifford Pinchot was named chief of the U.S. Department of Agriculture's Forestry Division. His tireless work led to its elevation to bureau status in 1903 and, two years later, consolidation of the government's several forestry-related bureaus and divisions into the Bureau of Forestry. Pinchot eliminated political appointees and professionalized the workforce, establishing regulations that included, beginning in 1906, user fees for grazing on public land. In 1907 forest reserves were renamed national forests, and by 1910 there were 150 of them, covering 172 million acres.

The floods and forest fires in the East that were the result of indiscriminate logging, which continued all over the nation, led to passage of the Weeks Act in 1911. Among other things it authorized federal grants for fire prevention, including additional fire-watch towers. In 1924 Congress passed the Clarke-McNary Act, which expanded the Weeks Act and authorized the Department of Agriculture to work with private property owners to reforest their land. The 1924 act also provided expanded grants-in-aid for state fire-protection programs. As a result, Georgia established its own system in 1925 and built fire towers, one of which was erected on Mount Yonah.[5]

The Weeks Act was a shift of focus from disposing of federal lands to buying more, particularly in the headwaters of navigable streams, which were being damaged by uncontrolled erosion. In 1911–12 the Forest Service identified eleven large tracts of land, which were called purchase units, each at least one hundred thousand acres encompassing the headwaters of significant watersheds. Two were in Georgia, one encompassing the Cohuttas in northwest Georgia and the other in eastern Rabun County, Georgia, and western Oconee County, South Carolina, around Ellicott Rock.

Most of the land had been partially cleared or culled for specific types of timber, especially yellow poplar and chestnut, but few of the early purchases were cut over completely. In contrast, many of the later purchases had been clear-cut. Almost without exception, the land was purchased from large timber, land, and investment companies. Enough

What had been a spruce and pine forest, logged and burned over. A large portion of the land acquired for the national forests in the Southern Appalachians was in a similar condition. Mastran and Lowerre, *Mountaineers and Rangers*, 38, fig. 40.

land had been acquired by October 1916 for President Wilson to proclaim Pisgah National Forest, the nation's first national forest made up almost entirely of purchases from private landowners. A national forest was proclaimed in Virginia in 1918, and two years later four more national forests were designated in the Southern Appalachians, including the Nantahala and Cherokee National Forests, which then included lands in Georgia. In the Southern Appalachians, 1,412,952 acres were initially proposed for purchase, but, by the end of 1930, over 4,000,000 acres had been acquired under the Weeks Act.

CHATTAHOOCHEE NATIONAL FOREST

As noted in the previous chapter, Andrew Gennett (1874–1941) and Nathaniel Gennett (1878–1958), sons of a prosperous Nashville merchant, formed the Gennett Lumber Company in 1901 and by World War I were among the largest landowners in the Southern Appalachians. In 1911, as the Byrd-Matthews Lumber Company was preparing to clear-cut tens of thousands of acres in White, Union, Towns, and Rabun Counties, the Forest Service made its first land purchase in Georgia under the Weeks Act when it bought 31,000 acres of timberland in Fannin, Gilmer, Union, and Lumpkin Counties from the Gennett brothers. That land became part of the Cherokee National Forest, while 7,500 acres in Rabun County that had also been acquired from the Gennetts became part of the Nantahala National Forest. The Forest Service began buying land in northern White County in the 1920s, and by 1930 much of that land was part of what was then called the Georgia National Forest. In 1936 the Forest Service reorganized the national forests to follow state boundaries and proclaimed the Chattahoochee National Forest, comprising all the national forest land in northern Georgia. Today more than 750,000 acres of public lands in northern Georgia are managed by the U.S. Forest Service, which includes 41,726 acres in White County, or nearly a quarter of the county's land area.[6]

As with national parks, historic sites, and other federally owned property, the congressional act establishing the national forests typically identified what were thought of as the ideal boundaries for the forest. Meeting those boundaries, however, depended on the willingness of individual property owners to sell their land, and many chose not to. As a result, a substantial part of the Chattahoochee National Forest authorized by Congress remains outside the jurisdiction of the federal government. As much as one third of the forest land in North Carolina, Georgia, and Tennessee was old-growth forests when it was acquired by the Forest Service, and many who worked to establish national forests expected those forests would be preserved. Some of it

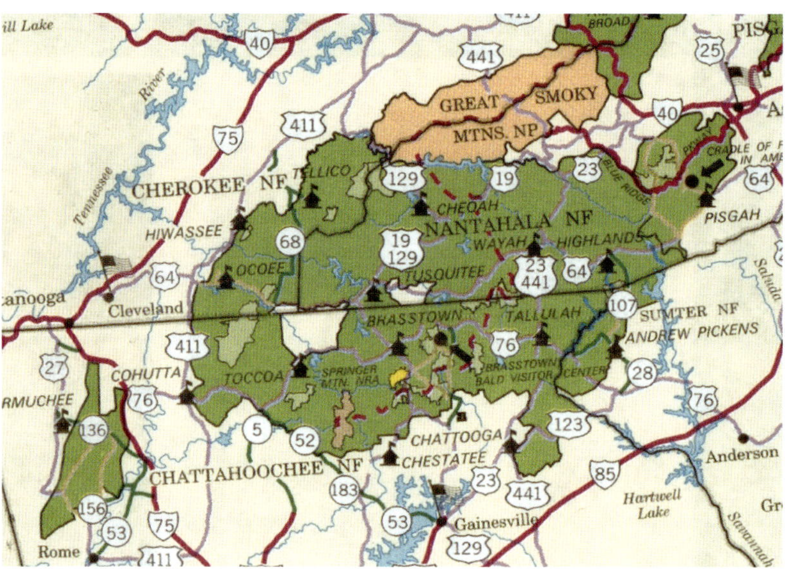

Map showing the extent of U.S. National Forest land in the Southern Appalachians, 1994. U.S. Forest Service.

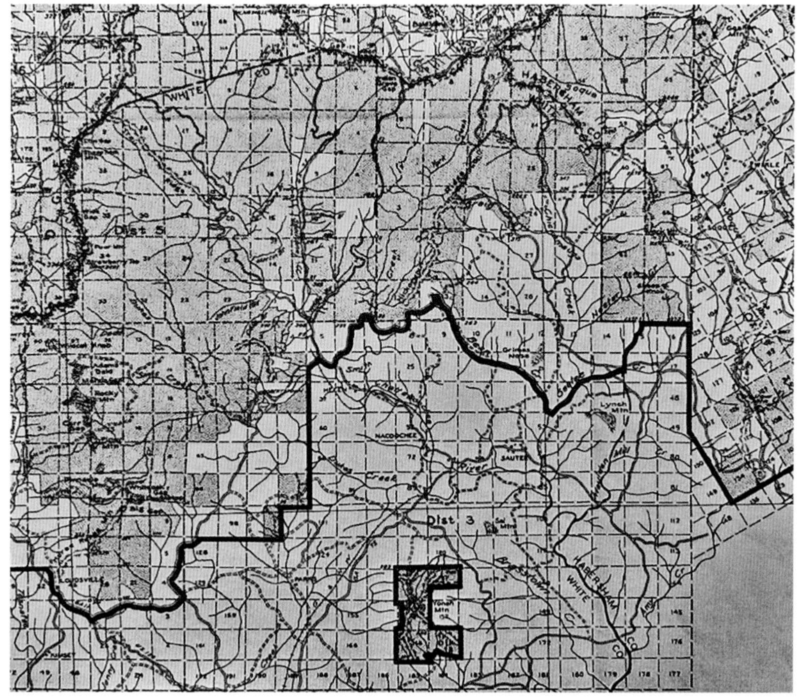

Part of the U.S. Forest Service map, showing the extent of land purchases (shaded land lots) within the authorized boundary of the Chattahoochee National Forest between 1934 and 1937. Georgia Secretary of State, Records of Surveyor General, Historical Map File, Chattahoochee National Forest, 1937, Georgia Archives, Morrow.

was later logged because of reserved timber rights, but much more was logged by the Forest Service itself in the 1950s. Today less than 2 percent of the Chattahoochee National Forest is virgin woodland.[7]

CIVILIAN CONSERVATION CORPS

Much of the land in the Chattahoochee National Forest and most other national forests in the Southern Appalachians had been clear-cut (and often burned over as well) prior to federal acquisition. Rehabilitating those lands through erosion and flood control, forest fire protection and suppression, and replanting decimated forests were among the primary objectives of the Civilian Conservation Corps (CCC), one of the first of the New Deal programs, enacted in March 1933. Ultimately the agency hired more than 250,000 unemployed men between the ages of seventeen and twenty-eight and paid them thirty dollars a month, of which twenty-five dollars were sent home to their families.

A CCC camp was established in White County in the summer of 1933, and, until funding for the CCC was eliminated in the first year of World War II, hundreds of young men resided at the camp, which was on Smith Creek where the lake at Unicoi State Park is now located. Most of their work in White County was in the national forest, where they planted seedlings, worked on erosion control, built roads, and fought fires. The crews also removed "undesirable buildings," which were the houses and outbuildings left when farmers sold their land to the government, and they maintained about a hundred miles of the Appalachian Trail as it ran through the national forest. CCC workers also built what is now known as the Walesi-Yi Interpretive Center at Neel's Gap.[8]

WILDERNESS ACT

Advocacy by the Sierra Club, the National Audubon Society, and others led Congress to pass the Wilderness Act, realizing a dream of John Muir and others to preserve natural areas for their own sake and not for the potential for commercial exploitation, as can happen in national forests. The act was signed into law in 1964 and allows presidential designation of certain federal lands as "an area where the earth and its community of life are untrammeled by man, where man himself is a visitor who does not remain."[9]

The Forest Service took issue with the designation of some of its lands, especially in the East, which it claimed were no longer "pristine" and so should not be designated wilderness. In response in 1974 President Ford signed the Eastern Wilderness Areas Act, which was explicit in rejecting "pristine" as a criterion for designation, since few parts of the world have not had some human intervention. Okefenokee Wilderness in South Georgia was designated as part of that act, and the following year a tract of 37,000 acres in Fannin and Murray Counties was designated the Cohutta Wilderness and a tract of 8,300 acres in Rabun County was designated Ellicott Rock Wilderness. In 1986 a large part of Tray Mountain in northeastern White County and southwestern Rabun County, Ravens Cliff in western White County, and Blood Mountain in Lumpkin County were designated national wilderness areas. Other areas were created in 1991, but by then resistance to additional wilderness designations had grown, primarily because of the restrictions on the use of motorized vehicles.[10]

Of the millions of bison that once roamed the Plains, perhaps as few as nine hundred remained in 1890, killed off for sport and as part of the government's campaign to destroy the Indians' way of life. The last passenger pigeon, which once darkened the skies with miles-long "rivers" of birds, died in the Cincinnati Zoo in 1914, and the last Carolina parakeet died at the same zoo in 1918, hunted to extinction mostly because of the damage the bird could do to orchards and other crops.[11] The dreadful exploitation of wading shore birds for their plumage is still evident in period photographs of women's fashions.

In the Southern Appalachians, clear-cut forests and unregulated hunting had even decimated some of the most common wildlife, including whitetail deer, bear, and turkey, which disappeared from all but the remotest parts of the mountains. Arthur Woody (1884–1946), one of the first two rangers in the Chattahoochee National Forest, claimed that, as a boy in 1895, he had seen his father kill the last deer in the Georgia mountains. While that fact would have been and remains difficult to verify, certainly by the 1890s the dwindling number of all sorts of wildlife was widely noted.

A hodgepodge of federal actions created a national system of wildlife refuges in the first decades of the twentieth century. Although none were established in northern Georgia, they formed a model for state action. In 1911 Georgia created a Department of Game and Fish, charged with the promotion and enforcement of game protection laws, which began with requirements for licenses to hunt and fish. In the 1920s federal agencies started working with the state to establish a series of state-controlled wildlife management areas in the national forests, the first of which was the Blue Ridge Wildlife Management Area, proclaimed in 1928 and now covering 20,900 acres in Lumpkin, Dawson, and Fannin Counties. In 1929 the Chattahoochee Wildlife Management Area was established on 25,000 acres in northern White County and southern Towns and Union Counties. In 2006 the state's first "youth-focused" wildlife management area was established on 582 acres of what had been Buck Shoals State Park on the Chattahoochee River, southeast of Cleveland.[12]

The first steps in restoring wildlife lost from logging and overhunting were taken by individuals, the most famous being Ranger Arthur

Ranger Woody, one of the first two rangers for what became the Chattahoochee National Forest. Mastran and Lowerre, *Mountaineers and Rangers*, 37, fig. 39.

Woody (1884–1946) of Suches. He had been hired as a ranger for what would become the Chattahoochee National Forest in 1918 and pursued the job with the dedication typical of many of the early rangers. About 1925 he used his own money to buy and transport a few deer from North Carolina to the national forest lands in Lumpkin County in a successful effort to reintroduce them to the Georgia mountains.[13] Woody also helped establish the state's first fish hatchery, near Rome, which produced 250,000 trout in 1928, its first year of operation. By World War II the state had restocked the state's lakes and rivers with over five million fish. Interest in these efforts caught the attention of Jesse Richardson Lumsden, and he is said to have "personally stocked" the headwaters of the Chattahoochee River with speckled trout.[14]

Recreation

With the widespread use of automobiles after World War I, there was a growing demand for better roads and more recreational opportunities, especially for camping, for an increasingly urban population. The conservation work begun in the 1920s was paying off even before World War II, as trout once again could be fished in the Chattahoochee River and deer could once again be hunted in the second-growth forests that were slowly recreating natural habitats in the mountains. The lack of paved roads in White County prior to World War II inhibited the development of recreational opportunities, but after the war more and more hunters and fishers were frequenting the mountains.

The idea of state parks for recreation was encouraged by the National Park Service and its director, Stephen Mather (1867–1930), who called a meeting of state representatives to discuss the concept in 1921. In 1927 Georgia took a first step toward a state-park system when it turned over jurisdiction of the Indian Springs Reserve, which it had owned since the Creek land cession in 1825, to the State Board of Forestry, as the state's first "forest park." Land to establish Vogel State Park, twenty miles northwest of Sautee Nacoochee, was donated that same year, but most of that park remained undeveloped until the CCC provided labor to construct the lake and facilities there beginning in 1933. By then the difficulty of combining recreation with traditional forestry led to the creation in 1931 of a separate department in the Forestry Board to manage recreational facilities.[15]

In 1937 Georgia created the Department of Natural Resources, which had jurisdiction over natural and cultural resources. Among its four divisions was the Division of State Parks, Historic Sites and Monuments, charged with, among other things, providing adequate recreational opportunities for the people of the state. By the beginning of World War II, there were eighteen Georgia state parks, including Unicoi and another state park at Black Rock Mountain in Rabun County. After the war the state park system expanded rapidly, as state policy sought to establish a park within fifty miles of every Georgian.

UNICOI STATE PARK

In 1908 Charles "Charlie" Maloof (1891–1971) immigrated from his native Lebanon and five years later opened a general store in Helen. In 1915 he married Nora Blanche Westmoreland (1896–1981), whose paternal ancestors had been among the pioneers who resettled the valleys in the 1820s. Maloof prospered with the town and became its biggest booster, as he served for over thirty years as mayor or city council member of Helen. When the sawmill closed in 1931, he bought part of the property and opened a hardwood-flooring factory. When much of the Mountain Ranch Hotel burned in 1946, what few tourists had been coming to Helen disappeared. Hoping to revive tourism and the town's fortunes, Maloof led a campaign to create a state park on the site of the old CCC camp on Smith Creek, northeast of Robertstown. With political connections in Atlanta, he pressured the state for the park, which was approved in 1950.[16]

Construction of a dam and lake on Smith Creek began in 1951, and the "White County Area Park" was opened to the public on 5 June 1954. The state announced the event: "White County Area is one of our latest State Park additions which has not yet been officially named. The Park is in White County near the towns of Helen and Robertstown and consists of 1800 acres which were leased from the Federal Government for the purpose of building a State Park. This Park is in about the same area as Vogel and recently a Dam and Spillway have been completed which will create a beautiful 49 acre lake. This lake will be stocked with fish and will provide boating, fishing and swimming."[17] Formally named Unicoi State Park in 1956, the park included 278 acres of state-owned land, with the rest leased from the Forest Service. Over the years the park has expanded to encompass 1,050 acres. In 1968 the North Georgia Mountains Authority was created by the state to manage the development of lodging and other facilities at designated state parks, and in 1973 a lodge opened at Unicoi. Now one of the state's most popular state parks, Unicoi has the state's largest concentration of park lodging, with a hundred rooms, thirty cabins, and eighty-two campsites. It remains one of the area's largest employers as well.[18]

SMITHGALL WOODS STATE PARK

Charles Augustus Smithgall Jr. (1911–2002) was born near Chipley, Florida, the son of a second-generation lumberman from Pennsylvania, who went to work in the piney woods of northern Florida before World War I. The family moved frequently before Charles Jr. graduated from Boys High School in Atlanta in 1929 and from the Georgia Institute of Technology four years later.[19] Smithgall began his career in radio while at Georgia Tech, where he was an announcer for the school's radio station, WGST. In 1936 he went to work for WSB Radio in Atlanta and was the first reporter on the scene after the dreadful tornado that destroyed downtown Gainesville in April of that year. In 1941 he was a founder of WGGA Radio in Gainesville as well as radio

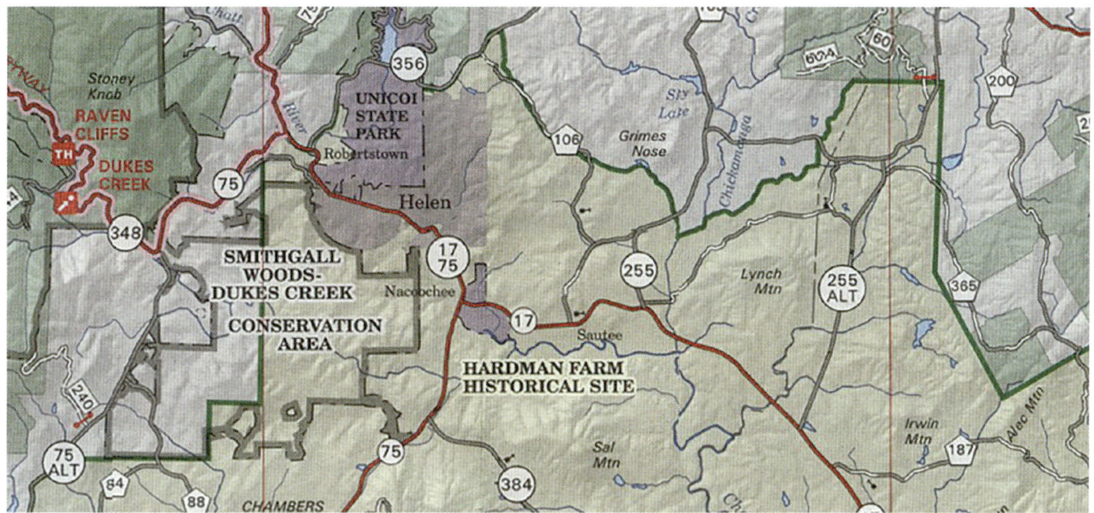

Detail from a map of Chattahoochee National Forest, showing adjacent state parks and reserves, which in combination have helped rehabilitate and maintain the forested mountains of the Blue Ridge in Georgia. The dark green line is the boundary of the authorized national forest; green-shaded areas are federally owned; gray areas are privately owned. U.S. Forest Service.

stations in Rome and Atlanta, Georgia, and Gadsden, Alabama. In 1947 he began publishing the *Gainesville Daily Times* and in the 1960s formed Georgia Community Papers, which published the *Gwinnett Daily News* and several small weekly papers in metropolitan Atlanta.

He inherited a love for the environment from his father, who in the 1920s had first developed the springs that are now at the center of Ponce De Leon Springs State Park in the Florida Panhandle. In the 1980s he sold his radio and newspaper interests and in eighty-one separate transactions between 1983 and 1994 acquired some 5,664 acres of much-abused land along Dukes Creek. Part of the land was inside the authorized boundary of the Chattahoochee National Forest, but much of it was not. Some of it was owned by the family of Benjamin Wilson Allison (1822–1905), whose father, uncle, and namesake grandfather had settled on the upper reaches of Dukes Creek in the early 1820s. Much of the land, which was known to some locals as "Old Ben's Fields," had been heavily damaged from hydraulic mining and from general neglect, but Smithgall spent considerable time and money beginning to undo the damage. In 1994 Smithgall conveyed the property to the State of Georgia in a part-gift and part-purchase arrangement, partially funded by a state grant. The park was dedicated as a state heritage preserve the following year and designated the Dukes Creek–Smithgall Woods Conservation Area.

After Governor Hardman's death in 1937, his children and grandchildren continued to maintain the farm at Nacoochee, using it mostly as a country retreat. By the late twentieth century, the cost of maintaining the house and the numerous fine outbuildings led the family to consider options that would ensure continued preservation of the historic buildings and landscapes. With the help of the Trust for Public Land and the Georgia Trust for Historic Preservation, a plan for the farm's preservation was developed in the mid-1990s, and in 1999 the house, nineteen outbuildings, and 183 acres of land were donated to the State of Georgia. After several years of work to rehabilitate the site, it was opened to the public as the Hardman Farm State Historic Site, operated and administered through Smithgall Woods State Park.

Tourism

Through much of the nineteenth century, the poor condition of most roads and general lack of public accommodations inhibited long-distance travel, but, as early as the 1830s, intrepid tourists were venturing away from "civilization" to search out natural wonders that were only then being "discovered." One of those was Tallulah Falls, which was already a tourist destination in 1837 when the Reverend Sherwood wrote, "Tallulah Falls attract thousands to view them every season. Rev. Mr. Hawthorne was bathing in the river in July, 1837, and was drowned! It is a pleasant ride out to them from Clarksville, and no one will forget the impression, while the awful grandeur of the scene steals upon him."[20]

With regular stagecoach service from Athens, where there was a railroad connection by 1841, Clarkesville was a gateway to other attractions in the northeastern Georgia mountains. Addison Richards already had an audience when he wrote in 1843, "Haply, these paragraphs may meet the eye of some who are dragging out the weary summer days in the city or in the 'dull town'; and if so, let us prevail with them to make a visit to Clarkesville and its noble 'lions'—Tallulah, Toccoa, Nacoochee, Currahee and Yonah!"[21]

The falls saw even more visitors after railroad service was established in 1882, and by 1900 several hotels had been built. There was also talk of making the area into a state park before it was ruined by "grasping capitalists," who by 1905 were already taking options on land and water rights around the gorge with the intent of generating electricity that could be used to power electric streetcars as far away as Atlanta. Vociferous opposition led by Helen Dortsch Longstreet (1863–1962), the widow of Confederate general James Longstreet, and her Tallulah Falls Conservation Association could not stop the development, and by 1913 the falls had been destroyed and the state's first hydroelectric plant was in operation. Most of the town of Tallulah Falls burned in 1921, and its tourist industry never really recovered.[22]

Meanwhile, the loggers were following the railroads into the mountains and laying waste to much that made the area attractive to tourists in the first place. All the forests around Sautee Nacoochee were not

Lithograph of Tallulah Falls by T. Addison Richards, from *American Scenery Illustrated*, 1854. *Tallulah Falls.*

clear-cut at once, since much of those in the immediate vicinity of the valleys had likely been cut early on, some by the Cherokees and much more by the nineteenth-century settlers, especially in areas that were being mined. In the early twentieth century, at least some second-growth forest was maturing, and people still came to enjoy the scenery. After World War I U.S. Forest Service management allowed the forests and streams to recover and, with support of the state's Department of Natural Resources, wildlife returned to the mountains, making the area attractive for hunting and fishing again.

With an improved natural environment, the creation of Unicoi State Park in the early 1950s formed the foundation for a resurgent tourist economy in White County in the last half of the twentieth century. Some of the few gristmills and general stores still in existence were reinvented to attract tourists, as was the entire town of Helen. In the 1990s Hardman Farm State Historic Site and Smithgall Woods State Park were created by the state, followed by the privately funded Folk Pottery Museum of Northeast Georgia. These and the relatively well-preserved rural landscape of the Sautee and Nacoochee Valleys have become tourist destinations in their own right.

HOTELS

Prior to the Civil War, travelers in the rural South often relied on residents along the way for food and overnight accommodations, usually in exchange for a little money. Poorer travelers still might beg for room and board or sleep in barns along the way, but those who could afford to do so appreciated the inns, hotels, and taverns that were sometimes available. Several hotels or boarding houses operated at Nacoochee in the nineteenth and early twentieth centuries, hosting transitory travelers, longer-term visitors, and year-round residents.

Nacoochee Hotel

As noted in chapter 4, James Rutherford Wyly, the contractor who built the Unicoi Turnpike, operated three inns on the road, one of them at Nacoochee. His earliest recorded purchase in the valley was Land Lot 74, which he acquired in November 1820, and that may have been the site of his first inn in the valley. The land lot encompasses the intersection of the Unicoi Turnpike and the ancient road from the Cherokee towns in western North Carolina that is now known as Rabun Road. It would have been an excellent location for an inn.

Research in the county's record of deeds and mortgages is incomplete, but Wyly is said to have bought a thousand acres (including Land Lot 75) from Benjamin Cleveland in April 1829. The notation of Wyly on the Moffat map suggests the possibility that he established an inn there before 1837. He reportedly sold 2,500 acres to Edwin Poore Williams in 1845, but he seems to have maintained the Nacoochee Hotel on Lot 76, east of Sautee Creek, at least until 1848, when he sold his remaining property at Nacoochee to Benjamin Cleveland and moved over the Blue Ridge into Union County. It is not clear if the hotel continued to operate, but it was closed before the Civil War.[23]

Green Hotel

Edwin Williams (1814–96) built his residence, Oak Hill, on the north side of the Unicoi Turnpike in Land Lot 75 in the 1840s, but it burned not long after the Civil War. Williams sold at least part of the land lot to Joseph I. Green (1841–85), who built a new hotel on the site in 1872. In 1896 his widow, Mary E. Green, who was a daughter of Charles L. Williams, sold "the dwelling house known as the Green Hotel, stables, barn, crib and tenant house" to William Henry Clay Alley (1843–1929).[24]

Alley House

Alley and his wife, Virginia Oakes Alley (1853–1919), christened the Greens' old hotel the Alley House and continued operating it as a hotel and boarding house until the building burned in 1898. The Alleys rebuilt about 1900, and their hotel remained popular until it closed

The view from Alley House, for years the main hotel at Sautee Nacoochee. Courtesy of Sautee Nacoochee Community Association.

The Glen House. Courtesy of Sautee Nacoochee Community Association.

in the 1930s. The property was sold in 1937, and the new owner tore down half of the building before the Alleys regained possession of the property in 1957.[25]

Glen House

In the early 1870s, Edwin Williams built a house on part of Land Lot 75, just east of the Green House, for James Glen (1836–1927) and his wife, Susan (1831–1909), the daughter of pioneer Abraham Littlejohn. They had at least three daughters. The original house had only three bedrooms, but around 1890 the Glens built a two-story addition that added four more bedrooms and began operating as a boarding house and hotel. In 1898 they constructed another addition that gave the house a total of eleven bedrooms. The Glens' daughters, Annie (1864–1936) and Elizabeth (1872–1938), did not marry, and after James Glen's death they continued to keep the boarding house until their own deaths in the late 1930s.[26]

View of Nacoochee Valley from the Alley House, circa 1900. Courtesy of Sautee Nacoochee Community Association.

Hotels in Helen

The Byrd-Matthews lumber town of Helen supported at least three hotels in the early twentieth century: the Marshall, the Commercial, and the Mitchell Mountain Ranch Hotel. The latter was built across the river from the sawmill by John Mitchell, a business associate of the Byrd-Matthews Company, who had bought property and sold lots that became the towns of Helen and Robertstown. His hotel was the first to have electric lighting and running water and was so busy that a fifteen-room addition was built in 1921. When the hotel burned in 1945, it dealt a serious blow to the local economy. John and Caroline Greear also operated a guest lodge that included cabin rentals. In 1952 the Chattahoochee Motel and Grill opened in Helen, hoping to capitalize on the new Unicoi State Park, which would not have non-camping lodging facilities until 1968.[27]

RICHARD B. RUSSELL SCENIC HIGHWAY

Unicoi State Park, which opened in 1954, quickly became one of the state's most popular state parks, bringing thousands of people into White County as tourists and having a significant impact on the local service economy. The editor and publisher of the *Cleveland Courier*, James Davidson (1897–1972), was a major booster of the nascent tourism industry and began editorializing in support for construction of a new road through the national forest between Helen and Blairsville. There was tremendous opposition from the Forest Service, conservationists, and many others who wished to preserve what the Forest Service had returned to wilderness. At a time when there was little consideration for environmental impacts, the support of Sen. Richard B. Russell ensured state and federal funding for the project. Construction began in 1962 on a route that left SR 75 Alt at Dukes Creek and ran along the mountains in a northwesterly direction to connect with the old Logan Turnpike at Tesnatee Gap. Completed in 1968 and designated the Richard Russell Scenic Highway, the road proved to be as popular as its proponents had promised, especially in the fall.[28]

ALPINE HELEN

In 1946 James Milligan Wilkins (1919–2007), a native of Chattanooga, moved to Helen and established the Wilco Company, manufacturer of hand-framed Argyle socks and employer of 125 mostly White County residents. He also joined Charlie Maloof in lobbying for Unicoi State Park, which for the first time brought large numbers of tourists through Helen, although few stopped except for gasoline and a snack.

Helen, before and after Kollock's remodel. Brooks and Greear, *Images of America*, 124.

In the spring of 1968, Wilkins and Helen businessmen Peter "Pete" Hodkinson III (1934–76) and Robert Fowler began discussing what could be done to make the town more attractive to tourists. They tasked local artist and historian John Kollock (1929–2014) to come up with a conceptual plan for a remodeled town, and in January 1969 his plan to remake Helen as a Bavarian village was approved. By September enough work had been done for the town to hold its first promotional event: the Chattahoochee Trout Festival and Alpine Hoedown. The following year Helen hosted the first of its wildly successful Oktoberfests, and quickly tourism became the engine that drove the county's economy. By 2010 as many as two million people were visiting the town each year. Many tourist towns, such as Gatlinburg, grew organically with little or no planning and were often subject to the worst excesses of commercialism. How well Helen has avoided that fate might be argued, especially while sitting in Oktoberfest traffic, but there can be no arguing that the redesign of Helen rejuvenated the local economy.[29]

SECOND HOMES

While today we think of the mountains as a way to escape the heat and humidity of summer, throughout the nineteenth century, escape from disease was equally important. Tropical and subtropical diseases such as dengue fever, malaria, and yellow fever regularly became epidemic and were especially feared by the planters in the Low Country of South Carolina and Georgia. Nearly all of them routinely abandoned their plantations and moved inland during the "fever months," which generally ran from late June until the frosts of autumn.

Almost as soon as the Southern Appalachians were opened to white settlement, wealthy planters were building summer homes in the cooler climate of the highlands. Clarkesville and Habersham County benefited from this trend, and there were enough summer residents in and around the town to organize an Episcopal church in 1838. George Walton Williams, whose primary residence remained in Charleston, appears to have been the first to build a summer home at Sautee Nacoochee when he built the Carpenter Gothic house he called Mountain Home in 1876. His son Henry built a summer house just to the east about 1890, and, like Mountain Home, it enjoyed running water and fountains fed by springs farther up the mountain. In 1896 George Walton Williams Jr. built Sautee Manor overlooking Sautee Valley just west of Mountain Home.

Automobiles and a rising middle class made possible another generation of second homes after World War I, but these were more modest and were built mostly a few miles to the northeast of Sautee Nacoochee on one of the new lakes on the Tallulah River: Rabun, Seed, and Burton Lakes. After its completion in 1957, a similar pattern was repeated at Lake Lanier, near Gainesville. In the last quarter of the twentieth century, there was an explosion of residential development across northeastern Georgia, some designed for retirees and some for vacation homes. The old summer camp at Sky Lake, just north of Sautee Valley, was redeveloped as a 1,400-acre gated residential community that today has over 450 houses or cabins and another ninety-nine vacant lots. Typical of late twentieth-century residential developments in the mountains, less than half of those are owner-occupied the year round. With lake-front property built out, the search for commanding views has also led some to construct on or very near ridgelines or prominent mountainsides, diminishing the aesthetics for everyone else. This trend has already scarred the historic views across Nacoochee Valley to Mount Yonah, and land has been cleared for a large residential development along the old Hardman Road as well.

Rural Conservation

The past fifty years have witnessed enormous changes in White County, as its population has nearly quadrupled, rising from 7,742 in 1970 to an estimated 28,319 in 2015. Helen has been an economic boon to the county, but at the same time the rapid growth has put tremendous pressure on the county's infrastructure and raised the specter of overdevelopment. In the decade following the birth of Alpine Helen, the county experienced its fastest rate of growth since the 1860s, and by 1980 it was clear that much of what makes White County attractive to tourists was in danger of being lost.

Lynch Mountain Road, one of the area's best-preserved early roads. Photograph by author, 2015.

Rural landscape on Lynch Mountain Road. Photograph by author, 2015.

MOUNTAIN AND RIVER PROTECTION ACTS

By the 1980s many people recognized the damage that unfettered development was doing to the state's resources. Gov. Zell Miller, who served from 1991 to 1999, was from Young Harris and was a vocal critic of mountaintop development. In 1991 the legislature passed the Georgia Mountain and River Corridor Protection Act, which required the state and local governments to establish minimum standards and coordinated planning procedures for land use development on mountains and along river corridors. Unfortunately, the act exempted much development from the legislation's requirements. In 2004 the White County Board of Commissioners found that "the protection of mountains and hillsides is an urgent matter given growing public concern about development on them" and amended the county's zoning code to adopt a new article that established minimum standards for "Mountain and Hillside Development." Applicable on any site located above 1,700 feet, its provisions were meant to meet or exceed state standards.[30]

SAUTEE NACOOCHEE COMMUNITY ASSOCIATION

In one of the first efforts to preserve something of the rural character of the Sautee and Nacoochee Valleys, which had survived by circumstance and not especially by design for 150 years, a group of local residents formed the Sautee Nacoochee Community Association (SNCA) in 1981. Among its first successes was obtaining a state grant to fund a landmark preservation study of the valleys, which was completed in November 1982 by Allen Stovall, now professor emeritus of landscape architecture at the University of Georgia.

In 1986 SNCA took possession of the old Nacoochee School, which they refurbished and converted into a thriving arts and community center. It includes an intimate theater that seats ninety-eight, two art galleries showcasing area artists, and a local history museum. By the 1990s the association had become the county's main voice for historic preservation, the environment, and the arts. In 1994 Sautee Nacoochee was included in John Villani's book *The 100 Best Small Art Towns in America: Where to Find Fresh Air, Creative People, and Affordable Living*. In the twenty-first century, the association has built on that foundation by expanding their interpretive program to include a major emphasis on local African American history and by the addition of the Folk Pottery Museum of Northeast Georgia, which was founded and funded by philanthropists Dean Swanson, a former board chair of the association, and his wife, Kay.

HISTORIC PRESERVATION

With the advocacy of the National Trust for Historic Preservation, the United States Conference of Mayors, and others, Congress passed and President Lyndon B. Johnson signed the National Historic Preservation Act in October 1966. The act precipitated the creation of a State Historic Preservation officer in each state and initiated decades-long efforts to survey each county and compile a list of old structures—which could include buildings, roads, bridges, dams, and canals—that were significant in local, state, or national history and ought to be preserved. The act also established the National Register of Historic Places, which is the official list of the nation's historic places "worthy of preservation." The register is meant to be "part of a national program to coordinate and support public and private efforts to identify, evaluate, and protect America's historic and archaeological resources."[31]

Listing in the National Register, however, generally requires a property owner's consent, which seriously limits the law's impact. Even when a property is listed, there are no restrictions on the owner's treatment of the historic building up to and including demolition, except when a project uses federal funds. Then a series of reviews, primarily through consultation required in Section 106 of the act, is meant to resolve or mitigate damage to or loss of the historic property. Far more effective in protecting historic resources from federal actions is the National Environmental Policy Act, passed in 1969, along with a variety of laws and regulations to protect archaeological resources. The state has a similar layer of review, but the only reliable guard against ill-considered actions of private property owners must come at the local level, typically through historic district zoning controls but also through historic preservation property easements.

White County has six National Register listings, two of which are in Cleveland (the courthouse and the jail); the rest are in Sautee

Nacoochee. The first listing was the Nacoochee National Register Historic District, which was nominated in 1980.[32] The district encompasses approximately 2,500 acres on both sides of the river, stretching from north of Nora Mill to east of Sautee Creek, its width generally defined by the 1,400-foot elevation contour line, which carries the district up Dukes Creek. Of the seven or eight dozen buildings located in the district today, which do not include the outbuildings at West End, nineteen are considered historically significant.

In 1986 the Sautee Valley Historic District was listed in the register. Beginning east of Chickamauga Creek, the district encompasses the southern half of the valley from the creek to Sautee, bounded on the east by Lynch Mountain Road and on the west by Bean Creek. The nomination notes that thirty-one buildings are considered historically significant, along with nine landscape features, a category not yet being considered when the Nacoochee Valley Historic District was listed in 1980. Archaeological resources are a major component of both National Register districts. Of the archaeological sites surveyed by Wauchope in 1938, sixteen are within the two districts, along with the Nacoochee Mound. In addition there are the ruins of Stovall Mill, at least two gold-mining sites on Bean Creek, and three historic archaeological sites.

Rural Preservation

The National Trust for Historic Preservation defines rural preservation as "protection of the countryside, including the preservation of buildings and villages of cultural significance, the protection of their surroundings, and the enhancement of the local economy and social institutions."[33] That, in essence, has been a primary mission of SNCA since its founding thirty-five years ago and was the reason that the organization commissioned the Stovall study. That study developed a comprehensive plan for preserving, conserving, and interpreting the Sautee and Nacoochee Valleys. The study is "about place," its author wrote. "It is place, molded by time and the events of time. A commitment to maintaining that sense of place is the focus of this study." Stovall's study has given SNCA a foundation on which to build and

Sautee Nacoochee Community Center.
Photograph by author, 2021.

continues to inform their work, especially in raising awareness of the tremendous cultural significance of the valleys.[34]

As noted earlier, White County has a mountain-protection ordinance and subdivision regulations that promote green space, but that has not prevented some development that has compromised the scenic qualities that make the valleys so special. In the summer of 2015, the Center for Community Design and Preservation at the University of Georgia completed a comprehensive historic resource survey of unincorporated White County, including Sautee Nacoochee. All the information and tools are now in place to ensure continued preservation of the rural landscape of Sautee Nacoochee. It is not yet clear, however, if there is the political will to establish historic district zoning, which could provide the comprehensive protection that the valleys so richly deserve.

Notes

Preface

1. Featherstonhaugh, *Canoe Voyage*, 177.

2. Lumsden's papers have recently been cataloged and can be found at the Hargrett Rare Book and Manuscript Library at the University of Georgia, collection number MS4204, accession number 2018-110.

Chapter 1. The Natural Environment

1. McPhee, *Annals*, 171.

2. See the U.S. Geological Survey's historical maps at "National Map Viewer."

3. McPhee, *Annals*, 36; Gore and Witherspoon, *Roadside Geology*, 190.

4. McPhee, *Annals*, 184, 227.

5. Gore and Witherspoon, *Roadside Geology*, 181–83, 316.

6. Gore and Witherspoon, 225.

7. McPhee, *Annals*, 218.

8. Gore and Witherspoon, *Roadside Geology*, 183, 187; McPhee, *Annals*, 223.

9. Soil Survey Staff, "Web Soil Survey."

10. S. P. Jones, *Second Report*, 27.

11. Fagan, *Complete Ice Age*, 47, 91.

12. Braun, "Glacial and Periglacial Erosion."

13. Carlson, "Younger Dryas Climate Event."

14. Russell et al., "Warm Thermal Enclave," 185; Gore and Witherspoon, *Roadside Geology*, 223.

15. Davis et al., *Southern United States*, 8–9.

16. Bartram, *Travels*, 56.

17. Voorhies, "Pleistocene Vertebrates."

18. Bartram, *Travels*, 62.

19. See Belue, *Long Hunt*.

20. Davis et al., *Southern United States*, 3–4.

Chapter 2. First People

1. Anderson, "Paleoindian Period."

2. M. White, *Archaeology and History*, 14, 24.

3. Ledbetter, "Georgia."

4. T. Lumsden, *Nacoochee Valley*, 3; M. White, *Archaeology and History*, 19.

5. Davis et al., *Southern United States*, 15–18.

6. Davis et al., 28; C. Hudson, *Southeastern Indians*, 44; M. White, *Archaeology and History*, 28.

7. Anderson and Hanson, "Early Archaic Settlement"; Baker, *Shadow of Yonah*, 16; T. Lumsden, *Nacoochee Valley*, 3–4.

8. M. White, *Archaeology and History*, 26, 38.

9. Davis et al., *Southern United States*, 29.

10. Davis et al., 30.

11. M. White, *Archaeology and History*, 32; C. Hudson, *Southeastern Indians*, 53; Elliot, *Live Oak Soapstone Quarry*.

12. Hopkins, *Report on the Asbestos*, 166–70.

13. Loubser and Frink, "Archaeological and Ethnohistorical Appraisal"; Loubser, "Recording and Interpretation."

14. See "Ancient Indian Petroglyph Boulders."

15. Davis et al., *Southern United States*, 35.

16. M. White, *Archaeology and History*, 28; Davis et al., *Southern United States*, 36; Cook, *Chattahoochee River User's Guide*, 33.

17. Davis et al., *Southern United States*, 37.

18. DeVorsey, "Indian Trails"; Myer, "Indian Trails of the Southeast," 739, 743.

19. A. Hudson, *Creek Paths*, 13.

20. Sassaman, "Stallings Island Site."

21. C. Hudson, *Southeastern Indians*, 55.

22. "Indigenous Ceramic Type Collection."

23. Loubser and Frink, "Archaeological and Ethnohistorical Appraisal."

24. Mooney, "Myths of the Cherokee," 332, 418–19.

25. M. White, *Archaeology and History*, 43–45; C. Hudson, *Southeastern Indians*, 62–69.

26. Blitz, "Adoption of the Bow," 131; Davis et al., *Southern United States*, 42.

27. Davis et al., *Southern United States*, 35; C. Hudson, *Southeastern Indians*, 19.

28. C. Hudson, *Southeastern Indians*, 60–62; Davis et al., *Southern United States*, 58–61.

29. C. Hudson, *Southeastern Indians*, 95.

30. Scarry, "Late Prehistoric Southeast," 29.

31. Davis et al., *Southern United States*, 51.

32. C. Hudson, *Southeastern Indians*, 81.

33. Davis et al., *Southern United States*, 59.

34. Richter, *Facing East*, 48.

35. C. Hudson, *Southeastern Indians*, 78; M. White, *Archaeology and History*, 91.

36. M. White, *Archaeology and History*, 64, 91; Anderson and Schuldenrein, "Mississippian Period Settlement."

37. Davis et al., *Southern United States*, 67–68.

38. M. White, *Archaeology and History*, 83; Davis et al., *Southern United States*, 37.

39. Widmer, "Structure of Southeastern Chiefdoms," 125–55; Shadburn, *Upon Our Ruins*, 143, 151n12.

40. Cobb, "Mississippian Chiefdoms"; Hudson and Tesser, *Forgotten Centuries*, 185.

41. M. White, *Archaeology and History*, 107.

42. Lanman, *Letters*, 25. One of Lanman's "mounds" was actually a natural feature.

43. The material in this section is based on C. Jones, *Antiquities*, 213–39.

44. C. Jones, *Antiquities*, 224.

45. C. Jones, *Antiquities*, 214–15, 225.

46. C. Jones, *Antiquities*, 226, 233.

47. C. Jones, *Antiquities*, 213.

48. C. Thomas, *Catalog of Prehistoric Works*, 7, 45–54.

49. "Johnson" and "Johnston" are often confused in historic documents. The reference here is probably to "Johnston" and not "Johnson."

50. C. Thomas, *Catalog of Prehistoric Works*, 54.

51. G. White, *Historical Collections of Georgia*, 487.

52. Pepper, "Museum of the American Indian."

53. Heye, Hodge, and Pepper, *Nacoochee Mound*, 4. Unless otherwise noted, all quotations in this section are taken from this report.

54. Heye, Hodge, and Pepper, *Nacoochee Mound*, 8.

55. Heye, Hodge, and Pepper, *Nacoochee Mound*, 8.

56. Heye, Hodge, and Pepper, *Nacoochee Mound*, 103.

57. Heye, Hodge, and Pepper, *Nacoochee Mound*, 99.

58. Middle American Research Institute, "Directors."

59. Wauchope, "Archaeological Survey," viii. Unless otherwise noted, all quotations in this section are taken from this report.

60. Wauchope, "Archaeological Survey," 440.

61. M. Williams, *Nacoochee Revisited.*

62. M. Williams, *Nacoochee Revisited*, 29.

Chapter 3. European Contact and Tribal America

1. Day, "John Day Letter."

2. "Giovanni Da Verrazzano."

3. Richter, *Facing East*, 11–19.

4. Davis et al., *Southern United States*, 81.

5. Richter, *Facing East*, 19.

6. Davis et al., *Southern United States*, 86.

7. See Rangel, "Account of the Northern Conquest"; and Elvas, "Relacam Verdadeira."

8. Swanton, *Final Report*; Baker, *Shadow of Yonah*, 24–25; T. Lumsden, *Nacoochee Valley*, 5.

9. Hudson and Tesser, *Forgotten Centuries*, 74–103; M. White, *Archaeology and History*, 94; Wauchope, "Archaeological Survey," 354–55.

10. G. White, *Historical Collections of Georgia*, 487.

11. Mooney, "Myths of the Cherokee," 29; M. White, *Archaeology and History*, 98; Baker, *Shadow of Yonah*, 26.

12. T. Lumsden, *Nacoochee Valley*, 5.

13. Waselkov, "Seventeenth-Century Trade."

14. Waselkov, 129; Davis et al., *Southern United States*, 90.

15. Hudson and Tesser, *Forgotten Centuries*, 258–59.

16. M. White, *Archaeology and History*, 94.

17. Hudson and Tesser, *Forgotten Centuries*, 265–72.

18. Davis et al., *Southern United States*, 91; Richter, *Facing East*, 62.

19. Saunt, "Creek Indians"; Adair, *American Indians*, 257, 268.

20. Hudson and Tesser, *Forgotten Centuries*, 377. See also Schnell, "Beginnings of the Creeks."

21. Hudson and Tesser, *Forgotten Centuries*, 385–86, 374, 385.

22. Adair, *American Indians*, 259.

23. M. White, *Archaeology and History*, 108–11.

24. Bartram, *Travels*, 296.

25. M. White, *Archaeology and History*, 104–5; Mooney, "Myths of the Cherokee," 39.

26. M. White, *Archaeology and History*, 120–21.

27. Warren, *Worlds the Shawnees Made*, 94.

28. Mooney, "Myths of the Cherokee," 15. Adair and others have given more fanciful accounts of the origin of the name.

29. Bartram, *Travels*, 367–68.

30. Bartram, *Travels*, 296–97.

31. Mooney, "Myths of the Cherokee," 492.

32. M. White, *Archaeology and History*, 94.

33. W. Sears, "Creek and Cherokee Culture," 148; D. King, *Cherokee Indian Nation*, 9.

34. Mooney, "Myths of the Cherokee," 16; Baker, *Shadow of Yonah*, 23.

35. D. King, *Cherokee Indian Nation*, 28.

36. Adair, *American Indians*, 226.

37. Bartram, *Travels*, 287, 297.

38. Riggs and Greene, *Cherokee Trail of Tears*, 105.

39. Heye, Hodge, and Pepper, *Nacoochee Mound*, 13–14.

40. Bartram, *Travels*, 297.

41. Bowen, *New & Accurate Map*.

42. Chicken, "Journal of the March," 336.

43. McLoughlin, *Cherokee Renascence*, 332–33; Perdue, *Cherokee Women*, loc. 1140 of 3423.

44. Mitchell, Kitchin, and Millar, *British and French Dominions*.

45. Corkran, *Cherokee Frontier*, 14.

46. D. King, *Cherokee Indian Nation*, 46–57. Nacoochee, Chota, Sautee, Tugaloo, and most other Cherokee names are rendered in a variety of generally phonetic spellings in the historic record, but for clarity modern place names are used here.

47. Wauchope, "Archaeological Survey," 347.

48. Chicken, "Journal of the March," 330.

49. Chicken, "Journal of the March," 332–33.

50. Chicken, "Journal of the March," 363.

51. Chicken, "Journal of the March," 327–41; Adair, *American Indians*, 231.

52. Vassar, "Some Short Remarks," 413–14.

53. Heye, Hodge, and Pepper note "a thin layer of charcoal," but that would not have related to the historic period since two feet had been removed from the top of the mound when the gazebo was built in the 1880s. *Nacoochee Mound*, 53.

54. Vassar, "Some Short Remarks," 408.

55. The data here is taken from "Cherokee Villages."

56. Mereness, "Colonel Chicken's Journal."

57. Adair, *American Indians*, 232–33.

58. Mooney, "Myths of the Cherokee," 38; Davis et al., *Southern United States*, 135.

59. Timberlake, quoted in Mooney, "Myths of the Cherokee," 39.

60. Adair, *American Indians*, 245–46, 251.

61. Mooney, "Myths of the Cherokee," 49.

62. T. Lumsden, *Nacoochee Valley*, 7; J. Lewis, "Cherokee Expedition."

63. Mooney, "Myths of the Cherokee," 50.

64. Baker, *Shadow of Yonah*, 28; Rogers, "General Andrew Pickens," 11.

65. U.S. Office of Indian Affairs. *Treaties*, 9.

66. McLoughlin, *Cherokee Renascence*, 30.

67. "Field Notebooks."

68. McLoughlin, *Cherokee Renascence*, 4; Adair, *American Indians*, 5.

69. McLoughlin, *Cherokee Renascence*, 4, 56, 163.

70. McLoughlin, 44, 68.

71. McLoughlin, 412–13.

72. "Cherokee Laws," *Cherokee Phoenix*, 13 March 1828, 1.

73. McLoughlin, 139–40, 161–62.

74. McLoughlin, *Champions of the Cherokee*, 93.

75. McLoughlin, *Cherokee Renascence*, 59.

76. Georgia Historical Society, *Letters of Benjamin Hawkins*, 16–17. See also Hemperley, "Benjamin Hawkins' Trip," 117. In 1939 Wauchope documented several "village" sites in Sautee Valley, which were all part of the larger town known as Sautee. Wauchope was unable to locate the town house, which would have been the town's focal point.

Chapter 4. Resettlement of the Valleys

1. The 1790 census schedules for Georgia have been lost, but see Lamar and Rothstein, *Reconstructed 1790 Census of Georgia*.

2. E. Davis, "Administrative Trail of Tears."

3. McLoughlin, *Cherokee Renascence*, 412.

4. C. Whitney, *Kansas City*, 104; Bailey and Cooper, *Biographical Directory*, 785–86.

5. Flowers, "Wofford Settlement"; "Wafford's [*sic*] Settlement."

6. Flowers, "Wofford Settlement," 259.

7. Flowers, 264.

8. McLoughlin, *Cherokee Renascence*, 128.

9. McLoughlin, 128.

10. McLoughlin, 129, 154–55.

11. Tolstoy, *War and Peace*, bk. 8, chap. 22.

12. McLoughlin, 180–81.

13. McLoughlin, 182, 203; Kappler, "Mar. 22, 1816," 2:124.

14. Kappler, "Mar. 22, 1816," 2:124–25.

15. Kappler, "Sept. 14, 1816," 2:133–34.

16. McLoughlin, *Cherokee Renascence*, 82; Ownby and Wharton, *Georgia's Old Federal Road*, 14; McLoughlin, *Cherokee Renascence*, 82.

17. McLoughlin, *Cherokee Renascence*, 85–86; A. Hudson, *Creek Paths*, 113.

18. Gonzalez, "Indian Sovereignty"; Bouman, *Traveler's Rest*, 90–91.

19. Lamar, *Compilation of the Laws*, 774–75.

20. Gedney, *Unicoi Road*, 19–21; Bouman, *Traveler's Rest*, 93.

21. Gedney, *Unicoi Road*, 21.

22. Bouman, *Traveler's Rest*, 93.

23. T. Lumsden, *Nacoochee Valley*, 14; Gedney, *Unicoi Road*, 23.

24. McLoughlin, *Cherokee Renascence*, 273.

25. McLoughlin, *Cherokee Renascence*, 236.

26. Kappler, "July 8, 1817."

27. McLoughlin, *Cherokee Renascence*, 236.

28. Kappler, "Feb. 27, 1819." Article 3 of the treaty offers the quoted characterization of successful applicants.

29. "Register of Cherokees." Note that the first name in the list of reserves in the 1819 list is "Yonah, aka Big Bear," a well-known Cherokee chief in western North Carolina.

30. Shadburn, *Upon Our Ruins*, 10.

31. Shadburn, app. B, 619–22.

32. T. Lumsden, *Nacoochee Valley*, 8–9.

33. Fields, "Between Two Cultures," 186–93; J. Martin, "Chief Justice John Martin," 33–35; Shadburn, *Upon Our Ruins*, 156–57.

34. Martin, "Chief Justice John Martin," 32–36.

35. Martin.

36. Chavez, "First CN Chief Justice."

37. Shadburn, *Upon Our Ruins*, 142–43.

38. "Habersham County, Georgia," *Deed Book*, D-45; T. Lumsden does not state the location of the Lynches' house, only the general location of the land. *Nacoochee Valley*, 9.

39. Shadburn, *Upon Our Ruins*, 144, 151.

40. Warren and Weeks, *Whites among the Cherokees*, 69.

41. Kimzey, *Genealogical and Historical Records*, 605–6.

42. Meserve, "Mayes," 56–57.

43. T. Lumsden, *Nacoochee Valley*, 8.

44. Journal of First Board or U.S. Commissioners at New Echota, Ga., Case 56. See Shadburn, *Upon Our Ruins*, 425–31.

45. "Register of Cherokee."

46. Gedney, *Unicoi Road*, 39–40; Shadburn, *Upon Our Ruins*, 429.

47. Shadburn, *Upon Our Ruins*, 428. "First beloved man" was an honorific bestowed on a leader, perhaps for a specific occasion like a treaty negotiation.

48. Shadburn, *Upon Our Ruins*, 425–41.

49. Shadburn, 427–30.

50. The Habersham County courthouse burned in 1923, but most public records were salvaged. Deed books and superior court minutes begin in July and September, respectively, 1819. Inferior court records are scattered prior to the commencement of the first minutes in 1824. Estate inventories start in 1863, and probate bonds begin in 1881.

51. Coleman and Gurr, *Dictionary of Georgia Biography*, 1:380–81.

52. Kimzey, *Genealogical and Historical Records*, 34; Lamar, *Compilation of the Laws*, 418.

53. Lamar, *Compilation of the Laws*, 226; Kimzey, *Genealogical and Historical Records*, 146; Coulter, "Tallulah Falls," pt. 2.

54. Lamar, *Compilation of the Laws*, 118.

55. Sherwood, *Gazetteer of the State* (1829), 94.

56. The federal census in 1850 and 1860 show his birthplace as Virginia, but Baker gives it as New York. His grave has not been located.

57. Broyles, *General Assembly*, 44–45.

58. Sherwood, *Gazetteer of the State* (1860), 75.

59. Baker, *Shadow of Yonah*, 64–66.

60. Sherwood, *Gazetteer of the State* (1860), 137.

61. Broyles, *General Assembly*, 45.

62. Baker, *Shadow of Yonah*, 71–72.

63. Baker, 71–74.

64. Hitz, "Georgia Militia Districts"; Baker, *Shadow of Yonah*, 69–71.

65. "Our History."

66. Lamar, *Compilation of the Laws*, 417–18.

67. Lamar.

68. Lamar, 419–20.

69. District plats of the survey and the field notebooks are online at http://cdm.georgiaarchives.org:2011/cdm/landingpage/collection/dmf. The individual land-lot surveys are on microfilm at the Georgia Archives in Morrow.

70. Lamar, *Compilation of the Laws*, 419.

71. Although the survey crews are identified by name in the records, there is no attached biographical information. What is provided here about the survey crew is based on circumstantial evidence in census data and cemetery records.

72. Lamar, *Compilation of the Laws*, 419.

73. Lamar, 421.

74. Lamar, 421.

75. Lucas, *Third and Fourth*. Lucas also compiled *Third or 1820 Land Lottery*.

76. "Habersham County, Georgia," *Deed Book C*.

77. Kimzey, *Genealogical and Historical Records*, 75–76.

78. Wishart, Ankron, and Zorick, "Settling Cherokee Georgia."

Chapter 5. The Pioneers

1. U.S. Census Bureau, *Abstract of the Returns*, 23–24.

2. U.S. Census Bureau, *1840 Census*, 48.

3. See Fischer, *Albion's Seed*.

4. Fischer, 608.

5. Baker, *Shadow of Yonah*, 42; Fischer, *Albion's Seed*, 608. See also Leyburn, *Scotch Irish*, for most of the supporting material on the Scots-Irish; and Sondley, *History of Buncombe County*, for the Scots-Irish in North Carolina.

6. Burke, *European Settlements in America*, 216.

7. Fischer, *Albion's Seed*, 615.

8. Sondley, *History of Buncombe County*, 370–71.

9. *Fecit* is Latin, meaning "he made [it]," and was once commonly inserted after an artist's name.

10. R. Moffat, *Moffat Genealogies*, 82.

11. Baker, *Shadow of Yonah*, 44; White County History, *History of White County*, 391.

12. In the family histories in this section, research was limited generally to cemetery, census, superior court, and probate records, with a goal of evaluating the veracity of traditional stories. Unless otherwise noted, the short descriptions of some of the families are from the traditional list of early settlers.

13. Lewis Richardson used 1823; Coulter seemed uncertain of the year.

14. Quoted in "History of Early Settlers"; see also Gedney, *Unicoi Road*, 29–31; and Kimzey, *Genealogical and Historical Records*, 5–10. The list is not mentioned in T. Lumsden's *Nacoochee Valley* or in Baker's *Shadow of Yonah*.

15. B. Huff, *First Permanent Settlers*.

16. Lewis Richardson notes Mrs. Evans being the last survivor. The handwritten note is at the end of McConnell, "Lawrence Graveyard." See also McConnell, "Lawrence Graveyard."

17. Quoted in "History of Early Settlers."

18. Gedney, *Unicoi Road*, 8.

19. U.S. Census Bureau, "Population Schedules." Also facsimiles of most census schedule pages are available through Ancestry at www.ancestry.com/search/categories/usfedcen/ancestry.com; White County History, *History of White County*, 116–17; T. Lumsden, *Nacoochee Valley*, 29; B. Huff, *First Permanent Settlers.*

20. The North Carolina county should not be confused with Burke County, Georgia, located just south of Augusta.

21. Hemperly, "Stagecoach Route"; Lucas, "Spartanburg District."

22. U.S. Department of Veterans Affairs. "War of 1812 Pension"; U.S. Census Bureau, "Population Schedules," 1820–50; and "Habersham County, Georgia" document the Holcombes.

23. Taylor, "Adam Poole Vandiver."

24. Richardson, "Move to Nacoochee."

25. The Brown and Williams families are well documented in the annals of Sautee Nacoochee. See, for example, Coulter, *George Walton Williams*, 1–15.

26. Gedney, *Unicoi Road*, 55–99, provides extensive documentation for the Conley family in White County.

27. T. Lumsden, *Nacoochee Valley*, 26; but see also B. Huff, *First Permanent Settlers.*

28. Coulter, *George Walton Williams*, 1–3, 274n1. Williams was not identified in a search of ancestry.com's index to veterans of the War of 1812.

29. Williams, *Nacoochee and Its Surroundings*, 412.

30. Tinius, "Description of Starlight"; Tinius, "Burning of Starlight." A recreation of the original Starlight was constructed by its owners, the Tinius family, who remained there in 2021.

31. "Maj. Edward Williams," obituary, *Southern Christian (S.C.) Advocate*, 20 March 1856, 4.

32. Gedney, *Unicoi Road*, 42–43; White County History, *History of White County*, 98.

33. Gedney, *Unicoi Road*, 59–61.

34. Gedney, 39–41.

35. Gedney, 44–52.

36. Kimzey, *Genealogical and Historical Records*, 148; Baker, *Shadow of Yonah*, 105.

37. Rigdon, *Historical Sketch & Roster.*

38. U.S. Census Bureau, "Population Schedules," 1790–1800, document the families' dispersal after the Revolution.

39. Baker, *Shadow of Yonah*, 66, 209; "Jehu Jasper Trammell."

40. Kimzey, *Genealogical and Historical Records.*

41. Williams Family File, SNCA.

42. "Habersham County, Georgia."

43. White County History, *History of White County*, 232.

44. Bishop, *Descendants of Bartholomew Stovall*, 58, 209.

45. T. Lumsden, *Nacoochee Valley*, 31.

46. See "Monroe Methodist Cemetery."

47. "Will/Braffitt/Johnson/Cook Genealogy."

48. Edmondson, "Ludlow L. Edmondson," cites "Habersham County, Georgia," *Deed Book O*, 355.

Chapter 6. Community Development

1. Lane, *Rambler in Georgia*, 148.

2. John R. Stilgoe, *Common Landscape*, 74.

3. Lanman, *Letters*, 21.

4. Baker, *Shadow of Yonah*, 102.

5. G. White, *Statistics of the State*, 299.

6. Norton, "White County History."

7. Baker, *Shadow of Yonah*, 113.

8. Historic transportation maps of White County are on file at Georgia Archives, and most of those can be accessed online at www.dot.ga.gov/DS/Maps. These maps are also useful because, until the 1970s, they depicted the location of rural residences, churches, stores, and so forth. The Georgia Department of Transportation has posted many of their historic highway maps online at www.dot.ga.gov/DS/Maps, although these do not include some of the earliest maps.

9. See U.S. Postal Service, *History*. Unless otherwise noted, the data in this section is documented by U.S. Post Office, "Record of Appointment of Postmasters," and U.S. Post Office, "Records of Site Locations."

10. U.S. Post Office, "Habersham County."

11. U.S. Post Office, "Record of Appointment of Postmasters."

12. The "continuously" quote is from the 1893 application to move the Nacoochee Post Office, M1126, roll 124, National Archives and Records Administration.

13. U.S. Post Office, "Record of Appointment of Postmasters."

14. U.S. Post Office, "Rural Free Delivery."

15. U.S. Census Bureau, *Seventh Census*, 386–93.

16. Lanman, *Letters*, 81–82.

17. Sherwood, *Gazetteer of the State* (1829), 247.

18. Sherwood, 245–47.

19. Kimzey, *Genealogical and Historical Records*, 387.

20. T. Lumsden, *Nacoochee Valley*, 19.

21. T. Lumsden, 19.

22. Knight, *Georgia's Roster*, 66.

23. "Our History."

24. "Georgia Baptist Association."

25. W. B. Lumsden's 1948 map of Nacoochee names the branch across from the church "Bristol Branch"; the U.S. Geological Survey maps call it "Ben Creek," and the Georgia DOT maps do not name it at all.

26. Baker, *Crescent Hill Church*, 26.

27. "Our History."

28. Caroline Crittenden to David Greear, personal correspondence, 20 July 2016.

29. Baker, *Crescent Hill Church*, 44–45.

30. Baker, 25–26; White County History, *History of White County*, 14.

31. "Nacoochee Institute," 2.

32. See Heyrman, *Southern Cross*, and Schmidt, *Holy Fairs*.

33. White County History, *History of White County*, 13.

34. Telford, "Early Methodists," 1, 2.

35. T. Lumsden, *Nacoochee Valley*, 20.

36. Transcript of deed by genealogy.com user, in "Re: Ebenezer Fain."

37. Telford, "Early Methodists," 4; T. Lumsden, *Nacoochee Valley*, 31; for notes on William and Elizabeth Crumley, see "Will/Braffitt/Johnson/Cook Genealogy." See also G. Smith, *History of Georgia Methodism*, 235–36. Emery and Warren, "Monroe Methodist Cemetery," contains a copy of the deed.

38. A framed copy of this deed hangs in the church today.

39. The original deed is framed and hanging in the Nacoochee Methodist church's sanctuary.

40. T. Lumsden, *Nacoochee Valley*, 20.

41. Lanman, *Letters*, 97; Coulter, *George Walton Williams*, 10.

42. "Nacoochee Methodist Church." The church's inventory of known burials is accessed at "Site Maps and Registry."

43. "White County, Georgia," *Deed Book A*, 166.

44. Giovino, "Mayor of Bean Creek," 20; Caroline Crittenden to David Greear, email message, 20 July 2016.

45. Levi Willard, "Early History of Decatur Written Many Years Ago," *DeKalb New Era*, 2 December 1920.

46. Baker, *Crescent Hill Church*, 34–35.

47. Williams, "Advice to Young Men."

48. G. White, *Statistics of the State*, 300.

49. Lynch, "Every Man Able."

50. LaBoone, "Public Education."

51. Coulter, *George Walton Williams*, 10.

52. Scotch-Irish Society, *Scotch-Irish in America*, 102.

53. Sherwood, *Gazetteer of the State* (1829), 95; G. White, *Statistics of the State*, 300.

54. LeBow, *Statistical View*, 371.

55. T. Lumsden, *Nacoochee Valley*, 23; E. Marsh, "Old Lumsden Place," 2.

56. Coulter, *History of Georgia*, 238–39.

57. Butchart, "Freedmen's Education."

58. Baker, *Shadow of Yonah*, 163.

59. Baker, 110–11, 309–11, the latter excerpting a 1938 report by John Benjamin Simmons titled "History of the Public Schools of White County, Georgia 1870 thru 1938."

60. White County History, *History of White County*, 25–26.

61. "Amelia Emma Starr Asbury."

62. "Methvin, John Jasper."

63. "White County, Georgia," *Deed Book G*, 112.

64. Baker, *Crescent Hill Church*, 27.

65. "Nacoochee Institute."

66. "Nacoochee Institute."

67. "History and Heritage."

68. "Nacoochee Institute"; George Walton Williams IV to Barbara Williams, email message, 12 July 2008.

69. The 1930 federal census of Nacoochee enumerates the Hollises and other individuals at the orphanage at that time.

70. See the federal censuses of 1920–30 for White County; "U.S., Social Security Death Index"; see also Giovino, "Mayor of Bean Creek," 20.

71. Baker, *Crescent Hill Church*, 26.

72. Richards, "Valley of Nacoochee," 49.

73. Cooke, "Sketches of Georgia," 775–76.

74. Keyes, "George Cooke."

75. Richards, "Valley of Nacoochee," 49.

76. Richards, 50.

77. Richards, *American Scenery*, 112.

78. Lanman, *Letters*, 25–30.

79. Lanman, 26–27.

80. Jackson, *Tallulah and Other Poems*.

81. L. Huff, "Samuel Jones Cassels."

Chapter 7. African Americans at Nacoochee

1. Grant, *Way It Was*, 68–70. The registers of free people of color in Habersham and White Counties were not indexed separately from other inferior court records and have not been searched during the course of the present study.

2. Inscoe, *Appalachians and Race*, 44–45.

3. U.S. Census Bureau, "Population Schedules," for the 1850 census in Habersham County: House 10, Family 10, for Mariah; House 41, Family 41, for Roman. The census taker mistakenly recorded Mariah's last name as "Inmate," a term apparently used for boarders, and which he recorded as part of listed occupations in other entries on the same schedule.

4. U.S. Census Bureau, "Population Schedules," for the 1860 census in White County: House 533, Family 536.

5. Baker, *Shadow of Yonah*, 76.

6. U.S. Census Bureau, "Population Schedules," for the 1860 census in White County: House 533, Family 536.

7. Phifer, "Slavery in Microcosm," 150.

8. Inscoe, *Appalachians and Race*, 158–59.

9. Inscoe, *Race, War, and Remembrance*, 172.

10. Phifer, "Slavery in Microcosm," 145.

11. F. Olmsted, *Journey in the Back Country*, 226.

12. Phifer, "Slavery in Microcosm,"136, 159.

13. "Habersham County, Georgia," *Deed Book*, N-163, and Q-283.

14. U.S. Census Bureau, "Other Schedules," 1820–60, and the antebellum tax digests for White County provide most of the local details regarding slavery. The schedules note the sex and age of slaves but omit names.

15. Author's tally derived from slave and population schedules for the 1860 census; see U.S. Census Bureau, "Population Schedules."

16. Phifer, "Slavery in Microcosm," 139, 160.

17. Pogue and Sandford, "Housing for the Enslaved."

18. Baker, *Shadow of Yonah*, 81–82. Hennion says that only nine voted for secession in White County.

19. G. Glover, "Horanto Hennion."

20. Baker, *Shadow of Yonah*, 82.

21. Coulter, *George Walton Williams*, 63.

22. Rigdon, "Historical Sketch & Roster."

23. Henderson, *Roster of Company B.*

24. Baker, *Shadow of Yonah*, 242–51, 85–88.

25. Henderson, *Roster of Company B.*

26. G. Glover, *Distribution of Salt*; Baker, *Shadow of Yonah*, 84.

27. Coulter, *George Walton Williams*, 78–79.

28. Coulter, 83.

29. An exhaustive search of White County tax digests and deed records, which was outside the scope of this project, might reveal significant information on the origins of the Bean Creek community.

30. Coulter, *George Walton Williams*, 78.

31. T. Lumsden, *Nacoochee Valley*,14; Coulter, *George Walton Williams*, 91. Edward Williams's father is thought to have operated a tannery at Easton, Massachusetts, in the last quarter of the eighteenth century.

32. G. Williams, *Nacoochee and Its Surroundings*, 37.

33. Inscoe, *Appalachians and Race*, 189.

34. Grant, *Way It Was*, 92.

35. Coulter, *History of Georgia*, 208.

36. White County, "Returns of Qualified Voters."

37. Grant, *Way It Was*, 103.

38. This grew out of General Sherman's special orders; see "Special Field Orders No. 15."

39. "Records of the Field Offices."

40. "State Records of Assistant Commissioners."

41. "State Records of Assistant Commissioners," December 1867; March, April, and May 1868; among others.

42. "Records of the Superintendent," August 1867. Note that these records include monthly reports, on printed forms, arranged chronologically. The many poorly indexed letters in the bureau records were not searched for the present work.

43. "Records of the Superintendent," February 1868.

44. "Records of the Superintendent," December 1867, February 1868.

45. "Records of the Superintendent," March 1868.

46. "Records of the Superintendent," May, June 1868.

47. Grant, *Way It Was*, 106.

48. Foner, *Reconstruction*, 602.

49. Baker, *Shadow of Yonah*, 97–98.

50. Baker, *Shadow of Yonah*, 94–98. See also Hebert, "Reconstruction-Era Violence."

51. Baudouin, *Ku Klux Klan*, 17.

52. Baker, *Shadow of Yonah*, 306–7.

53. Moore, "1895 Atlanta Exposition Speech."

54. Plessy v. Ferguson, 163 U.S. 537 (1896).

55. Grant, *Way It Was*, 170.

56. Mixon, "Atlanta Race Riot."

57. Grant, *Way It Was*, 170. For a full account of the terror in Forsyth County, see Phillips, *Blood at the Root*.

58. Grant, *Way It Was*, 293–94.

59. The definitive history of the migration of African Americans out of the South is Wilkerson, *Warmth of Other Suns*.

60. M. King, *Daddy King*, 14.

61. U.S. Census Bureau, "Population Schedules," 1870–1970.

62. Brown v. Board of Education of Topeka, 347 U.S. 483.

63. Baker, *Shadow of Yonah*, 163–64.

64. Giovino, "Mayor of Bean Creek," 20.

65. "White County, Georgia," *Deed Book B*, 170, *Deed Book O*, 355.

66. White County, "Returns of Qualified Voters." The federal census provides most details of the lives of the Welches and of the other nineteenth-century residents of Bean Creek.

67. S. Jones, "Stovall House," 60.

68. There are Habersham County deeds referencing "Burnett's gold lots" in Land Lots 23 and 24, but no other documentation for these has yet been located.

69. Coulter, *George Walton Williams*, 83.

70. "Our History."

71. Inscoe, *Appalachians and Race*, 220–34.

72. Inscoe, 228.

73. "Records of the Education Division."

74. S. Jones, "Stovall House," 60; U.S. Census Bureau, "Population Schedules," 1870 census in White County, Georgia.

75. Brandee A. Thomas, "Descendants Reluctantly Keep Slave History Alive in White County Community," *Gainesville Times*, 19 February 2012.

76. The discussion of employment here is derived from the U.S. Census Bureau, "Population Schedules," census for the Nacoochee District, which includes Sautee Nacoochee and Helen.

77. Occupations shown in federal census, 1870–1940.

Chapter 8. Agriculture and Commerce

1. G. White, *Statistics of the State*, 301.

2. Bonner, *History of Georgia Agriculture*, 50.

3. Davis et al., *Southern United States*, 141–43.

4. Coulter, *George Walton Williams*, 12, 14.

5. Wigginton, *Foxfire 3*, 300–301.

6. D. Lewis, *Transactions*, 19.

7. U.S. Census Bureau, *1840 Census*, 48–51.

8. Bonner, *History of Georgia Agriculture*, 60, 64–65.

9. Lane, *Rambler in Georgia*, 207.

10. Range, *Century of Georgia Agriculture*, 28.

11. Catron-Sullivan, "Jarvis Van Buren."

12. Van Buren, "Fruit at the South," 226.

13. Coulter, *George Walton Williams*, 14.

14. G. Williams, *Sketches of Travel*, 330–31.

15. D. Lewis, *Transactions*, 322, 334.

16. Lanman, *Letters*, 56–57; D. Lewis, *Transactions*, 329–31; Bonner, *History of Georgia Agriculture*, 136–37.

17. McInvale, "James Hall Nichols."

18. White County History, *History of White County*, 39, 43–44.

19. "Calvin Hunnicutt."

20. Hackbart-Dean, "Georgia's Renaissance Governor."

21. Dorsey, "Lamartine Griffin Hardman."

22. Ball, "Notes on the Use."

23. Bishir et al., *Architects and Builders*, 196–201.

24. G. Williams, *Nacoochee and Its Surroundings*, 59–60.

25. G. Williams, 59–60.

26. T. Lumsden, *Nacoochee Valley*, 14.

27. T. Lumsden, 30.

28. T. Lumsden, 25–26.

29. J. Glover, "Water-Powered Mills," 6.

30. "White County, Georgia," *Deed Book E*, 244–45.

31. Kephart, *Our Southern Highlanders*, 123, 154.

32. Moss, *Southern Spirits*, 40.

33. Moss, 67.

34. G. Williams, *Nacoochee and Its Surroundings*, 47.

35. "White County, Georgia," *Deed Book N*, 125.

36. Moss, *Southern Spirits*, 144.

37. See Stewart, *Moonshiners and Prohibitionists*.

38. Baker, *Shadow of Yonah*, 132.

39. Developed in the 1830s, the balloon frame depended on standardized lumber dimensions and mass-produced wire nails. It was the precursor to modern wood-frame construction.

40. Sherwood, *Gazetteer of the State* (1829); "Nacoochee Valley."

41. G. Williams, *Sketches of Travel*, 431.

42. T. Lumsden, *Nacoochee Valley*, 21–22; Coulter, *George Walton Williams*, 15; G. Williams, *Sketches of Travel*, 463.

43. 1860 federal agricultural census of White County.

44. National Park Service, "Harshaw-Stovall House."

45. T. Lumsden, *Nacoochee Valley*, 16; federal census of White County, 1910–40. Businesses, including filling stations, are noted on the State Highway Board maps of White County beginning in 1940.

46. Sautee Valley Historic District nomination; Walter Lumsden's map, *Nacoochee Valley*, in 1948.

Chapter 9. Mining, Manufacturing, and Timber

1. Hopkins, *Report on the Asbestos*, 194.

2. Hopkins, 270.

3. T. Lumsden, *Nacoochee Valley*, 32.

4. Haseltine, *Iron Ore Deposits*, 133.

5. 1840 federal census of manufactures, Habersham County; Swank, *History of the Manufacture*, 279.

6. See Baker, *Shadow of Yonah*, 241–51, for Hennion's memoir, which provides most of the details of his life cited here.

7. Baker, *Shadow of Yonah*, 241–51.

8. Hopkins, *Report on the Asbestos*, 97, 77.

9. Baker, *Shadow of Yonah*, 115.

10. Hopkins, *Report on the Asbestos*, 107.

11. Hopkins, 113, 169; Baker, *Shadow of Yonah*, 116.

12. Hopkins, *Report on the Asbestos*, 116, 269.

13. T. Lumsden, *Nacoochee Valley*, 5; Yeates, McCallie, and King, *Preliminary Report*, 27–28.

14. S. P. Jones, *Second Report*, 19.

15. Nitze and Wilkens, *Gold Mining*, 35.

16. See chapter 6 in Gedney, *Unicoi Road* for excellent details on the history of gold mining in and around Helen.

17. Jefferson, *Notes on the State of Virginia*, 25.

18. D. Williams, *Georgia Gold Rush*, 11.

19. D. Olmsted, "Gold Mines," 5.

20. D. Williams, *Georgia Gold Rush*, 11.

21. S. P. Jones, *Second Report*, 27–28.

22. D. Williams, *Georgia Gold Rush*, 12.

23. Baker, *Shadow of Yonah*, 53; Yeates, McCallie, and King, *Preliminary Report*, 33, 35.

24. Newspaper quoted in D. Williams, *Georgia Gold Rush*, 24.

25. J. Whitney, *Metallic Wealth*, 118.

26. Baker, *Shadow of Yonah*, 53–54.

27. Quotation from *Georgia Journal*, 1 August 1829.
28. Gilmer, quoted in D. Williams, *Georgia Gold Rush*, 26.
29. Parks, quoted in D. Williams, 54.
30. Baker, *Shadow of Yonah*, 58; Yeates, McCallie, and King, *Preliminary Report*, 34. A Troy ton of gold today would be worth nearly $100 million.
31. Yeates, McCallie, and King, *Preliminary Report*, 30.
32. Whitney, *Metallic Wealth*, 119.
33. Tischendorf, "British Enterprise."
34. Nitze and Wilkens, *Gold Mining*, 79.
35. S. P. Jones, *Second Report*, 205.
36. Edom Hamby is the only person with that surname in the antebellum federal censuses of Habersham County.
37. S. P. Jones, *Second Report*, 221; T. Lumsden, *Nacoochee Valley*, 14–15.
38. Yeates, McCallie, and King, *Preliminary Report*, 52.
39. Richards, "Valley of Nacoochee," 50.
40. G. White, *Statistics of the State*, 303.
41. Gedney, *Story of Helen*, loc. 512 of 2188.
42. Gedney, loc. 525 of 2188; Coulter, *George Walton Williams*, 11–12. See Cole and Foley, *Collett Leventhorpe*, for the Deans' life at Nacoochee in the late 1850s.
43. Yeates, McCallie, and King, *Preliminary Report*, 54; S. P. Jones, *Second Report*, 218; G. Williams, *Sketches of Travel*, 423.
44. The federal censuses of 1900 to 1920 document the date of William Crammond Martin's immigration. Some local sources suggest that he moved to Nacoochee in the late 1870s, but he cannot be located in the census before 1900.
45. S. P. Jones, *Second Report*, 218.
46. Newspaper quotation from Baker, *Shadow of Yonah*, 113.
47. Yeates, McCallie, and King, *Preliminary Report*, 79; G. Williams, *Sketches of Travel*, 423.
48. S. P. Jones, *Second Report*, 225; T. Lumsden, *Nacoochee Valley*, 30.
49. S. P. Jones, *Second Report*, 227.
50. S. P. Jones, 227.
51. S. P. Jones, 227; Yeates, McCallie, and King, *Preliminary Report*, 39; Nitze and Wilkens, *Gold Mining*, 108.
52. G. Williams, *Sketches of Travel*, 461.
53. T. Lumsden, *Nacoochee Valley*, 14.
54. Wigginton, *Foxfire* 3, 55–78.
55. Wigginton, 55–78.
56. G. Williams, *Nacoochee and Its Surroundings*, 59–60.

57. T. Lumsden, *Nacoochee Valley*, 22, states that Glen came to Nacoochee from Clarke County, Georgia, but he is found in the 1850 census of Henry County and not in that of Clarke County.

58. See the 1880 mortality census schedule reporting the death of John Glen.

59. T. Lumsden, *Nacoochee Valley*, 22.

60. "Felt."

61. Burrison, *Brothers in Clay*, 120.

62. D. Davis, *Where There Are Mountains*, 139, 155.

63. Davis et al., *Southern United States*, 168.

64. Baker, *Shadow of Yonah*, 130–31; H. Thomas, *Digest of the Railroad*, 279.

65. D. Davis, *Where There Are Mountains*, 166–67; Mastran and Lowerre, *Mountaineers and Rangers*, 1–5.

66. "Gennett Lumber Company"; Mastran and Lowerre, *Mountaineers and Rangers*, 5. See also Hayler, *Sound Wormy*.

67. Mastran and Lowerre, *Mountaineers and Rangers*, 3.

68. Gedney, *Story of Helen*, loc. 696–98 of 2188.

69. Robert Pinckney Tucker to Herbert Wythe, 16 December 1901, quoted in Giltner, "Art of Serving," n451.

70. Gedney, *Story of Helen*, loc. 831 of 2188.

71. Gedney, loc. 668–69, 686 of 2188.

72. Gedney, loc. 864–85 of 2188.

73. Baker, *Shadow of Yonah*, 133–35; Brooks and Greear, *Images of America*, chap. 2; T. Lumsden, *Nacoochee Valley*, 27–29.

74. Gedney, *Story of Helen*, loc. 778, 819 of 2188.

75. Gedney, loc. 941–79, 1530 of 2188.

76. Byars is buried at the Nacoochee Methodist church cemetery.

77. Gedney, *Story of Helen*, locs. 1534–61, 1545–50 of 2188.

78. Brooks and Greear, *Images of America*, 8.

79. Baker, *Shadow of Yonah*, 120.

80. Gedney, *Story of Helen*, loc. 1662 of 2188.

81. D. Davis, *Where There Are Mountains*, 176–77.

82. D. Davis, 177.

83. Mastran and Lowerre, *Mountaineers and Rangers*, 25; D. Davis, *Where There Are Mountains*, 179.

Chapter 10. Conservation, Recreation, and Tourism

1. "Gifford Pinchot."

2. G. Marsh, *Man and Nature*, iii.

3. Yosemite Act of June 30, 1864, ch. 184, §§ 1, 2; 13 Stat. 325[1].

4. The legislation's formal title is *Sundry Civil Appropriations Act of 1897.*

5. Mastran and Lowerre, *Mountaineers and Rangers*, 36.

6. Baker, *Shadow of Yonah*, 146–48.

7. B. Martin, "Forest Removal."

8. Mastran and Lowerre, *Mountaineers and Rangers*, 78–79; Baker, *Shadow of Yonah*, 149.

9. Public Law 88-577 (16 U.S.C. 1131–36) 88th Cong., 2d Sess., 3 September 1964.

10. Mastran and Lowerre, *Mountaineers and Rangers*, viii.

11. Department of Vertebrate Zoology, "Passenger Pigeon."

12. "Youth-Focused Wildlife Management Area."

13. Mastran and Lowerre, 36.

14. White County History, *History of White County*, 247.

15. "Indian Spring State Park," 22.

16. Baker, *Shadow of Yonah*, 164–65.

17. Department of Parks, "White County Area," 8; Brooks and Greear, *Images of America*, 92–93.

18. Townsend, *Georgia State Parks*, 77–78; Brooks and Greear, *Images of America.*

19. "Charles Smithgall."

20. Sherwood, *Gazetteer of the State* (1837).

21. Richards, "Valley of Nacoochee," 41.

22. Coulter, "Tallulah Falls," pt. 1.

23. T. Lumsden, *Nacoochee Valley*, 14; G. White, *Statistics of the State*, 301; White County History, *History of White County*, 388.

24. White County History, *History of White County*, 38; T. Lumsden, *Nacoochee Valley*, 17.

25. T. Lumsden, *Nacoochee Valley*, 17.

26. White County History, *History of White County*, 40; T. Lumsden, *Nacoochee Valley*, 17.

27. Baker, *Shadow of Yonah*, 137; Brooks and Greear, *Images of America*, 78.

28. Baker, *Shadow of Yonah*, 171–72.

29. Baker, 173–76. Brooks and Greear, *Images of America*, provides an excellent chronicle of the town's transformation.

30. "Mountain and Hillside Protection."

31. "National Register of Historic Places."

32. Nomination is tantamount to listing.

33. Murtagh, *Keeping Time*, 120.

34. Stovall, *Sautee and Nacoochee Valleys*, 2.

Bibliography

Published Works

Adair, James. *The History of the American Indians; Particularly Those Nations Adjoining to the Mississippi, East and West Florida, Georgia, South and North Carolina, and Virginia*. London: Dilly, 1775. 11 May 2024. https://hdl.handle.net/2027/umn.31951002290818e.

"Amelia Emma Starr Asbury." Find a Grave. Accessed 5 June 2024. www.findagrave.com/memorial/87449781/amelia_emma-asbury.

"Ancient Indian Petroglyph Boulders." Waymarking. Accessed 11 May 2024. www.waymarking.com/waymarks/wm6NBN_Ancient_Indian_Petroglyph_Boulders_UGA_Campus_Athens_GA.

Anderson, David G. "Paleoindian Period: Overview." *New Georgia Encyclopedia*. Last modified 8 June 2017. www.georgiaencyclopedia.org/articles/history-archaeology/paleoindian-period-overview/.

Anderson, David G., and Glen T. Hanson. "Early Archaic Settlement in the Southeastern United States: A Case Study from the Savannah River Valley." *American Antiquity* 53, no. 2 (April 1988): 262–86.

Anderson, David G., and Joseph Schuldenrein. "Mississippian Period Settlement in the Southern Piedmont: Evidence from the Ruckers Bottom Site, Elbert County, Georgia." *Southeastern Archaeology* 2, no. 2 (Winter 1983): 98–117.

Audubon, John James. *The Birds of America*. 2 vols. New York: J. J. Audubon, 1840.

Audubon, John James, and Rev. John Bachman. *The Viviparous Quadrupeds of North America*. 2 vols. New York: J. J. Audubon, 1844.

Bagley, Garland C. *The History of Forsyth County, Georgia, 1832–1932*. Vol. 1. Reprint, Milledgeville, Ga.: Boyd, 1996.

Bailey, N. Louise, and Elizabeth Ivey Cooper, eds. *Biographical Directory of the South Carolina House of Representatives*. Vol. 3, *1775–1790*. Columbia: University of South Carolina Press, 1981.

Baker, Garrison. *A History of Crescent Hill Church*. Cleveland, Ga.: Fern Creek, 1997.

———. *In the Shadow of Yonah: A History of White County, Georgia*. Cleveland, Ga.: Brasstown Creek, 2005.

Ball, Donald B. "Notes on the Use of Tubmills in Southern Appalachia." *Material Culture* 40, no. 2 (2008): 1–20. www.jstor.org/stable/29764465.

Bartram, William. *Travels of William Bartram*. Edited by Mark Van Doren. 1928. Reprint, Mineola, N.Y.: Dover, 1955.

Baudouin, Richard, ed. *Ku Klux Klan: History of Racism and Violence*. 6th ed. Montgomery, Ala.: Southern Poverty Law Center, n.d. www.splcenter.org/sites/default/files/Ku-Klux-Klan-A-History-of-Racism.pdf.

Becker, George F. *Reconnaissance of the Gold Fields of the Southern Appalachians*. Washington, D.C.: Department of the Interior, 1895.

Belue, Ted Franklin. *The Long Hunt: Death of the Buffalo East of the Mississippi*. Mechanicsburg, Pa.: Stackpole, 1996.

Berlin, Ira, Steven F. Miller, and Leslie S. Rowland. "Afro-American Families in the Transition from Slavery to Freedom." *Radical History Review* 42 (1988): 89–121.

Bishir, Catherine, Charlotte Brown, Carl Lounsbury, and Ernest H. Wood III. *Architects and Builders in North Carolina: A History of the Practice of Building*. Chapel Hill: University of North Carolina Press, 1990.

Bishop, Donald E. *Descendants of Bartholomew Stovall (1655–1722)*. N.p.: Stovall Family Association, 1999.

Blitz, John H. "Adoption of the Bow in Prehistoric North America." *North American Archaeologist* 9, no. 2 (1988):123–45.

Bonner, James C. *A History of Georgia Agriculture, 1732–1860*. Athens: University of Georgia Press, 1964.

Botkin, Benjamin A., ed. *Slave Narratives: A Folk History of Slavery in the United States from Interviews with Former Slaves*. Washington, D.C., 1941.

Bouman, Robert Eldridge. *Traveler's Rest and the Tugaloo Crossroads*. Atlanta: State of Georgia Department of Natural Resources, Parks, Recreation, and Historic Sites, Historic Preservation Section, 1980.

Boyd, Kenneth W. *The Historical Markers of North Georgia*. Marietta, Ga.: Cherokee, 1993.

Braun, Duane D. "Glacial and Periglacial Erosion of the Appalachians." *Geomorphology* 2, nos. 1–3 (September 1989): 233–56.

Brooks, Chris, and David Greear. *Images of America: Helen*. Charleston, S.C.: Arcadia, 2012.

Broyles, Edwin N., ed. *Acts of the General Assembly . . . November and December 1857*. Columbus, Ga.: State Printer, 1858.

Burke, Edmund. *European Settlements in America*. Vol. 2. London, 1760.
Burrison, John A. *Brothers in Clay: The Story of Georgia Folk Pottery*. Athens: University of Georgia Press, 1989.
Butchart, Ronald E. "Freedmen's Education during Reconstruction." *New Georgia Encyclopedia*. 13 April 2016. www.georgiaencyclopedia.org/articles/history-archaeology/freedmens-education-during-reconstruction/.
Bynum, William B., ed. *The Heritage of Rutherford County, North Carolina*. Vol. 1. Winston-Salem, N.C.: Genealogical Society of Old Tryon County, 1984.
"Calvin Hunnicutt, Pioneer, Is Dead." *Atlanta Constitution*, 21 January 1915.
Carlson, Anders E. "The Younger Dryas Climate Event." In *Encyclopedia of Quaternary Science*, 126–34. 2nd ed. Amsterdam: Elsevier, 2013.
Cashin, Edward L., ed. *A Wilderness Still the Cradle of Nature: Frontier Georgia*. Savannah: Library of Georgia, 1994.
Cassels, Samuel J. *America Discovered: A Poem*. Savannah, 1851.
———. *Liberty Poems*. New York, 1851. https://archive.org/details/libertypoemsoocass.
———. *Providence and Other Poems*. Savannah, 1838.
Catron-Sullivan, Staci. "Jarvis Van Buren: A Brief History of Georgia Horticulturist, Writer, Nurseryman, and Builder." *Magnolia: Bulletin of the Southern Garden History Society* 16, no. 1 (Fall 2000): 1–9.
Center for Community Design and Preservation. *Historic Resources Survey Report, White County, Georgia*. Athens: University of Georgia, College of Environment and Design, 2015.
Chapin, George. *Health Resorts of the South: Containing Numerous Engravings Descriptive of the Most Desirable Resorts of the Southern States, Together with Some Representative Northern Resorts*. Boston, 1892.
"Charles Smithgall, 91, Preserve's Namesake." Obituary. *Atlanta Constitution*, 21 August 2002.
Chavez, Will. "First CN Chief Justice Honored with Monument." *Cherokee (Okla.) Phoenix*, 20 June 2013.
"Cherokee Villages in the Nacoochee Valley." Human Family: Kinship, Social Organization. 18 August 2015. https://consanguinityandaffinity.wordpress.com/2015/08/18/nacooche-mound.
Chicken, George. "Journal of the March of the Carolinians into the Cherokee Mountains in the Yemassee Indian War, 1715–16." In *Charleston Year Book, 1894*, edited by Langdon Cheves, 314–52. 12 May 2024. https://ia601906.us.archive.org/1/items/yearbookcityofchoounse_6/yearbookcityofchoounse_6.pdf.
Church, Mary L. *The Hills of Habersham*. Clarkesville, Ga.: Self-published, 1988.

Cobb, Charles R. "Mississippian Chiefdoms: How Complex?" *Annual Review of Anthropology* 32 (October 2003): 63–84.

Cole, Timothy, and Bradley R. Foley. *Collett Leventhorpe, the English Confederate: The Life of a Civil War General, 1815–1889*. Jefferson, N.C.: McFarland, 2006.

Coleman, Kenneth, ed. *A History of Georgia*. 2nd ed. Athens: University of Georgia Press, 1991.

Coleman, Kenneth, and Stephen Gurr, eds. *Dictionary of Georgia Biography*. 2 vols. Athens: University of Georgia Press, 1983.

Conley, Robert J. *A Cherokee Encyclopedia*. Albuquerque: University of New Mexico Press, 2007.

Cook, Joe. *Chattahoochee River User's Guide*. Athens: University of Georgia Press, 2014.

Cooke, George. "Sketches of Georgia." *Southern Literary Messenger* 6, no. 10 (1840): 775–77.

Corkran, David H. *The Cherokee Frontier: Conflict and Survival, 1740–1762*. 1962. Reprint, Norman: University of Oklahoma Press, 2016.

Cornell, Nancy J. *1864 Census for Re-organizing the Georgia Militia*. Baltimore: Genealogical, 2000.

Coulter, E. Merton. *George Walton Williams: The Life of a Southern Merchant and Banker, 1820–1903*. Athens, Ga.: Hibriten, 1976.

———. *Georgia: A Short History*. Chapel Hill: University of North Carolina Press, 1947.

———. "The Georgia-Tennessee Boundary Line." *Georgia Historical Quarterly* 35, no. 4 (December 1951): 269–306.

———, ed. *A History of Georgia*. Athens: University of Georgia Press, 1977.

———. "Tallulah Falls, Georgia's Natural Wonder, from Creation to Destruction, Part I." *Georgia Historical Quarterly* 47, no. 2 (June 1963): 121–57.

———. "Tallulah Falls, Georgia's Natural Wonder, from Creation to Destruction, Part II." *Georgia Historical Quarterly* 47, no. 3 (September 1963): 249–75.

Cummings, William P. *The Southeast in Early Maps*. 3rd ed. Chapel Hill: University of North Carolina Press, 1986.

Cyriaque, Jeanne. "Interpreting African American Life in the Sautee-Nacoochee Valley: The Bean Creek History Project." *Reflections* 6, no. 2 (July 2006): 1–3.

Davis, Donald Edward. *Where There Are Mountains: An Environmental History of the Southern Appalachians*. Athens: University of Georgia Press, 2003.

Davis, Donald Edward, Craig E. Colten, Megan Kate Nelson, Barbara L. Allen, and Mikko Saikku. *Southern United States: An Environmental History*. Santa Barbara, CA: ABC-CLIO, 2006.

Davis, Etha. "An Administrative Trail of Tears: Indian Removal." *American Journal of Legal History* 50, no. 1 (January 2008–10): 49–100.

Day, John. "John Day Letter to the Lord Grand Admiral, Winter 1497/8." *The Smugglers' City*. University of Bristol. Accessed 14 May 2024. www.bristol.ac.uk/Depts/History/Maritime/Sources/1497johnday.htm.

Debo, Angie. *The Road to Disappearance: A History of the Creek Indians*. Norman: University of Oklahoma Press, 1993.

Department of Parks, Historic Sites, and Monuments. "White County Area." *From the Hills of Habersham to Tybee Light: 8*. Georgia Department of Natural Resources. 1952. Accessed 11 May 2024. https://dlg.galileo.usg.edu/data/dlg/ggpd/pdfs/dlg_ggpd_y-ga-bp300-b-ps1-bd5-b1952.pdf.

Department of Vertebrate Zoology. "The Passenger Pigeon." National Museum of Natural History/Public Inquiry Services, Smithsonian Institution. Accessed 18 May 2024. www.si.edu/spotlight/passenger-pigeon.

DeVorsey, Louis. "Indian Trails." *New Georgia Encyclopedia*. Last modified 29 September 2020. www.georgiaencyclopedia.org/articles/history-archaeology/indian-trails/.

Dittmer, John. *Black Georgia in the Progressive Era, 1900–1920*. Urbana: University of Illinois Press, 1980.

Dorsey, James. "Lamartine Griffin Hardman." In Coleman and Gurr, *Dictionary of Georgia Biography*, 1:392–93. Athens: University of Georgia Press, 1983.

Dunaway, Wilma A. *Slavery in the American Mountain South*. New York: Cambridge University Press, 2003.

Dunn, Durwood. *An Abolitionist in the Appalachian South*. Knoxville: University of Tennessee Press, 1997.

Edelman, Steven H., Angang Liu, and Robert D. Hatcher Jr. "The Brevard Zone in South Carolina and Adjacent Areas: An Alleghenian Orogen-Scale Dextral Shear Zone Reactivated as a Thrust Fault." *Journal of Geology* 95, no. 6 (November 1987): 793–806.

Edmondson, S. W. "Ludlow L. Edmondson of Habersham County, GA, 1840." Genealogy. 31 January 2009. www.genealogy.com/forum/surnames/topics/edmondson/2206/.

Elliot, Daniel. *The Live Oak Soapstone Quarry, DeKalb County, Georgia*. 1986. 11 May 2024. www.academia.edu/2962715/The_Live_Oak_Soapstone_Quarry_Dekalb_County_Georgia.

Elvas, Fidalgo de. "Relacam Verdadeira." *Early Visions Bucket*. Digital Commons, University of South Florida. https://digitalcom.

Emery, Sylvia Crumley, and Hazel Crumley Warren. "Monroe Methodist Cemetery." History Paper 212 (previously 215). Cleveland, Ga.: White County Historical Society, n.d.

Evans, Richard T., and Helen M. Frye. *History of the Topographic Branch (Division).* USGS Circular 1341. 2009. 11 May 2024. https://pubs.usgs.gov/circ/1341/.

Fagan, Brian, ed. *The Complete Ice Age: How Climate Change Shaped the World.* New York: Thames and Hudson, 2009.

——— *The Little Ice Age: How Climate Made History, 1300–1850.* New York: Basic, 2001.

Featherstonhaugh, G. W. *A Canoe Voyage Up the Minnay Sotor; with an Account of the Lead and Copper Deposits in Wisconsin; of the Gold Region in the Cherokee Country; and Sketches of Popular Manners; &c.* Vol. 2. Accessed 11 May 2024. http://www.loc.gov/resource/lhbum.6643b.

Federal Writers' Project. "Born in Slavery: Slave Narratives from the Federal Writers' Project, 1936 to 1938." Library of Congress. Accessed 13 May 2024. www.loc.gov/collections/slave-narratives-from-the-federal-writers-project-1936-to-1938/about-this-collection/.

"Felt." Wikipedia. 17 April 2024. https://en.wikipedia.org/wiki/Felt.

Fewkes, J. Walter, ed. *Forty-Second Annual Report of the Bureau of American Ethnology to the Secretary of the Smithsonian, 1924–1925.* Washington, D.C.: Government Printing Office, 1928. Accessed 12 May 2024. www.biodiversitylibrary.org/item/107718#page/5/mode/1up.

Fields, Elizabeth Arnett. "Between Two Cultures: Judge John Martin and the Struggle for Cherokee Sovereignty." In *The Southern Colonial Backcountry: Interdisciplinary Perspectives on Frontier Communities,* edited by David Colin Crass, Steven D. Smith, Martha A. Zierden, and Richard D. Brooks, 186–93. Knoxville: University of Tennessee Press, 1998.

Fischer, David Hackett. *Albion's Seed: Four British Folkways in America.* New York: Oxford University Press, 1989.

Flatt, William. "Agriculture in Georgia: Overview." *New Georgia Encyclopedia.* Last modified 17 October 2016. www.georgiaencyclopedia.org/articles/business-economy/agriculture-in-georgia-overview/.

Flowers, Carl, Jr. "The Wofford Settlement on the Georgia Frontier." *Georgia Historical Quarterly* 61, no. 3 (Fall 1977): 258–67.

Foner, Eric. *Reconstruction: America's Unfinished Revolution, 1863–1877.* New York: Harper and Row, 1988.

Frothingham, Earl Hazeltine. *Timber Growing and Logging Practice in the Southern Appalachian Region.* Technical Bulletin 250. Washington, D.C.: Government Printing Office, 1931.

Garrett, Franklin M. *Atlanta and Environs: A Chronicle of Its People and Events*. 2 vols. Athens: University of Georgia Press, 1969.
Gedney, Matt. *Living on the Unicoi Road: Helen's Pioneer Century and Tales from the Georgia Gold Rush*. Marietta, Ga.: Littlestar, 2008.
———. *The Story of Helen, Georgia: Desperately Seeking Helen*. Marietta, Ga.: Littlestar, 2014. Kindle.
"Gennett Lumber Company." Homepage. Accessed 5 June 2024. www.gennettlumber.com/about.php.
Georgia Department of Transportation. "Archive Maps." Accessed 5 June 2024. www.dot.ga.gov/GDOT/pages/Maps.aspx.
Georgia Historical Society. *Letters of Benjamin Hawkins, 1796–1806: Collections of Georgia Historical Society*. Vol. 9. Savannah, Ga.: Morning News, 1916.
Georgia Secretary of State. "Vanishing Georgia." Georgia Archives. Accessed 13 May 2024. https://dlg.usg.edu/collection/dlg_vang.
Gerrell, Pete. *Old Trees: The Illustrated History of Logging the Virgin Timber of the Southeastern United States*. Tallahassee, Fla.: Southern Yellow Pine, 2000.
"Gifford Pinchot." U.S. Forest Service History. 2015. 12 May 2024. https://foresthistory.org/research-explore/us-forest-service-history/people/chiefs/gifford-pinchot-1865-1946/.
Gilmer, George R. *Sketches of Some of the First Settlers of Upper Georgia, of the Cherokees, and the Author*. 1855. Reprint, Baltimore: Genealogical Publishing Company, 1970.
Giltner, Scott Edward. "The Art of Serving Is with Them Innate." Master's thesis, Hiram College, 1996, University of Pittsburgh, 1998. 12 May 2024. https://d-scholarship.pitt.edu/8423/1/GiltnerETD2005.pdf.
"Giovanni Da Verrazzano." Carolana. Accessed 14 May 2024. www.carolana.com/Carolina/Explorers/giovanniverrazano.html.
Giovino, Heather. "The Mayor of Bean Creek." *Foxfire* 48, nos. 1–2 (Spring–Summer 2014): 19–24.
Glover, George. *Distribution of Salt in White County, 1862–1863*. Cleveland, Ga.: White County Historical Society, 1999.
———. "Horanto Hennion and the Mossy Creek Iron Works (a Unionist in Confederate White County)." History Paper 72. Cleveland, Ga.: White County Historical Society, n.d.
Glover, J. Porter. "Water-Powered Mills in White County." History Paper 19. Cleveland, Ga.: White County Historical Society, 1973.
Goff, John H. *Georgia Early Roads and Trails, circa 1730–1850*. Atlanta: Georgia Department of Transportation, [1980?].

Gonzalez, Gerald T. E. "Indian Sovereignty and the Tribal Right to Charter a Municipality for Non-Indians: A New Perspective for Jurisdiction on Indian Land." *New Mexico Law Review* 7 (1977): 194–95.

Gore, Pamela J. W., and William Witherspoon. *Roadside Geology of Georgia*. Missoula, Mont.: Mountain, 2013.

Grant, Donald L. *The Way It Was in the South: The Black Experience in Georgia*. New York: Carol, 1993.

Griffin, Clarence W. *History of Old Tryon and Rutherford Counties, North Carolina, 1730–1936*. Asheville: Miller, 1937.

Griffin, Rebecca J. " 'Goin' Back Over There to See That Girl': Competing Social Spaces in the Lives of the Enslaved in Antebellum North Carolina." *Slavery & Abolition: A Journal of Slave and Post-slave Studies* 25, no. 1 (2004): 94–113.

"Habersham County, Georgia History, Early History and First Settlers." *Georgia Genealogical Trails*. Accessed 5 June 2024. http://genealogytrails.com/geo/habersham/history2.htm.

Hackbart-Dean, Pamela. "Georgia's Renaissance Governor: Lamartine Hardman." *Georgia Historical Quarterly* 79, no. 2 (1995): 441–52.

Hahn, Steven. *The Roots of Southern Populism: Yeoman Farmers and the Transformation of the Georgia Upcountry, 1850–1890*. New York: Oxford University Press, 1983.

Haseltine, Raymond Holden. *Iron Ore Deposits of Georgia*. Bulletin 41. Atlanta: Geological Survey of Georgia, 1924. https://epd.georgia.gov/outreach/publications/georgia-geologic-survey-bulletins.

Hayler, Nicole, ed. *Sound Wormy: Memoir of Andrew Gennett, Lumberman*. Athens: University of Georgia Press, 2002.

Hebert, Keith S. "Reconstruction-Era Violence in North Georgia: The Mossy Creek Ku Klux Klan's Defense of Local Autonomy." In Slap, *Reconstructing Appalachia*, 49–70.

Hemperley, Marion R. "Benjamin Hawkins' Trip across Georgia in 1796." *Georgia Historical Quarterly* 55, no. 1 (Spring 1971): 114–37.

———. *Historic Indian Trails of Georgia*. Athens: Garden Club of Georgia, 1989.

Henderson, Lillian. *Roster of Company B, 52d Regiment, Georgia Volunteer Infantry, White County, Georgia, "Cleveland Volunteers."* 12 May 2024. http://tn-roots.com/Georgia/White/civilwar/index.htm.

Heye, George G., Fredrich Webb Hodge, and George H. Pepper. *The Nacoochee Mound in Georgia*. New York: Museum of the American Indian, Heye Foundation, 1918.

Heyrman, Christine. *Southern Cross: The Beginnings of the Bible Belt*. Chapel Hill: University of North Carolina Press, 1998.

Highsmith, Betty Telford. "Rock Springs Campground." History Paper 49. Cleveland, Ga.: White County Historical Society, 1992.

"History and Heritage." Rabun Gap–Nacoochee School. Accessed 8 June 2024. www.rabungap.org/alumni/history—heritage.

"History of Early Settlers of Nacoochee Valley." *Georgia Pioneers Genealogical Magazine* 3, no. 2 (May 1966). https://genealogytrails.com/geo/habersham/early_settlers.html.

"History of the Georgia State Parks and Historic Sites Division." Georgia Department of Natural Resources. Accessed 18 May 2024. https://gastateparks.org/sites/default/files/parks/pdf/HistoryOfGSPHSD.pdf.

Hitz, Alex M. *Authentic List of All Land Lottery Grants Made to Veterans of the Revolutionary War by the State of Georgia*. Atlanta: Secretary of State of Georgia, 1966.

———. "Georgia Militia Districts." *Georgia Bar Journal* 18, no. 3 (February 1956).

Hooker, Richard J., ed. *The Carolina Backcountry on the Eve of the Revolution: The Journal and Other Writings of Charles Woodmason, Anglican Itinerant*. Chapel Hill: University of North Carolina Press, 1953.

Hopkins, Oliver B. *A Report on the Asbestos, Talc and Soapstone Deposits of Georgia*. Bulletin 29. Atlanta: Byrd, 1914. https://epd.georgia.gov/sites/epd.georgia.gov/files/related_files/site_page/B-29.pdf.

Hough, Franklin B. *On the Duty of Governments in the Preservation of Forests*. [Salem: Salem Press, 1873.] www.loc.gov/item/12018878/.

———. *Report upon Forestry*. U.S. Department of Agriculture. 1878. www.loc.gov/item/12029209/.

Hudson, Angela Pulley. *Creek Paths and Federal Roads: Indians, Settlers, and Slaves and the Making of the American South*. Chapel Hill: University of North Carolina Press, 2010.

Hudson, Charles. *The Southeastern Indians*. Knoxville: University of Tennessee Press, 1976.

Hudson, Charles, and Carmen Chaves Tesser, eds. *The Forgotten Centuries: Indians and Europeans in the American South, 1521–1704*. Athens: University of Georgia Press, 1994.

Huff, Bill. *The First Permanent Settlers and the Brown and Williams Families*. Cleveland, Ga.: White County Historical Society, n.d.

Huff, Lawrence. "Samuel Jones Cassels: A Pioneer Georgia Poet." *Georgia Historical Quarterly* 47, no. 4 (December 1963): 408–19.

Hurst, Vernon J. *Gold in East-Central Georgia*. Bulletin 112. Atlanta: Georgia Geologic Survey, 1990.

Hurst, Vernon J., Thomas Kremer, and Parshall B. Bush. *A Geochemical Reconnaissance for Gold in East-Central Georgia*. Information Circular 83. Atlanta: Georgia Geologic Survey, 1990.

"Indian Spring State Park." History of the Georgia State Parks and Historic Sites Division. Accessed 18 May 2024. https://gastateparks.org/sites/default/files/parks/pdf/HistoryOfGSPHSD.pdf.

"Indigenous Ceramic Type Collection." University of Georgia. Accessed 13 May 2024. https://archaeology.uga.edu/ceramic-type-collection.

Ingersoll, Ernest. *Gold-Mining in Georgia: A Visit to Dahlonega, Georgia in 1879*. Dahlonega, Ga.: Porter Springs Media, 2015.

Inscoe, John C., ed. *Appalachians and Race: The Mountain South from Slavery to Segregation*. Lexington: University Press of Kentucky, 2001.

———. *Race, War, and Remembrance in the Appalachian South*. Lexington: University Press of Kentucky, 2010.

Jackson, Henry R. *Tallulah and Other Poems*. Savannah, Ga.: Cooper, 1850.

Jefferson, Thomas. *Notes on the State of Virginia*. 1782. Reprint, Richmond, Va.: J. W. Randolph, 1853.

"Jehu Jasper Trammell." Find a Grave. Accessed 5 June 2024. www.findagrave.com/memorial/19448135/jehu_jasper-trammell.

Johnson, J. G. "The Spaniards in Northern Georgia during the Sixteenth Century." *Georgia Historical Quarterly* 9, no. 2 (June 1925): 159–68. www.jstor.org/stable/40575817.

Jones, Charles C., Jr. *Antiquities of the Southern Indians, Particularly of the Georgia Tribes*. New York: Appleton, 1873. Reprint, Tuscaloosa: University of Alabama Press, 1989.

———. *The History of Georgia*. Boston: Houghton and Mifflin, 1883.

Jones, S. Percy. *Second Report of Gold Deposits of Georgia*. Atlanta: Byrd; State Printer, 1909.

Jones, Stephanie. "The History of the Stovall House." *Foxfire* 48, nos. 1–2 (Spring–Summer 2014).

Kappler, Charles J., ed. *Indian Affairs: Laws and Treaties*. Vol. 2. Washington, D.C.: Government Printing Office, 1904.

———. "Treaty with the Cherokee, Feb. 27, 1819." In Kappler, *Indian Affairs*, 2:177–82.

———. "Treaty with the Cherokee, July 8, 1817." In Kappler, *Indian Affairs*, 2:140–44.

———. "Treaty with the Cherokee, Mar. 22, 1816." In Kappler, *Indian Affairs*, 2:125–26.

———. "Treaty with the Cherokee, Oct. 24, 1804." In Kappler, *Indian Affairs*, 2:73–74

———. "Treaty with the Cherokee, Sept. 14, 1816." In Kappler, *Indian Affairs*, 2:133–34.

Kephart, Horace. *Our Southern Highlanders: A Narrative of Adventure in the*

Southern Appalachians and a Study of Life among the Mountaineers. New York: MacMillan, 1922.

Keyes, Donald. "George Cooke." *New Georgia Encyclopedia*. Last modified 20 January 2017. www.georgiaencyclopedia.org/articles/arts-culture/george-cooke-1793-1849/.

Kimzey, Herbert B. *Early Genealogical and Historical Records, Habersham County, Georgia*. Athens, Ga.: N. K. Dempsey, 1988.

King, Duane H., ed. *The Cherokee Indian Nation: A Troubled History*. Knoxville: University of Tennessee Press, 1979.

King, Martin Luther. *Daddy King: An Autobiography*. New York: Morrow, 1980.

Kitchen, Lionel. *A New Map of the Cherokee Nation*. New York Public Library Digital Collections. Accessed 21 May 2024. https://digitalcollections.nypl.org/items/ed2125b0-1f49-0133-cc6d-58d385a7b928.

Knight, Lucian Lamar. *Georgia's Landmarks, Memorials and Legends*. 2 vols. Atlanta: Byrd, 1914.

———. *Georgia's Roster of the Revolution*. Atlanta: Index, 1920.

———. *A Standard History of Georgia and Georgians*. Vol. 5. Atlanta: Lewis, 1917.

LaBoone, John. "Public Education." *New Georgia Encyclopedia*. 16 May 2024. www.georgiaencyclopedia.org/articles/education/public-education-prek-12/.

Lamar, Lucius Q. C. *A Compilation of the Laws of the State of Georgia: Passed by the Legislature since the Year 1810 to the Year 1819, Inclusive*. Augusta, Ga.: Hannon, 1821. http://digitalcommons.law.uga.edu/ga_code/22/.

Lamar, Marie de, and Elisabeth Rothstein. *The Reconstructed 1790 Census of Georgia*. Baltimore: Genealogical, 1985.

Lane, Mills, ed. *Neither More Nor Less Than Men: Slavery in Georgia*. Savannah, Ga.: Beehive, 1993.

———. *The Rambler in Georgia*. Savannah, Ga.: Beehive, 1973.

Lanman, Charles. *Letters from the Alleghany Mountains*. New York: Putnam, 1849.

Ledbetter, Jerald. "Georgia: Paleoindian and Early Archaic Projectile Point Data." Paleoindian Database of the Americas. 12 May 2024. http://pidba.org/ga_pics.htm.

Lewis, David W. *Transactions of the Southern Central Agricultural Society, from Its Organization in 1846 to 1851*. Macon, Ga.: Griffin, 1852.

Lewis, J. D. "Cherokee Expedition, 1776: 'Rutherford's Campaign.'" American Revolution in North Carolina. 12 May 2024. www.carolina.com/NC/Revolution/revolution_cherokee_expedition_1776.html.

Leyburn, James. G. *The Scotch-Irish: A Social History*. Chapel Hill: University of North Carolina Press, 1962.

Lothrop, Samuel Kirkland. "George Gustav Heye, 1874–1956." *American Antiquity* 23, no. 1 (July 1957): 66–67.

Loubser, Johannes (Jannie). "The Recording and Interpretation of Two Petroglyph Locales, Track Rock Gap and Hickorynut Mountain." Report 6776. Georgia Archaeological Site File. 12 May 2024. https://archaeology.uga.edu/gasf/report/6776.

Loubser, Johannes (Jannie), and Douglas Frink. "An Archaeological and Ethnohistorical Appraisal of a Piled Stone Feature Complex in the Mountains of North Georgia." *Early Georgia* 38, no. 1 (2010): 29–50.

Lucas, Silas Emmett, ed. "Spartanburg District." *Mills' Atlas of the State of South Carolina, 1825*. Greenville, S.C.: Southern Historical Press, 1980.

———. *The Third and Fourth or 1820 and 1821 Land Lotteries of Georgia*. Easley, S.C.: Georgia Genealogical Reprints/Southern Historical Press, 1973.

———. *The Third or 1820 Land Lottery of Georgia*. Greenville, S.C.: Southern Historical Press, 2005.

Lumsden, Thomas N. *Nacoochee Valley: Its Times and Its Places*. Sautee Nacoochee, Ga.: Sautee Nacoochee Community Association, 1989.

Lynch, Jack. "Every Man Able to Read." Colonial Williamsburg. Accessed 12 May 2024, www.history.org/Foundation/journal/Winter11/literacy.cfm.

Mann, Charles C. *1491: New Revelations of the Americas before Columbus*. New York: Vintage Books, 2006.

———. "The Clovis Point and the Discovery of America's First Culture." *Smithsonian Magazine*, November 2013. www.smithsonianmag.com/history/the-clovis-point-and-the-discovery-of-americas-first-culture-3825828/.

"Marker Monday: Hernando de Soto in Georgia." Georgia Historical Society. 17 October 2016. www.georgiahistory.com/marker-monday-hernando-de-soto-in-georgia/.

Marley, Timothy. *Southward Ho! Notes of a Tour to and through the State of Georgia in the Winter of 1885–6*. 14 May 2024. https://dlg.galileo.usg.edu/georgiabooks/do-pdf:gb0120.

Marsh, Elizabeth. "The Old Lumsden Place." History Paper 4. Cleveland, Ga.: White County Historical Society, n.d.

Marsh, George Perkins. *Man and Nature, or Physical Geography as Modified by Human Action*. New York: Scribner, 1864.

Martin, Brent. "Forest Removal in the Georgia Mountains." *New Georgia Encyclopedia*. 11 February 2013. www.georgiaencyclopedia.org/articles/geography-environment/forest-removal-georgia-mountains.

Martin, J. Matthews. "Chief Justice John Martin and the Origins of Westernized Tribal Jurisprudence." *Elon Law Review* 4, no. 1 (May 2012): 31–53.

Mastran, Shelley Smith, and Nan Lowerre. *Mountaineers and Rangers: A History of Federal Forest Management in the Southern Appalachians, 1900–1981*. Washington, D.C.: U.S. Forest Service, 1983. https://foresthistory.org/wp-content/uploads/2017/11/MountaineersAndRangers-chaps1-2.pdf.

Mathews, Donald G. *Religion in the Old South*. Chicago: University of Chicago Press, 1979.

McConnell, Marianne S. "The Lawrence Graveyard." History Paper 39. White County Historical Society.

McInvale, Morton R. "James Hall Nichols." In Coleman and Gurr, *Dictionary of Georgia Biography*, 2: 743–44. Athens: University of Georgia Press, 1983.

McLoughlin, William G. *Champions of the Cherokee: Evan and John B. Jones*. Princeton, N.J.: Princeton University Press, 1990.

———. *Cherokee Renascence in the New Republic*. Princeton, N.J.: Princeton University Press, 1986.

McPhee, John. *Annals of the Former World*. New York: Farrar, Straus and Giroux, 2000.

Mereness, Newton D., ed. "Colonel Chicken's Journal to the Cherokees, 1725." In *Travels in the American Colonies*, 95–174. New York: Macmillan, 1916. https://dn790005.ca.archive.org/0/items/travelsinameric01goog/travelsinameric01goog.pdf.

Meserve, John Bartlett. "The Mayes." *Chronicles of Oklahoma*, Spring 1937. https://gateway.okhistory.org/ark:/67531/metadc2192042/.

"Methvin, John Jasper (1846–1941)." *Encyclopedia of Oklahoma History and Culture*. Oklahoma Historical Society. Accessed 5 June 2024. www.okhistory.org/publications/enc/entry?entry=ME022.

Middle American Research Institute. "Directors." Tulane University. Accessed 11 May 2024. http://mari.tulane.edu/directors.html.

Mixon, Gregory. "Atlanta Race Riot of 1906." *New Georgia Encyclopedia*. 23 September 2005. www.georgiaencyclopedia.org/articles/history-archaeology/atlanta-race-riot-of-1906/.

Moffat, Alistair. *The Borders: A History of the Borders from Earliest Times*. Edinburgh, U.K.: Birlinn, 2007.

Moffat, R. Burnham. *Moffat Genealogies: Descent of Rev. John Moffat of Ulster County*. New York: Privately printed, 1909.

"Monroe Methodist Cemetery." Find a Grave. Accessed 19 May 2024. www.findagrave.com/cemetery/2437290/monroe-methodist-cemetery.

Mooney, James. "Myths of the Cherokee." In *Nineteenth Annual Report of the Bureau of American Ethnology*, edited by John Wesley Harding, 3–548. Washington, D.C.: Government Printing Office, 1902.

Moore, Jacqueline. "Booker T. Washington's 1895 Atlanta Exposition Speech (1908 Recreation)." National Registry. Accessed 17 May 2024. www.loc.gov/static/programs/national-recording-preservation-board/documents/BookerT.pdf.

Morley, Joe A. *The Way We Lived in North Carolina*. Chapel Hill: University of North Carolina Press, 2006.

Moss, Robert F. *Southern Spirits: Four Hundred Years of Drinking in the American South, with Recipes*. Berkeley, Ca: Ten Speed, 2015.

"Mountain and Hillside Protection." Community and Economic Development. Accessed 18 May 2024. www.whitecountyga.gov/ced/page/mountain-and-hillside-protection.

Murtagh, William J. *Keeping Time: The History and Theory of Preservation in America*. New York: Wiley and Sons, 2006.

Myer, William Edward. "Indian Trails of the Southeast." In Fewkes, *Forty-Second Annual Report*, 727–847.

"Nacoochee Institute." History Paper 12. Cleveland, Ga.: White County Historical Society, n.d.

"Nacoochee Methodist Church African American Cemetery Assessment Project Report." Georgia Mountains Regional Commission Planning Department. 2011. www.snca.org/snc/museums/intro/NUMCcemeteryRpt.pdf.

"Nacoochee Valley." *Augusta Sentinel-Chronicle*, 8 October 1846.

Nitze, Henry B. C., and H. A. J. Wilkens. *Gold Mining in North Carolina and Adjacent South Appalachian Regions*. Raleigh: North Carolina Geological Survey, 1897.

Norton, Eliza Kenimer. "White County History, Articles from the Gainesville Eagle." History Paper 11. Cleveland, Ga.: White County Historical Society, n.d.

Olmsted, Denison. "On the Gold Mines of North Carolina." *American Journal of Science and Arts* 9 (June 1825): 5–15.

Olmsted, Frederick Law. *A Journey in the Back Country*. New York: Mason, 1863.

———. *A Journey in the Seaboard Slave States, with Remarks on Their Economy*. New York: Dix and Edwards, 1856.

"Our History." Bean Creek Missionary Baptist Church. 12 May 2024. http://beancreekmbc.org/history.

Owens, Jesse, and Ethan Phillips, eds. "The Sautee Nacoochee Community." *Foxfire* 48, nos. 1–2 (Spring–Summer 2014).

Ownby, Ted, and David Wharton. *Georgia's Old Federal Road: Phase I, Development of a Historical Context for the Federal Road in North Georgia, Parts I and II*. Lafayette: Center for the Study of Southern Culture, University of Mississippi, 2006. https://dlg.galileo.usg.edu/data/dlg/ggpd/pdfs/dlg_ggpd_s-ga-bt700-pr4-bm1-b2006-bf4.pdf.

"Paleoindian Database of the Americas." University of Tennessee. 13 May 2024. http://pidba.utk.edu/.

Pepper, George H. "The Museum of the American Indian, Heye Foundation." *Geographical Review* 2, no. 6 (1916): 401–18.

Perdue, Theda. *Cherokee Women: Gender and Culture Change, 1700–1835*. Lincoln, Nebr.: Bison, 1999. Kindle.

Phifer, Edward W., Jr. *Burke County, a Brief History*. Raleigh, N.C.: Department of Cultural Resources, Division of Archives and History, 1979.

———. "Slavery in Microcosm: Burke County, North Carolina." *Journal of Southern History* 28, no. 2 (1962): 145–50.

Phillips. Patrick. *Blood at the Root: A Racial Cleansing in America*. New York: Norton, 2016.

Pogue, Dennis, and Douglas Sanford. "Housing for the Enslaved in Virginia." In *Encyclopedia Virginia*. Virginia Humanities. 2020. https://encyclopediavirginia.org/entries/slave-housing-in-virginia/.

Pomerantz, Gary. *Where Peachtree Meets Sweet Auburn: A Saga of Two Families and the Making of Atlanta*. New York: Lisa Drew Book/Scribner, 1996.

Powell, John Wesley, ed. *Fifth Annual Report of the Bureau of Ethnology to the Secretary of the Smithsonian Institution, 1883–1884*. Washington D.C.: Government Printing Office, 1887.

Prince, Oliver H. "1822 Prince's Digest." In *Historic Georgia Digests and Codes*. Bk. 6. Milledgeville, Ga.: Grandland and Orme, 1822.

Pringle, Heather. "The First Americans: Mounting Evidence Prompts Researchers to Reconsider the Peopling of the New World." *Scientific American*, 12 October 2011.

Rabbitt, Mary C. *The United States Geological Survey: 1879–1989*. U.S. Geological Survey Circular 1050. Washington, D.C.: Department of the Interior, 1989.

Range, Willard. *A Century of Georgia Agriculture, 1850–1950*. 1954. Reprint, Athens: University of Georgia Press, 1969.

Rangel, Rodrigo. "Account of the Northern Conquest and Discovery of Hernando de Soto." In *Historia general y natural de las Indias*. By Gonzalo Fernándz de Oviedo y Valdés. Translated by John E. Worth. Madrid: Real Academia de la Historia, 1851.

Rasmussen, M., Sarah L. Anzick, Michael R. Waters, Pontus Skoglund, Michael DeGiorgio, Thomas W. Stafford Jr., Simon Rasmussen, et al., eds.

"The Genome of a Late Pleistocene Human from a Clovis Burial Site in Western Montana." *Nature* 506 (2014): 225–29.

"Records of the Georgia Baptist Association and Convention." 2015. Georgia Baptist History Depository. Accessed 22 May 2024. https://ursa.mercer.edu/bitstream/handle/10898/13592/Assns-GA-2015_yellow_.pdf.

"Re: Ebenezer Fain." Genealogy Forum. Accessed May 29, 1998. www.genealogy.com/forum/surnames/topics/fain/22/.

Richards, T. Addison. *American Scenery, Illustrated*. New York: Leavitt and Allen, 1854. https://archive.org/details/ americansceneryi00rich.

———. "The Valley of Nacoochee, with the Legend of the 'Evening Star.'" *Orion: A Monthly Magazine of Literature and Art* 3, no. 2 (October 1843): 49–507. https://ia903406.us.archive.org/27/items/sim_orion-a-monthly-magazine-of-literature-and-art_1843-10_3_2/sim_orion-a-monthly-magazine-of-literature-and-art_1843-10_3_2.pdf.

Richards, T. Addison, and William Cary. *Georgia Illustrated in a Series of Views: Embracing Natural Scenery and Public Edifices*. Penfield, Ga.: W. and W. C. Richards, 1842.

Richardson, Lewis W. "The Move to Nacoochee: 1823." History Paper 65. Cleveland, Ga.: White County Historical Society, n.d.

Richter, David K. *Facing East from Indian Country: A Native History of Early America*. Cambridge, Mass.: Harvard University Press, 2001.

Rigdon, John C. *Historical Sketch & Roster of the Georgia 24th Infantry Regiment*. Clearwater, S.C.: Eastern Digital Resources, 2015. Kindle.

Riggs, Brett, and Lance Greene. *The Cherokee Trail of Tears in North Carolina: An Inventory of Trail Resources in Cherokee, Clay, Graham, Macon, and Swain Counties*. Santa Fe, N.M.: National Park Service, 2006.

Rogers, Hugh I. "General Andrew Pickens: Backcountry Warrior." Kettle Creek Battlefield Association. Accessed 14 May 2024. www.kettlecreekbattlefield.com/.

Royce, Charles C. "The Cherokee 'Nation' of Indians: A Narrative of Their Official Relations with the Colonial and Federal Governments." Accessed 12 May 2024. https://repository.si.edu/handle/10088/91635.

Russell Dale A., Fredrick J. Rich, Vincent Schneider, and Jean Lynch-Stieglitz. "A Warm Thermal Enclave in the Late Pleistocene of the South-Eastern United States." *Biological Reviews of the Cambridge Philosophical Society* 84, no. 2 (2009): 173–202. doi: 10.1111/j.1469-185x.2008.00069.x. PMID: 19391200.

Sassaman, Kenneth. "Stallings Island Site." *New Georgia Encyclopedia*. Last modified 8 June 2017. www.georgiaencyclopedia.org/articles/history-archaeology/stallings-island-site/.

Saunt, Claudio. "Creek Indians." *New Georgia Encyclopedia*. Last modified 25 August 2020. www.georgiaencyclopedia.org/articles/history-archaeology/creek-indians/.

"Sautee Nacoochee Center." Sautee Nacoochee Community Association. Accessed 13 May 2024. http://snca.org/snc/home.php.

Sawyer, Gordon. *Northeast Georgia: A History*. Arcadia, 2001.

Schmidt, Leigh Eric. *Holy Fairs: Scottish Communions and American Revivals in the Early Modern Period*. Princeton, N.J.: Princeton University Press, 1989.

Schnell, Frank T. "The Beginnings of the Creeks: Where Did They First 'Sit Down?'" *Early Georgia* 17, nos. 1–2 (1989): 25–29.

Scotch-Irish Society of America. *The Scotch-Irish in America: Proceedings and Addresses of the Fourth Congress at Atlanta, Ga., April 28 to May 1, 1892*. Nashville: Methodist Episcopal Church South, 1892.

Sears, Joan N. "Town Planning in White and Habersham County." *Georgia Historical Quarterly* 54, no. 1 (Spring 1970): 20–40.

Sears, William H. "Creek and Cherokee Culture in the 18th Century." *American Antiquity* 21, no. 2 (October 1955): 144–49.

Shadburn, Don L. *Upon Our Ruins: A Study in Cherokee History and Genealogy*. With John D. Strange III. Cumming, Ga.: Cottonpatch, 2011.

Sherwood, Adiel. *A Gazetteer of the State of Georgia*. 2nd ed. Philadelphia: J. W. Martin & W. K. Boden, 1829.

———. *A Gazetteer of the State of Georgia*. 3rd ed. Washington City, Ga.: Force, 1837.

———. *A Gazetteer of the State of Georgia*. 4th ed. Macon, Ga.: S. S. Boykin, 1860.

Sholes, A. E., and J. H. Estill. *Georgia State Gazetteer, Business and Planters, 1886–1887*. Savannah, Ga.: Morning News, 1886.

Simon, Rachel Ellen. "Whitewashing Reality: Diversity in Appalachia." *Appalachian Voices*, 7 February 2014.

"Site Maps and Registry." NMC Historic Registry. Accessed 6 June 2022. https://sauteewebsites.com/cemeteryWebsite/cemetery/cemetery.php.

Slap, Andrew L. *Reconstructing Appalachia: The Civil War's Aftermath*. Lexington: University Press of Kentucky, 2010.

Smith, George Gilman. *The History of Georgia Methodism from 1786 to 1866*. Atlanta: Caldwell, 1913.

Society for American Archaeology. "The New Archaic." *SSA Archaeological Record* 8, no. 5 (November 2008): 6–40.

Soil Survey Staff. "Web Soil Survey." Natural Resources Conservation Service. Accessed 22 March 2024. http://websoilsurvey.sc.egov.usda.gov/.

Sondley, Foster Alexander. *A History of Buncombe County, North Carolina*. Spartanburg, S.C.: Reprint Company, 1977.

Spalding, Phinizy, ed. *Georgia: The WPA Guide to Its Towns and Countryside*. Columbia: University of South Carolina Press, 1990.

"Special Field Orders No. 15." Wikipedia. 7 April 2024. https://en.wikipedia.org/wiki/Special_Field_Orders_No._15.

Stewart, Bruce E. *Moonshiners and Prohibitionists: The Battle over Alcohol in the Southern Appalachians*. Lexington: University Press of Kentucky, 2011.

Stilgoe, John R. *Common Landscape of America, 1580–1845*. New Haven: Yale University Press, 1982.

Stovall, Allen D. *The Sautee and Nacoochee Valleys: A Preservation Study*. Sautee Nacoochee, Ga.: Sautee Nacoochee Community Association, 1982.

Stromberg, Joseph. "Ancient Migration Patterns to North America Are Hidden in Languages Spoken Today." *Smithsonian Magazine*, 14 March 2014. www.smithsonianmag.com/science-nature/ancient-migration-patterns-north-america-are-hidden-languages-spoken-today-180950053/.

Sullivan, Lynn, and Susan C. Prezzano, eds. "Cherokee Settlement Patterns." In *Archaeology of the Appalachian Highlands*. Knoxville: University of Tennessee Press, 2001.

Swank, James Moore. *History of the Manufacture of Iron in All Ages, and Particularly in the United States from Colonial Times to 1891*. Philadelphia: American Iron and Steel Association, 1892.

Swanton, John Reed. *Early History of the Creek Indians and Their Neighbors*. Washington, D.C.: Government Printing Office, 1922.

———. *Final Report of the United States DeSoto Expedition Commission*. 1939. Reprint, Washington D.C.: Smithsonian Institution Press, 1985.

Tallulah Falls. Wikipedia Commons. Accessed 5 June 2024. https://commons.wikimedia.org/wiki/File:Tallulah_Falls_7151.jpg.

Taylor, Jen. "Adam Poole Vandiver." Find a Grave. Accessed 5 June 2024. www.findagrave.com/memorial/143262573/adam-poole-vandiver#source.

Telford, Nell Kenimer. "Early Methodists in N.E. Georgia." History Paper 3. Cleveland, Ga.: White County Historical Society, n.d.

Tell, Bernice. "'Separate Yet One': Booker T. Washington's Atlanta Compromise Displayed at Library." *Library of Congress Information Bulletin* 55, no. 3 (1996). www.loc.gov/loc/lcib/9603/booker.html.

Thomas, Cyrus. *Catalog of Prehistoric Works East of the Rocky Mountains*. Washington, D.C.: Government Printing Office, 1891. https://archive.org/stream/ catalogueprehis01thomgoog#page/n12/ mode/2up.

———. *The Cherokees in Pre-Columbian Times*. New York: Hodges, 1890. www.jstor.org/stable/1767655.

Thomas, Henry Walker. *Digest of the Railroad Laws of Georgia*. Atlanta: Franklin, 1895.

Thurmond, Michael. *Freedom: An African-American History of Georgia, 1733–1865*. Gainesville, Ga.: Longstreet, 2003.

Tinius, Mary Bell. "The Burning of Starlight." History Paper 16. Cleveland, Ga.: White County Historical Society, n.d.

———. "Description of Starlight." History Paper 6. Cleveland, Ga.: White County Historical Society, n.d.

Tischendorf, Alfred. "British Enterprise in Georgia, 1865–1907." *Georgia Historical Quarterly* 42, no. 2 (1958): 170–75.

Tolstoy, Leo. *War and Peace*. Oxford: Oxford University Press, 2017.

Townsend, Billy. *History of the Georgia State Parks and Historic Sites Division*. Last modified 2001. http://gastateparks.org/ content/georgia/parks/75th _Anniv/parks_ history.pdf.

Turck, John A., Mark Williams, and John F. Chamblee. "Examining Variation in the Human Settlement of Prehistoric Georgia." 2011. www.academia .edu/1074369/Examining_Variation_in_the_Human_Settlement_of _Prehistoric_Georgia.

Turner, William H., and Edward J. Cabbell, eds. *Blacks in Appalachia*. Lexington: University Press of Kentucky, 1985.

Utley, Francis Lee, and Marion R. Hemperley. *Place Names of Georgia: Essays of John H. Goff*. 1975. Reprint, Athens: University of Georgia Press, 2007.

Van Buren, Jarvis. "Fruit at the South." *Horticulturalist and Journal of Rural Art and Rural Taste* 13 (1858): 226.

Vassar, Rena, ed. "Some Short Remarks on the Indian Trade to the Charikees and in Management Thereof since the Year 1717." *Ethnohistory* 8, no. 4 (Autumn 1961): 401–23.

Villani, John. *The 100 Best Small Art Towns In America: Where to Find Fresh Air, Creative People, and Affordable Living*. Santa Fe: Muir, 1994.

Voorhies, Michael R. "Pleistocene Vertebrates with Boreal Affinities in the Georgia Piedmont." *Quaternary Research* 4, no. 1 (March 1974): 85–93.

Warren, Mary Bondurant, and Eve. B. Weeks. *Whites among the Cherokees*. Danielsville, Ga.: Heritage, 1987.

Warren, Stephen. *The Worlds the Shawnees Made: Migration and Violence in Early America*. Chapel Hill: University of North Carolina Press, 2014.

Waselkov, Gregory A. "Seventeenth-Century Trade in the Colonial Southeast." *Southeastern Archaeology* 8, no. 2 (Winter 1989): 117–33.

Watson, Thomas L. *A Preliminary Report on a Part of the Granites and Gneisses of Georgia*. Atlanta: Franklin, 1902.

Wauchope, Robert. "Archaeological Survey of Northern Georgia: With a Test of Some Cultural Hypotheses." *Memoirs of the Society for American Archaeology* 21 (1966): iii–482.

Webb, Althea. "African Americans in Appalachia." Oxford African American Studies Center. Accessed 12 May 2024. https://oxfordaasc.com/page/2527.

Weeks, Stephen Beauregard, ed. *Letters of Benjamin Hawkins, 1796–1806*. Savannah: Georgia Historical Society, 1916.

White, George. *Historical Collections of Georgia: Containing the Most Interesting Facts, Traditions, Biographical Sketches, Anecdotes, Etc.* New York: Pudney & Russell, 1855.

———. *Statistics of the State of Georgia: Including an Account of Its Natural, Civil, and Ecclesiastical History . . . and a Correct Map of the State*. Savannah, Ga.: Williams, 1849.

White, Max E. *The Archaeology and History of the Native Georgia Tribes*. Gainesville: University Press of Florida, 2002.

White County History Book Committee. *A History of White County, 1857–1980*. Cleveland, Ga.: White County History Book Committee, 1981.

Whitney, Carrie Westlake. *Kansas City, Missouri: Its History and Its People, 1800–1908*. 3 vols. Kansas City: Clarke, 1908.

Whitney, Josiah Dwight. *The Metallic Wealth of the United States, Described and Compared with That of Other Countries*. Philadelphia: Lippincott, Grambo, 1854.

Wigginton, Eliot, ed. *Foxfire 3*. New York: Anchor, 1975.

Wilkerson, Isabel. *The Warmth of Other Suns: The Epic Story of America's Great Migration*. New York: Random House, 2010.

"Will/Braffitt/Johnson/Cook Genealogy: Person Sheet." Accessed 4 June 2024. www.willbraffitt.org/roots/ps02/ps02_479.html.

Williams, David. *The Georgia Gold Rush: Twenty-Niners, Cherokees, and Gold Fever*. Columbia: University of South Carolina Press, 1993.

Williams, George Walton. "Advice to Young Men, and a Sketch of Nacoochee, Georgia and Its Surroundings." Digital Library of Georgia. 16 May 2024. https://dlg.galileo.usg.edu/georgiabooks/do-pdf:gb0294.

———. *Nacoochee and Its Surroundings*. Charleston, S.C.: George W. Williams, 1874.

———. *Sketches of Travel in the Old and New World*. Charleston, S.C.: Walker, Evans & Cogswell, 1871.

Williams, George Walton, and George Sherwood Dickerman. *History of Banking in South Carolina from 1712 to 1900*. Charleston, S.C.: Walker Evans & Cogswell, 1903.

Williams, Mark. *Nacoochee Revisited: The 2004 Project*. Lamar Institute 72.

Box Springs, Ga.: Lamar Institute, 2004. www.thelamarinstitute.org/images/PDFs/publication_72.pdf.

Wilms, Douglas C. "Cherokee Settlement Patterns in Nineteenth-Century Georgia." *Southeastern Geographer* 14, no. 1 (1974): 46–53.

Wishart, David M., Jeff A. Ankron, and Wendy H. Zorick. "Settling Cherokee Georgia: Land Grab, Gold Rush, or Both?" Paper presented at the Fourteenth International Economic History Congress, Helsinki, Finland, 2006.

Wood, Curtis W., Jr. *From Ulster to Carolina: The Migration of the Scotch-Irish to Southwestern North Carolina*. Raleigh: North Carolina Office of Archives and History, 1998.

Woodard, Colin. *American Nations: A History of the Eleven Rival Regional Cultures of North America*. New York: Penguin, 2011.

Woodward, C. Vann. *The Strange Career of Jim Crow*. Oxford, U.K.: Oxford University Press, 1963.

Wooley, James E., and Vivian Wooley. *Rutherford County, North Carolina, Wills and Miscellaneous Records, 1783–1868*. Greenville, S.C.: Southern Historical Press, 1983.

Yarnell, Susan L. *The Southern Appalachians: A History of the Landscape*. General Technical Report SRS-18. Asheville, NC: U.S. Department of Agriculture, Forest Service, 1998.

Yeates, William Smith, Samuel Washington McCallie, and Francis P. King. *A Preliminary Report on a Part of the Gold Deposits of Georgia*. Geological Survey of Georgia Bulletin 4-A. Atlanta: State Printing Office, 1896.

Public Records

FEDERAL RECORDS

LeBow, J. D. B., ed. *Statistical View of the United States: Compendium of the Seventh Census*. Washington, D.C.: U.S. Census Bureau, 1854.

National Park Service. "Harshaw-Stovall House." *National Register of Historic Places Inventory: Nomination Form*. 1984. https://npgallery.nps.gov/GetAsset/262c8c74-d12a-4c09-9580-c25b2172d1ab.

"National Register of Historic Places." National Park Service. Accessed 18 May 2024. www.nps.gov/subjects/nationalregister/index.htm.

"Records of the Education Division of the Bureau of Refugees, Freedmen, and Abandoned Lands, 1865–1871." M803. 35 rolls. National Archives and Records Administration, Washington, D.C.

"Records of the Field Offices for the State of Georgia." Record Group 105. U.S. Bureau of Refugees, Freedmen, and Abandoned Lands, 1865–72. National

Archives and Records Administration. Accessed 22 May 2024. www.archives.gov/files/research/microfilm/m1903.pdf.

"Records of the Superintendent of Education for the State of Georgia Bureau of Refugees, Freedmen and Abandoned Lands, 1865–1870." National Archives and Records Service. Accessed 17 May 2024. www.archives.gov/files/research/microfilm/m799.pdf.

U.S. Bureau of Indian Affairs. "Register of Cherokees Who Wished to Remain in the East." 1817–1819. Microfilm A21. National Archives and Records Administration, Washington, D.C.

U.S. Census Bureau. *1840 Census: Compendium of the Enumeration of the Inhabitants and Statistics of the United States*. Washington, D.C.: U.S. Census Bureau, 1840.

———. *Abstract of the Returns of the Fifth Census*. Washington, D.C.: U.S. Census Bureau, 1832.

———. "Population Schedules, 1880–1940." Accessed 5 June 2024. www.census.gov/prod/ www/decennial.html.

———. "Population, Slave, Manufacturing, Agricultural and Other Schedules, 1790–1950." Accessed 5 June 2024. www.census.gov/prod/ www/decennial.html.

———. *The Seventh Census of the United States: 1850, an Appendix*. Washington, D.C: U.S. Census Bureau, 1853.

U.S. Department of Veterans Affairs. "War of 1812 Pension Application Files Index, 1812–15." Accessed 5 June 2024. www.ancestry.com/search/collections/1133/.

U.S. Geological Survey. "National Map Viewer." 2005. http://viewer.nationalmap.gov/viewer/.

U.S. Office of Indian Affairs. *Treaties between the United States of America and the Several Indian Tribes, from 1778 to 1837*. Washington, D.C.: Langtree and O'Sullivan, 1837.

U.S. Post Office. *History of the United States Postal Service, 1775–1993*. Publication 100. Washington, D.C.: U.S. Post Office, May 2007.

———. "Nacoochee Post Office Application." 1893. M1126. Roll 124. National Archives and Records Administration, Morrow, Ga.

———. "Record of Appointment of Postmasters, 1832–1971." October 1789–1832. M1131. Roll 3. Habersham County. National Archives and Records Administration, Morrow, Ga.

———. "Records of Site Locations." 1837–1950. M1126. Rolls 112 and 124. Habersham and White Counties. National Archives and Records Administration, Morrow, Ga.

———. "Rural Free Delivery." Postal History. 2013. https://about.usps.com/who-we-are/postal-history/rural-free- delivery.pdf.

"U.S., Social Security Death Index, 1935–2014." Ancestry. Accessed 5 June 2024. www.ancestry.com/search/collections/3693/.

STATE RECORDS

"Georgia's First Youth-Focused Wildlife Management Area." Wildlife Resources Division, Georgia DNR. Accessed 18 May 2024. https://georgiawildlife.blog/2016/09/22/georgias-first-youth-focused-wildlife-management-area/.
Secretary of State. *District Plats of Survey, 1819–20*. Georgia Archives. Accessed 13 May 2024. https://dlg.usg.edu/collection/gaarchives_dmf?page=2andper_page=100andsort=score+desc%2C+year+ascandview=list.
———. "District Survey Field Notebooks, 1819–20." Records of the Surveyor General. Georgia Archives. Accessed 13 May 2024. https://vault.georgiaarchives.org/digital/collection/fieldnotes.
"State Records of Assistant Commissioners and Superintendents of Education." January 1868. Microfilm 799. Rolls 16–26. Bureau of Refugees, Freedmen, and Abandoned Lands, 1865–1872. National Archives, Morrow, Ga.

COUNTY RECORDS

"Habersham County, Georgia." Clerk of Superior Court. Record of Deeds and Mortgages, 1819–1856. Courthouse in Clarkesville. Microfilm. Georgia Archives, Morrow.
White County. "Returns of Qualified Voters, 1867–1868." Microfilm 297. Roll 25. Georgia Archives, Morrow.
"White County, Georgia." Clerk of Superior Court. Record of Deeds and Mortgages, 1857–1889. Courthouse in Cleveland. Microfilm. Georgia Archives, Morrow.

Maps

Barker, William, and Mathew Carey. *Georgia, from the Latest Authorities*. [Philadelphia: M. Carey, 1795.]
Bonne, Rigobert, and Peter André. *Carte de la Louisiane, et de la Floride*. [Geneva, 1780]. www.loc.gov/ item/73697603/.
Bowen, Emanuel. *A New & Accurate Map of the Provinces of North & South Carolina, Georgia, &c.* In *A Complete Atlas or Distinct View of the Known World*. [London: Innys and Joseph Richardson, 1752.]
Bradley, Abraham, Jr., and Jedidiah Morse. *A Map of the Southern Parts of the United States of America*. [1800]. Historic Maps. Surveyor General. RG 3-8-65. Georgia Archives, Morrow.

Brazier, Robert H. B. *A New Map of the State of North Carolina*. Fayetteville, N.C.: Published under the patronage of the legislature by John MacRae; Philadelphia: H. S. Tanner, 1833.

Burr, David H. *Map of Georgia & Alabama Exhibiting the Post Offices, Post Roads, Canals, Rail Roads &c.* [London, 1839.] https://lccn.loc.gov/98688462.

Carey, Matthew. *Carey's General Atlas, Improved and Enlarged; Being a Collection of Maps of the World and Quarters, Their Principal Empires, Kingdoms, &c.* Philadelphia, 1814. https://digitalcollections.nypl.org/items/24e82000-c603-012f-a87b-58d385a7bc34.

Claflin, Fred F. [*National Forest Service Land Map*.] Atlanta, 1937. Shows land lots to document Forest Service purchases, 1934–37. Georgia Archives, Morrow.

Cram, George Franklin. *Railroad and County Map of Georgia*. [Chicago, 1883]. www.loc.gov/item/98688460/.

Crisp, Edward, Thomas Nairne, John Harris, Maurice Mathews, and John Love. *A Compleat Description of the Province of Carolina in 3 Parts: 1st, the Improved Part from the Surveys of Maurice Mathews & Mr. John Love; 2ly, the West Part by Capt. Tho. Nairn: 3ly, a Chart of the Coast from Virginia to Cape Florida*. [London: Edw. Crisp?, 1711]. www.loc.gov/item/2004626926/.

Early, Eleazer. *Map of the State of Georgia Prepared from Actual Surveys and Other Documents for Eleazer Early*. David Rumsey Map Center, Stanford Libraries. 1818. www.davidrumsey.com/luna/servlet/detail/RUMSEY~8~1~240502~5512285.

Fry, Joshua, and Peter Jefferson. *A Map of the Most Inhabited Part of Virginia Containing the Whole Province of Maryland with Part of Pensilvania, New Jersey and North Carolina*. London: Jefferys, 1755.

Gaston, Samuel N. *The Campaign Atlas for 1861*. New York: Gaston, 1861. www.loc.gov/item/2008626958/.

Georgia Department of Mines, Mining, and Geology. *Mineral Resources of Union, Towns, Lumpkin, and White Counties, Georgia*. Map RM-2. Tennessee Valley Authority. 1950. https://dlg.galileo.usg.edu/data/dlg/ggpd/pdfs/dlg_ggpd_s-ga-bm500-b-pm1-b1950-bu5.pdf.

Georgia Department of Transportation. *General Highway Map, White County*. Atlanta: State Highway Department of Georgia, 1951–1985.

Gridley, Enoch G., and Mathew Carey. *The State of Georgia*. Library of Congress. 1818. www.loc.gov/item/2006635240/.

Hall Brothers. *Hall's Original County Map of Georgia, Showing Present and Original Counties and Land Districts, Completed from State Records*. Georgia Archives. 1895. https://dlg.usg.edu/record/gyca_mmap_gen-027#item.

Hammerton, William. *Map of the Southeastern Part of North America.* New Haven, Conn.: Yale Center for British Art, 1721.

Hammond, C. S. *Hammond's Complete Map of Georgia.* New York: C. S. Hammond & Company, 1926.

Hemperley, Marion. *Stagecoach Route, Georgia, circa 1825.* 1985. Georgia Archives, Morrow.

Herbert, John. *A New Mapp of His Majesty's Flourishing Province of South Carolina Showing Settlements of Ye English, French, and Indian Nation.* 1725. Reprint, London, 1744.

Hollis, Victor R. *Nacoochee Valley, 1837.* Redrawing of Moffat map, with added notes and topographical features. Self-published, [1922?].

Hopkins, Oliver B. *Map Showing Distribution of Asbestos, Talc, Soapstone, and a Part of the Basic, Magnesian Rocks in Georgia. Geological Survey of Georgia.* Bulletin 29. Atlanta: Byrd, State Printer, 1914.

Johnson, Alvin Jewett, and Joseph Hutchens Colton. *Johnson's New Illustrated Steel Plate Family Atlas, with Descriptions, Geographical, Statistical, and Historical.* New York: Johnson & Ward, 1862.

Kitchin, Thomas. *Map of the United States in North America.* London, 1783.

———. *A New Map of North & South Carolina & Georgia, Drawn from the Latest Authorities.* London, 1765.

———. *A New Map of the Cherokee Nation with the Names of the Towns and Rivers.* London, [1760]. Lionel Pincus and Princess Firyal Map Division. Library of Congress. Accessed 14 May 2024. www.loc.gov/item/2021586055/.

Lattré, Jean. *Carte des Etats-Unis de l'Amerique suivant le Traité de Paix de Paris.* Paris, Chez Lattré. 1784. Library of Congress. Accessed 14 May 2024. https://lccn.loc.gov/73691628.

Le Moyne De Morgues, Jacques. *Floridae Americae provinciae recens & exactissima descriptio.* [1591.] www.loc.gov/item/2003623393/.

Lewis, Samuel, and Harry Schenck Tanner. *Georgia.* Baltimore: Fielding Lucas, 1817.

L'Isle, Guillaume de, and Nicholas Guérard. *L'Amerique septentrionale. Dressée sur les observations de Mrs. de l'Academie Royale des Sciences & quelques autres, & sur les memoires les plus recens.* 1700. Reprint, [Paris, 1718]. www.loc.gov/item/gm71005434/.

Lucas, Silas, Jr., ed. *Atlas of the State of South Carolina.* 1825. Reprint, Greenville, S.C.: Southern Historical Press, 1994.

Lumsden, Walter B., Jr. *Nacoochee Valley.* Self published, 1948.

Map by Which the Creek Indians Gave Their Statement at Fort Strother on the 22nd Jany, 1816: [Alabama and Georgia]. Library of Congress. Accessed 13 May 2024. www.loc.gov/item/2007626786/.

McCallie, Samuel W. *Map Showing the Distribution of a Part of the Gold Deposits of North Georgia*. In Yeates, McCallie, and King, *Preliminary Report*.

Merrill, William Emery, N. Finegan, H. Riemann, and U.S. Army of the Cumberland. *Map of Northern Georgia*. Chattanooga, Tenn.: Topographical Engineers Office, U.S. Army of the Cumberland, 1864. www.loc.gov/item/2006458675/.

Mitchell, John, Thomas Kitchin, and Andrew Millar. *A Map of the British and French Dominions in North America, with the Roads, Distances, Limits, and Extent of the Settlements, Humbly Inscribed to the Right Honourable the Earl of Halifax, and the Other Right Honourable the Lords Commissioners for Trade & Plantations*. [London: Millar, 1755.]

Mitchell, S. Augustus. *County Map of the States of Georgia and Alabama*. Philadelphia, 1879.

———. *A New Map of Georgia with Its Roads and Distances*. Philadelphia, 1846.

Moffat, Reuben Curtis. *Nacoochee Valley, 1837*. Sautee Nacoochee, Ga.: Sautee Nacoochee Community Association, 1891.

Moll, Herman. *A New and Exact Map of the Dominions of the King of Great Britain on Ye Continent of North America, Containing Newfoundland, New Scotland, New England, New York, New Jersey, Pensilvania, Maryland, Virginia and Carolina*. [London, 1731.]

Morse, Sidney E., and Samuel Breese. *Georgia*. In *Morse's North American Atlas*. New York: Harpers, 1842.

Mouzon, Henry, Robert Laurie, and James Whittle. *An Accurate Map of North and South Carolina with Their Indian Frontier*. [London: Laurie & Whittle, 1794.] www.loc.gov/item/2006626006/.

Myer, William Edward. "The Trail System of the Southeastern United States in the Early Colonial Period." In *Forty-Second Annual Report*, 727–857.

A New and Accurate Map of the Province of Georgia in North America. [London?, 1779]. Library of Congress. Accessed 13 May 2024. www.loc.gov/item/2008625108/.

Nicholson, Francis. *Map of the Several Nations of Indians to the Northwest of South Carolina*. Library of Congress. 1724. www.loc.gov/item/2005625337/.

Ortelius, Abraham, Geronimo Chaves, Diego Hurtado De Mendoza, Christophe Plantin, and Abraham Ortelius. *Pervviae avriferae regionis typvs/La Florida/auctore Hieron. Chiaues: Gvastecan reg*. [Antwerp: Christophe Plantin?, 1584]. www.loc.gov/item/84696980/.

Pickering, Samuel M., Jr., and J. B. Murray. *Geologic Map of Georgia*. 1950. Reprint, Atlanta: Georgia Geological Survey, Geologic and Water Resources Division, 1976.

Romans, Bernard. *A General Map of the Southern British Colonies in America Comprehending North and South Carolina, Georgia, East and West Florida, and the Neighboring Indian Countries*. London, 1776.
Royce, Charles C. *Map of the Former Territorial Limits of the Cherokee "Nation of" Indians: Map Showing the Territory Originally Assigned Cherokee "Nation of" Indians*. [1884?]. www.loc.gov/item/99446145/.
Salley, Samuel Alexander. *George Hunter's Map of the Cherokee Country and the Path Thereto in 1730*. Columbia: South Carolina State Library, 1917.
Society for the Diffusion of Useful Knowledge. *Georgia with Parts of North & South Carolina, Tennessee, Alabama & Florida*. North America Sheet 12. London: Baldwin & Craddock, 1833.
State Highway Board of Georgia. *Map of White County*. Atlanta, 1934.
———. *Map of White County*. Atlanta, 1940.
———. *State of Georgia, System of State Roads*. Atlanta: Foote and Davies, 1938.
State of Georgia. *White County*. Atlanta, 1867.
U.S. Geological Survey. *Clarkesville NE, Ga*. 1:24,000. 7.5 Minute Series. Reston, Ga.: U.S. Geological Survey, 2014.
———. *Georgia: Dahlonega Sheet*. 1:125,000. 1885 Survey. Reston, Ga.: U.S. Geological Survey, 1892.
———. *Helen, Ga*. 1:24,000. 7.5 Minute Series. Reston, Ga.: U.S. Geological Survey, 1957.
———. *Helen Quadrangle, Georgia*. 1:24,000. 7.5 Minute Series. Reston, Ga.: U.S. Geological Survey, 1957, 1985, 2014.

Index